ENVIRONMENTAL LEAD

ECOTOXICOLOGY AND ENVIRONMENTAL QUALITY SERIES

Series Editors: *Frederick Coulston*
and
Freidhelm Korte

OTHER SERIES VOLUMES

Water Quality: Proceedings of an International Forum
F. Coulston and E. Mrak, editors

Regulatory Aspects of Carcinogenesis and Food Additives: The Delaney Clause
F. Coulston, editor

ENVIRONMENTAL LEAD

Edited by

Donald R. Lynam
Lillian G. Piantanida
Jerome F. Cole

International Lead Zinc Research Organization
New York, New York

ACADEMIC PRESS 1981

A Subsidiary of Harcourt Brace Jovanovich, Publishers

New York London Toronto Sydney San Francisco

Academic Press Rapid Manuscript Reproduction

Proceedings of the Second International Symposium on Environmental Lead Research, held in Cincinnati, Ohio, December 5–7, 1978.

ACADEMIC PRESS, INC.
111 Fifth Avenue, New York, New York 10003

United Kingdom Edition published by
ACADEMIC PRESS, INC. (LONDON) LTD.
24/28 Oval Road, London NW1 7DX

Library of Congress Cataloging in Publication Data

International Symposium on Environmental Lead Research, 2d, University of Cincinnati, 1978.
Environmental lead.

(Ecotoxicology and environmental quality)
Includes index.
1. Lead-poisoning--Congresses. 2. Lead--Physiological effect--Congresses. 3. Lead--Environmental aspects--Congresses. I. Lynam, Donald R. II. Piantanida, Lillian. III. Title. IV. Series. [DNLM: 1. Environmental pollution--Congresses. 2. Lead poisoning--Congresses. 3. Nervous system diseases--Etiology--Congresses. W3 IN917MH 2d 1978e [QV 292 I63 1978e]
RA1231.L4I54 1978 615.9'25688 80-29487
ISBN 0-12-460520-6

PRINTED IN THE UNITED STATES OF AMERICA

81 82 83 84 9 8 7 6 5 4 3 2 1

CONTENTS

III. Epidemiologic Studies on Lead

IV. Dynamics of Lead in Adults

V. Effects of Lead on the Kidney

VI. Biochemical Studies on Lead

VII. Air and Water Pollution Studies

CONTRIBUTORS

Numbers in parentheses indicate the pages on which the authors' contributions begin.

F. Behse (69), *University Hospital (Rigshospitalet), Research Building, Section 41-1-2, Room 1220, Blegdamsvej 9, 21000 Copenhagen O, Denmark*

R. Bornschein (49), *Department of Environmental Health, University of Cincinnati Medical Center, 3223 Eden Avenue, Cincinnati, Ohio 45267*

F. Buchthal (69), *University Hospital (Rigshospitalet), Research Bldg., Section 41-1-2, Room 1220, Blegdamsvej 9, 2100 Copenhagen O, Denmark*

A. C. Chamberlain (175), *Harwell Environmental and Medical Sciences Division, AERE Harwell, Oxfordshire, OX11 ORA England*

J. J. Chisolm Jr. (1), *Department of Pediatrics, Baltimore City Hospital, G Building/Room 224, Baltimore, Maryland 21224*

W. C. Cooper (111), *The Great Western Building, 2150 Shattuck Avenue, Suite 401, Berkeley, California 94704*

G. Craun (199), *U.S. Environmental Protection Agency, Cincinnati, Ohio*

P. J. Craig (333), *Fresh Water Biological Institute, University of Minnesota, Navarre, Minnesota 55392*

G. W. Crockford (95), *London School of Hygiene and Tropical Medicine, TUC Centenary Institute of Occupational Health, Keppel Street (Gower Street), London WC1E 7HT England*

B. D. Culver (253), *University of California, Irvine, California*

E. DeVos (199), *Harvard Graduate School of Education, Cambridge, Massachusetts*

J. W. Embree (231), *Rincon Annex, University of California, PO Box 7604, San Francisco, California 94120*

M. Fugăs (145), *Yugoslav Academy of Science and Arts, Institute for Medical Research, 158, Mose Pijade, 41000 Zagreb 1, Yugoslavia*

P. Garibaldi (9), *Assoreni, 20097, S. Donato Milanese, Milano, Italy*

G. Gilli (9), *Institute of Hygiene Turin University, Turin, Italy*

H. C. Gonick (253), *University of California Center for the Health Sciences, s Los Angeles, California 90024*

J. Gorczynski (291), *New York University Medical Center, Institute of Environmental Medicine, 550 First Avenue, New York, New York 10016*

S. Hall (231), *Rincon Annex, University of California, PO Box 7604, San Francisco, California 94120*

P. Hammond (227), *University of Cincinnati, Department of Environmental Health, Kettering Laboratory 3223 Eden Avenue, Cincinnati, Ohio 45267*

L. Hastings (49), *Department of Environmental Health, University of Cincinnati Medical Center, 3223 Eden Avenue, Cincinnati, Ohio 45267*

M. J. Heard (175), *Harwell Environmental and Medical Sciences Division, AERE Harwell, Oxfordshire, OX11 ORA, England*

C. H. Hine (231), *Rincon Annex, University of California, San Francisco, California 94120*

P. Karelekas (199), *U.S. Environmental Protection Agency, Cincinnati, Ohio*

E. King (169), *New York University Medical Center, Institute of Environmental Medicine, New York, New York 10016*

M. T. Kleinman (291), *New York University Medical Center, Institute of Environmental Medicine, 550 First Avenue, New York, New York 10016*

T. Kneip (291), *New York University Medical Center, Institute of Environmental Medicine, 550 First Avenue, New York, New York 10016*

D. Lawrence (269), *Department of Microbiology and Immunology, The Neil Hellman Medical Research Building, The Albany Medical College, Albany, New York 12208*

H. A. Lewis (231), *Rincon Annex, University of California, San Francisco, California 94120*

M. Lieberman (199), *Harvard Graduate School of Education, Cambridge, Massachusetts*

M. Lippman (291), *New York University Medical Center, Institute of Environmental Medicine, 550 First Avenue, New York, New York 10016*

R. K. Loch (49), *Department of Environmental Health, University of Cincinnati Medical Center, 3223 Eden Avenue, Cincinnati, Ohio 45267*

I. A. Michaelson (49), *Department of Environmental Health, University of Cincinnati Medical Center, 3223 Eden Avenue, Cincinnati, Ohio 45267*

E. Mitran (95), *Institute of Hygiene, Bucharest, Romania*

A. Matranga (199), *Tufts Medical School, Newton, Massachusetts 02158*

J. Northrup (231), *Rincon Annex, University of California, San Francisco, California 94120*

L. S. Rafales (49), *Department of Environmental Health, University of Cincinnati Medical Center, 3223 Eden Avenue, Cincinnati, Ohio 45267*

S. R. V. Raghavan (253), *University of California Center for Health Sciences, Los Angeles, California*

C. Ryan (199), *U.S. Environmental Protection Agency, Boston, Massachusetts*

H. K. Sachs (41), *182 La Pier Street, Glencoe, Illinois 60022*

M. Sărîc (145), *Yugoslav Academy of Science and Arts, Institute for Medical Research, 158, Mose Pijade, 41000 Zagreb 1, Yugoslavia*

J. Sayre (23), *Department of Pediatrics, The University of Rochester Medical Center, Rochester, New York 14642*

G. Vanini (9), *Institute of Hygiene Turin University, Turin, Italy*

L J. Warren (309), *CSIRO, Division of Mineral Chemistry, Institute of Earth Resources, Port Melbourne, Australia 3207*

J. M. Wood (333), *Fresh Water Biological Institute, University of Minnesota, Navarre, Minnesota 55392*

D. Worth (199), *Tufts Medical School, Newton, Massachusetts 02158*

PREFACE

The Second International Symposium on Environmental Lead Research, held in Cincinnati, Ohio, December 5–7, 1978, was attended by approximately 250 persons from throughout the world who came to hear ILZRO lead research grantees speak on all major facets of the ILZRO lead environmental health program. These proceedings of that symposium contain the papers presented as well as the discussions that followed. The conference provided an opportunity for the exchange of scientific data on key topics of concern to the scientific community as well as to policy decision makers.

The proceedings follow the format of the symposium, which was divided into seven major topical groupings: Dynamics of Lead in Children; Neurologic Effects of Lead on the Kidney; Biochemical Studies on Lead; and Air and Water Pollution Studies. Anyone with an interest in the environmental aspects of exposure to lead, such as industrial hygienists, toxicologists, physicians, and policy decision makers, will find the book of interest.

The International Lead Zinc Research Organization, Inc. (ILZRO), cosponsor of the conference, along with the Department of Environmental Health, University of Cincinnati, is the research arm of the world-wide lead and zinc industry. Since its inception, ILZRO, alone and in cooperation with other industrial and governmental groups, has sponsored a broad program of research in order to further our understanding of the potential effects of lead on health, to both the general population and the working population. This research has concentrated on critical information needs required for realistic evaluation of potential health effects.

Donald R. Lynam *

Lillian G. Piantanida[+]

Jerome F. Cole
International Lead Zinc Research Organization, Inc.
New York, New York

*Present affiliation: Ethyl Corporation, Baton Rouge, Louisiana

[+]Present affiliation: Department of Industrial Relations, Division of Occupational Safety and Health, Berkeley, California

DOSE-EFFECT RELATIONSHIPS FOR LEAD IN YOUNG CHILDREN: EVIDENCE IN CHILDREN FOR INTERACTIONS AMONG LEAD, ZINC, AND IRON[1]

J. J. Chisolm, Jr.

Johns Hopkins University and John F. Kennedy Institute
Baltimore, Maryland

A growing body of evidence in experimental animals indicates that interactions among essential and non-essential trace elements have significant toxicological and nutritional implications. In rats, induced dietary deficiencies of calcium, iron, copper, zinc, and selenium are reported to enhance absorption and retention of lead. Zinc is said to be protective against some of the neurotoxic effects of lead in horses. Delta-aminolevulinic acid dehydratase (ALAD) is a zinc-dependent enzyme, and lead-induced inhibition of this enzyme can be reversed *in vivo* by zinc in animals. Experimentally, lead is a competitive inhibitor of the utilization of iron for heme formation. These experimental observations point to a strong possibility that nutritional factors may well account for some of the variability in biologic responses to lead in man. This report provides the first evidence in man of significant interactions among lead, zinc, and iron.

Children were selected for in-patient studies on the Pediatric Clinical Research Unit of The Johns Hopkins Hospital on the basis of various combinations of blood lead (PbB) and FEP, such as low PbB - low FEP, low PbB - high FEP, high PbB - low FEP, and high PbB - high FEP, as determined in the lead clinic of The John F. Kennedy Institute, to which children are referred for evaluation and management. In the Pediatric Clinical Research Unit, serial measurements of PbB and FEP are made and urine is collected quantitatively daily for ALAU, catecholamine metabolites, creatinine, and trace metals (Pb, Zn, Cd, Fe, Mn, Co, Cr, and Cu). Routine

[1]*A detailed version of this paper is being submitted to Pediatric Research for publication.*

ISBN 0-12-460520-6

urinalysis, complete blood counts, hemoglobin typing, serum iron, total serum iron binding capacity, serum ferritin, lactose tolerance test, and serum zinc and copper are also determined.

After a four-day control period, CaEDTA (1,000 $mg/m^2/day$) is injected intramuscularly in two divided doses at 12 hour intervals. Since CaEDTA causes a diuresis not only of lead, but also of zinc, iron, cadmium, and manganese, this test can serve as a "chemical biopsy" of the mobile stores of these trace metals in the tissues. The amounts of these metals excreted following CaEDTA may be called the "chelatable zinc," "chelatable iron," "chelatable lead," etc. To date, 66 children have been studied under this protocol. Laboratory determinations (except catecholamine metabolites) and statistical analyses have been completed on data from 45 children. Admission data on these 45 children who are the subject of this report, are given in Table 1.

RESULTS

According to the design of the study, no statistical significant relationship between PbB and FEP was expected and none was found. A modest relationship was found between PbB and chelatable lead ($r = 0.407$, $p < .02$, $> .01$), but there was no relationship between FEP and chelatable lead in these children. However, highly significant statistical relationships were found when ALAU ($mg/m^2/24$ hr) was taken as the metabolic indicator of interference in the biosynthesis of heme. Most impressive is the highly significant linear relationship between ALAU and chelatable lead ($r = .831$, $p < .0001$). Conversely, no relationship between ALAU and chelatable iron or chelatable zinc was found. However, polynomial regression analysis revealed highly significant curvilinear relationships between ALAU and the chelatable zinc - lead and chelatable iron - lead ratios. The best fit with the smallest variance was found with an intersecting lines approach.

Statistically well-defined points of intersection were found. The data indicate that increase in ALAU above the normal range (1.11 ± 0.37 $mg/m^2/24$ hr) is significantly associated with a decrease in chelatable zinc - lead ratio to < 18.45 and a decrease in chelatable iron / lead ratio to < 0.593. This occurs as PbB rises through the 45 to 60 μg Pb range. Multiple regression analysis of the data indicates that decrease in labile tissue iron

TABLE 1. Admission Data on the 45 Children in the Study

	Mean	*Range*
Age (mo.)	*36*	*16 - 66*
PbB (μg Pb)	*50*	*26 - 66*
FEP (μg/dl rbc)	*523*	*138 - 1094*

stores, as measured by the chelatable iron - lead ratio, contributes significantly to the ALAU effect. The multiple regression equation for the iron-lead interaction involving ALAU is as follows:

$$\text{ALAU (mg/m}^2\text{/24 hr)} = 0.3917 + 1.5658 \text{ (chelatable lead)} - 0.4246 \text{ (chelatable iron)}^2 \pm 0.975$$

$$(r = .853).$$

Biologically, accumulation and excretion of ALA may be viewed as the consequence of the relative activities of δ-aminolevulinic acid synthetase (ALAS), which controls the formation of ALA, and δ-aminolevulinic acid dehydratase (ALAD), which catalyzes the conversion of ALA to porphobilinogen, the next step in the biosynthesis of heme. The chelatable zinc - lead data are readily explained by a zinc - lead interaction involving δ-aminolevulinic dehydratase. The iron - lead data probably reflect compensatory increase in the activity of ALAS secondary to reduced formation of heme, which, in turn, can result from reduced bioavailability of iron for heme formation. Failure to find significant relationships between FEP and the other parameters measured, as well as nutritional and other environmental implications of the data will be discussed.

DISCUSSION OF PAPER BY DR. J. JULIAN CHISOLM

Vincent N. Finelli, University of Cincinnato, Kettering Laboratory: I would like to know if you have ever considered treatment of lead exposed children or adults, with zinc therapy in addition to EDTA chelation. We have done some animal studies, which show that if zinc is administered to the patient concurrently with EDTA therapy, there is a much quicker elevation in the red blood cell ALAD and, therefore, a quick, drastic drop in urinary ALA. I agree with you that urinary ALA is one of the most important biological indicators of lead poisoning. I believe that ALAD, therefore, becomes a very important biological indicator of lead exposure, lead poisoning. Therefore, this increase in ALAD in the erythrocytes of the zinc treated patients might be a good indication of quick reversal of lead poisoning.

Dr. J. Julian Chisolm, Baltimore City Hospital: Although I have thought about this possibility, I have not done any studies on zinc supplementation in children as yet.

Harold Petering, University of Cincinnati, Kettering Laboratory: First of all, I'd like to congratulate you, on a very interesting and provocative experimental approach as well as a very nice presentation. We've done some work with respect to the ratio of cadmium and lead in hair of people of many ages, including children, and we find a very high correlation. So, there is evidence that there must be some kind of association here either metabolically or by intake. The other point is that if you want to study the relationship of copper, I suggest you might use penicillamine which greatly increases copper as well as zinc, lead, and cadmium excretion.

Chisolm: In answer to your last question, I have only presented the results on the first five days of a six day protocol. I have given BAL on the sixth day to a few patients to evaluate copper status. The number studied is insufficient for any conclusions. In answer to your first question, we have not found any relationship between either spontaneous urinary cadmium or "chelatable" cadmium and blood and urine lead measurements in this group of children. What this means I can not say, although the data suggest that, as a group, the children do not have a common source for Cadmium and lead. We have, however, noted an apparent positive relationship between chelatable lead and chelatable manganese.

Hammond: Was there an association between zinc and iron excretion, that is, was the iron-ALAU relationship a distinct relationship apart from the zinc-ALAU relationship? If, indeed, that were the case, then you might have a better basis for invoking two separate mechanisms, one of them depression of ALAS, in the case of iron, and the other nonutilization of ALAD. Did the children with the high zinc levels also have the high iron levels?

Chisolm: First, the range in output for iron is really rather narrow. It's much, much greater for zinc. In terms of having lower zincs and lower irons, they are, in general, in the same children.

Kathryn R. Mahaffey, The Food and Drug Administration: The importance of this is quite great, because the nutritional status of pre-school children, especially from low-income groups for nutrients such as Zinc and iron, is often inadequate. In fact, there is some estimate that as high as 30 percent of pre-school children in low income groups are iron deficient, to varying degrees. Do you have any plans for supplementing the therapeutic regimes that you've talked about with administration of trace elements?

Chisolm: Yes, I do give prophylactic amounts of iron, zinc and manganese to children on long term d-penicillamine therapy. We propose to study the issue of essential trace metal depletion during chelation therapy more formally in the future. I might add, parenthetically, that I have been a little heavy-handed on the iron on not very good scientific grounds for the past 25 years. Recent experimental studies would appear to support this approach.

Emil A. Pfitzer, Hoffmann-LaRoche, Incorporated: I'd like to thank you for a very stimulating discussion on interactions and their importance. I have three questions. First, are you telling us that when you're looking for sensitive and subtle changes of heme synthesis as related to lead exposure that FEP is not a very good measure, particularly as compared to ALAU? Second, you mentioned that about one-quarter of your children had iron deficiency anemia problems. Did you have any confounding effect on the statistical analysis or were these individuals sort of scattered throughout the group? Third, can you help me understand the significance of your chelatable metal as a function of the EDTA administered? I'm used to seeing the amount of lead in 24 hours or 48 hours; I'm having some trouble understanding the units you used.

Chisolm: With regard to FEP versus ALAU, I'm not saying that FEP is not a good measure of changes in heme synthesis. We're really doing short-term studies that last only about six days. We're dealing with a part of the pathway where enzymatic turnover is relatively rapid, whereas zinc protoporphyrin apparently persists throughout the life span of the red cell. What does surprise me is the fact that we can certainly answer the question which is a better predictor of chelatable lead, blood lead or FEP. On the basis of this study, it's the blood lead although it's not very good. Therefore, I don't think that what we have would support the statement that the administration of chelating agents should be based upon the FEP. But I'm not saying that FEP is not important. It would take a longer study to show it.

As to your second question, we have plotted percent saturation (serum iron/total iron binding capacity) and serum ferritin against all of these parameters. We find no relationship between serum ferritin and percent saturation on the one hand and chelatable iron on the other hand. In short, measurements made in serum do not show any relationship with urinary measurements of iron. In answer to another of your questions, we do have low hemoglobin values spread throughout the group, although anemic children are perhaps more numerous among those with blood lead concentrations greater than 50 μg. Thirdly, you ask about expressing the data as amount of metal excreted in relation to dose of EDTA administered. This is done to compensate for the different sizes of the children which results in different doses being administered to each child according to his surface area.

Pfitzer: I guess I'm still having a little trouble appreciating the significance of the amount of excretion relative to the actual amount administered, but I'll think about that.

Chisolm: Well, I think it's dose-related. The amount of metal, certainly of lead, removed is dose-related. We just have to make sure we have a high enough dose so that we can saturate the "chelatable" pool.

Pfitzer: One further comment about ALAU. The relationships you showed were very striking. How do you feel about using ALAU measurements to a greater extent than they are currently?

Chisolm: In a clinical research unit setting, where 24 hour urine collections can be made in children, ALAU measurements are very useful. However, in children it's a virtually useless measurement if one has to depend upon spot samples. In previous studies we have found ALAU/creatinine ratios and ALAU measurements adjusted for specific gravity relatively useless in children. Also, in a study such as the one reported here, we have successive 24 hour collections which greatly increase our ability to detect collections that are incomplete. Incomplete collections, of course, have to be rejected.

Vostal: Did you measure the concentration of EDTA in the urine, or are you using only the dose which was administered for your ratio?

Chisolm: No, we did not measure EDTA in the urine. We do give a standard dose of CaEDTA, which is 1,000 mg/m^2. I rely on the studies of others which indicate that virtually 100% of EDTA is removed by the kidney through glomeruler filtration.

Vostal: The reason why I've asked is that it has been shown in the literature that people who have higher levels of lead might have a slight delayed excretion of EDTA, so you might not get the complete amount of EDTA in the urine during the first day.

Chisolm: I see what you mean. I have, of course, observed in the past what you describe as slightly delayed excretion of lead in children with blood lead concentrations greater than 100 μg/Pb/dl whole blood. Most of these children had acute symptoms of lead poisoning, including encephalopathy. In such children, some of whom also had impaired renal function, we often saw the highest excretion of lead on the second or third day of EDTA administration. In the group of asymptomatic children just reported, we are dealing with a far lower blood lead concentration range; namely, 26 to 66 µg/Pb/dl whole blood. In 13 of these children, we have been able to get complete urine collections on the first three days of CaEDTA administration. In this subgroup we find that the second day's lead excretion is 60% of the first day's lead excretion and the third day's lead excretion is 60% of the second day's lead excretion. There is a highly significant correlation between the first day's output of lead, zinc and iron and the three day total outputs for these three metals. I conclude, therefore, that the dosage of CaEDTA provides a sufficient molar excess of CaEDTA so that the first day's output of these metals does provide a representative "chemical biopsy" for these metals.

ISOTOPIC TRACING OF LEAD INTO CHILDREN FROM AUTOMOBILE EXHAUST

P. Garibaldi

Assoreni, Milan, Italy

G. Vanini
G. Gilli

Hygiene Institute of Turin University
Turin, Italy

INTRODUCTION

The present study of children is intended as a contribution to the solution of the problem represented by the absorption of lead into the human body in its early development stages from atmospheric particulates. In particular, this research aims at defining the magnitude of the automotive source in contributing to the total lead content in children's blood. This section is sponsored by ILZRO and is part of a broader research effort, sponsored jointly by EEC and EPA, directed to the same object, which studies the adult population.

This report, therefore, contains a description of the current research work in broad outline, followed by the results obtained up to now in the section devoted to children. It should also be noted that this work is in progress and consequently the results given in this paper are subject to modifications due both to constant improvements of the analytical techniques used (particularly as far as the isotopic composition determination is concerned) and to the amount of experimental data likely to be obtained in the next years. This presentation represents a Progress Report, and the reader is expected to evaluate it accordingly.

ISBN 0-12-460520-6

THE OVERALL EXPERIMENT

The Isotopic Lead Experiment (ILE) Project, based on the large-scale use of isotopically differentiated lead (stable natural isotopes) as additive to commercial gasolines, aims at assessing the contribution of automobile lead exhaust to the environment. The experiments are designed to provide answers to the following points:

1) Determination of the contribution of road traffic to atmospheric lead pollution in selected urban and rural areas.
2) Evaluation of the contribution of automotive lead to absorption in man.
3) Estimation of the distribution of automotive lead in different environmental areas.

In order to differentiate automotive lead from other lead sources, it was decided to control the production of alkyl lead for use as an additive in most gasolines in Italy (100% in the experimental region) by using only lead from the Broken Hill mine in Australia, having an isotopic composition (Pb206/Pb207 = 1.04). This is significantly different from the lead normally present in the environment in Italy and different from lead coming from other sources.

In Table 1, the isotopic composition of lead is reported for rock samples gathered in Italy. Table 2 contains environmental

TABLE 1. Pb 206/Pb 207 Ratio in Different Samples of Ores Collected in Italy

Goni Mine (Sardinia)	*1.18*
Monteponi (Sardinia)	*1.14-1.17*
Alpes Ivrea Area (Piedmont)	*1.17*
Alpes Swiss Area (Sempione)	*1.19*
Val d'Antrona (Piedmont)	*1.20*
Val d'Anzasca (Piedmont)	*1.20*
Traversella (Piedmont)	*1.19*
Serravezza (Tuscany)	*1.15*

TABLE 2. Pb 206/Pb 207 Ratio in Different Environmental Samples Collected in Piedmont Area Before July 1975

Soil	*1.17*
Vegetation	*1.16*
Gasolines	*1.18*
Blood	*1.17*

samples taken in 1974 in the course of the feasibility study. Isotopic composition data for lead samples coming from mines located in several parts of the world are reported in Table 3.
3. A feasibility study has shown that one region is particular, the Piedmont (Fig. 1), offered interesting aspects for carrying out this work. The main reasons were the following:

- The output of the local refineries exceeds the regional consumption (Fig. 2).
- Most of the refineries in the area were already supplied with lead alkyl by one company only (the Società Italiana Additivi per Carburanti (SIAC)), and they agreed to cooperate in the experiment.
- It has a high traffic density.
- In the city of Turin, both dense traffic and very high industrial activity are present.
- An efficient atmospheric particulate sampling network was already in operation in Turin.

As a consequence, the study is being conducted in the Piedmont region, in particular in the city of Turin and in 7 villages belonging to the surrounding district, all located within 40 km of the city itself. The experiment was scheduled to start during 1975. SIAC had to import an amount of lead of Australian origin sufficient to cover the requirements of the area selected for the duration of the study. Since SIAC did not supply the entire market, appropriate exchanges of alkyl lead with the other competing company, ETHYL, had to be arranged without affecting in any way their respective share of the market.

Unfortunately, the introduction of the Australian lead into the test region did not follow the original schedule, and the exchange of alkyl lead between the two companies in the study area began in January, 1977. Between October, 1975 and January,

TABLE 3. Isotopic Ratio of Lead Samples Originating from Different Rocks and Mining Deposits

Origin	*Pb 206/207*
Rudny-Altai	*1.15*
Kazakistan Central	*1.12*
Samo Island	*1.21*
Bulgaria	*1.15-1.18*
High Slesia	*1.17*
Boheno Mountain	*1.15-1.18*
Australia (Broken Hill)	*1.04*
South Africa	*1.16*
Terranova	*0.934*
Cerro de Pasco - Peru	*1.20*
Mexico	*1.21*

Fig. 1. Oil refineries in the study area.

1977, the two companies went on supplying their respective customers with the usual alkyl lead product. This fact caused the isotopic ratio of lead in gasoline to fluctuate around an intermediate value.

Finally, in May 1977, 100% of the gasoline distributed in the northwestern part of Italy could be said to contain Australian lead exclusively. Therefore, based on the distribution of the latter in the gasoline present in this region (Fig. 3), the experiment can be subdivided into the following four periods.

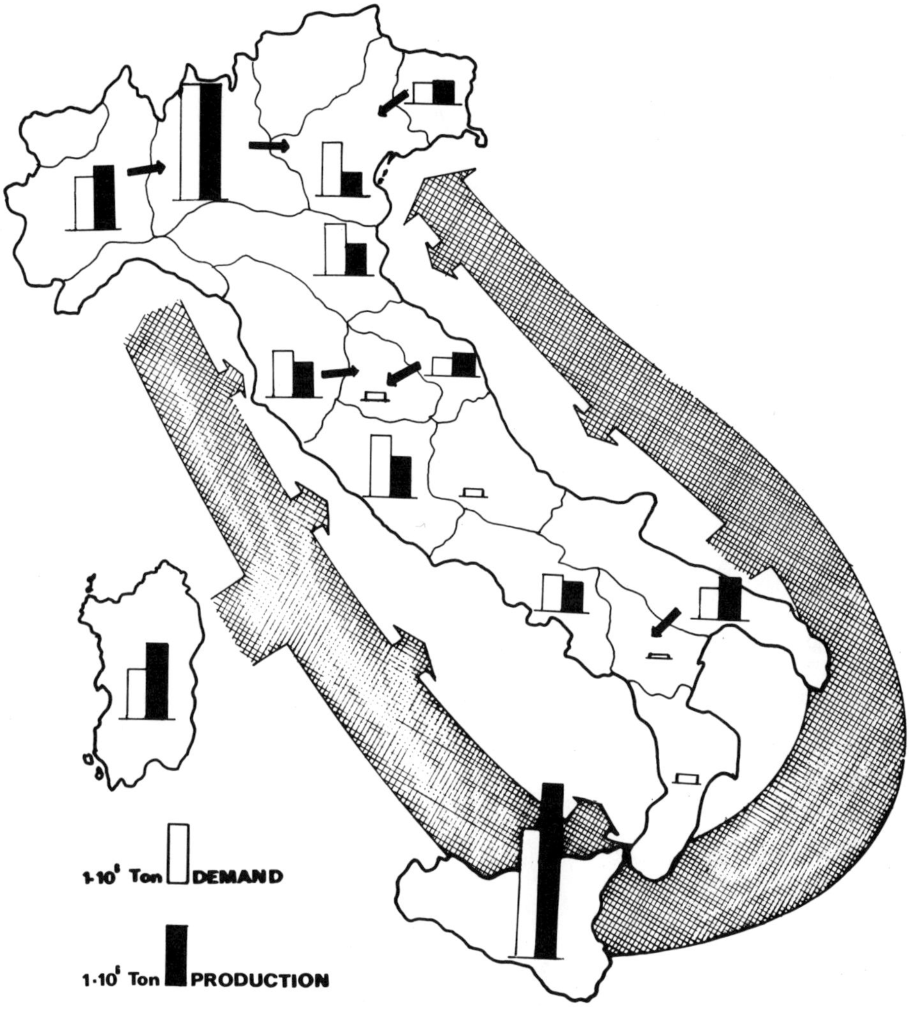

Fig. 2. Regional internal movement of gasoline.

- Phase 0: background definition - July 1974-July 1975
- Phase 1: transition period from the old lead to the Australian lead in gasoline - August 1975-April 1977
- Phase 2: use of 100% Australian lead (Pb-206/Pb-207=1.04) as additive in the gasoline - May 1977-December 1978
- Phase 3: restoration of the initial isotopic conditions - January 1979-December 1979

A decision on whether to lengthen Phase 2 is still to be made, as it depends on the isotopic composition findings concerning blood samples collected during the last months of 1978. The intention

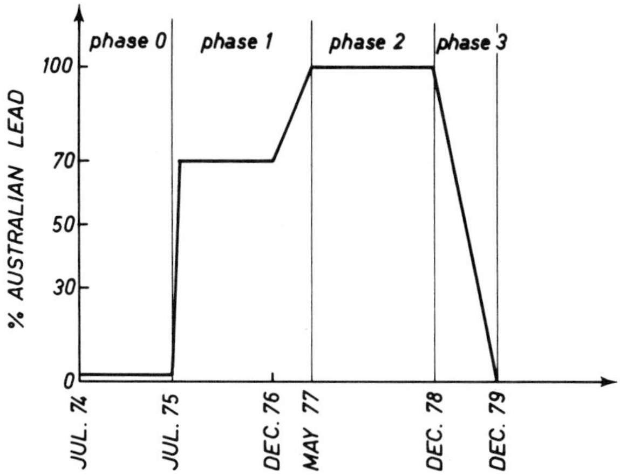

Fig. 3.

is to extend, if possible, the use of Australian lead for at least 12 more months.

The general object of this research is to define a correlation between lead in gasoline and lead in the environment. In particular, the study represents an attempt to verify the actual existence and validity of the relation:

Gasoline Pb : airborne particulate Pb : human blood Pb

Therefore, as shown in Figure 4, an accurate control of the pollu-

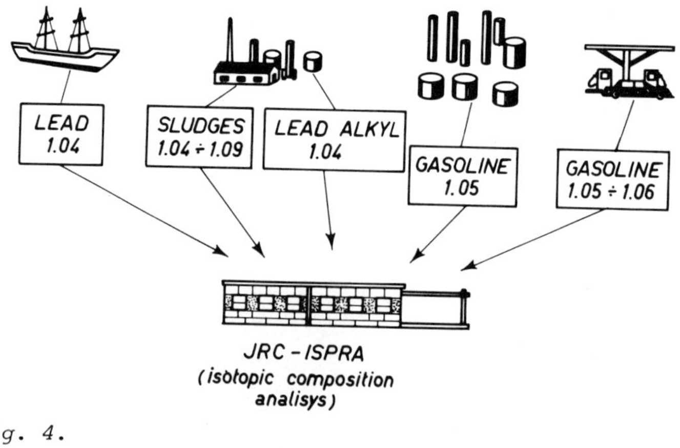

Fig. 4.

tion source represented by gasoline is being carried out with the aim of determining its effects. This is made by following the entire production and distribution process, starting from metallic lead through to the gasoline sold at gas pumps. Collection of the entire series of samples is to be performed in a continuous manner:

- Total airborne particulate
- Size-distributed airborne particulate
- Soil
- Vegetation
- Surface and drinking water
- Precipitation
- Blood

To give some idea of the size and complexity of the research being carried out, the total number of samples collected by October 1978, divided in the various phases is reported in Table 4. In particular, it might be interesting to describe briefly the collecting program used to take blood samples from adults.

More than 800 subjects, divided into two large groups, one living in town and the other in the countryside, are sampled periodically in order to follow, over the course of time, the possible variation of lead isotope composition in the blood of each subject (Table 5). It is thus possible to obtain curves of the type reported in Figure 5, which should allow one to deduce for each subject the contribution from automotive lead.

If one wishes to take into account the effect of a possible variation of concentration with time, the following empirical formula for each subject could be used:

$$\text{Contribution to total Pb content in blood due to automotive Pb} = \frac{r_0 - r_i}{r_o - 1.04}\,(c_i/c_o) \times 100$$

TABLE 4. Number of Samples Collected from July 1974 to October 1978

Sample	*Ph.0*	*Ph.1*	*Ph.2*	*Total*
Metallic Pb	*13*	*20*	*26*	*59*
Alkyl Pb	*1*	*42*	*31*	*74*
Refinery gas	--	*23*	*93*	*116*
Service station gas	*47*	*210*	*1430*	*1687*
Airborne particulate	*491*	*1897*	*1116*	*3504*
Soil+veg.+water etc.	*279*	*189*	*100*	*568*
Blood (adults)	*751*	*1180*	*487*	*2418*
Blood (children)	--	*59*	*162*	*221*
Total	*1582*	*3620*	*3445*	*8647*

(Ph. = Phase)

TABLE 5. Distribution of Adult Subjects by Number of Blood Samplings during Transition and Steady State Phase Through August 1978

Place	Subj.	Number of Persons Submitted to Successive Sampling in Piedmont								
		1	2	3	4	5	6	7	8	9
Turin										
City	198	17	82	99						
Country	634	288	132	87	48	34	16	18	9	2
Total	832	305	214	186	48	34	16	18	9	2

where

r_o is the initial isotopic ratio (Phase 0)
r_i is the isotopic ratio at time i
c_o is the initial concentration of lead in blood
c_i is the lead concentration in blood at time i
1.04 is the 206/207 ratio of the Australian lead.

Most samples have already been analyzed for the concentration of lead. However, the isotope composition has been measured on only a small number of samples. The recent availability of new instrumentation should make it possible to complete within the next year the isotope analysis of the most significant samples.

On the basis of the available data in Figure 6, the time trend of the isotopic composition of alkyl lead observed at the production stage, in gasoline taken at gas pumps, and finally in the atmospheric particulate of the city of Turin and the surrounding agricultural district has been reported. A better definition of these trends will undoubtedly be possible when new data of isotope composition are available.

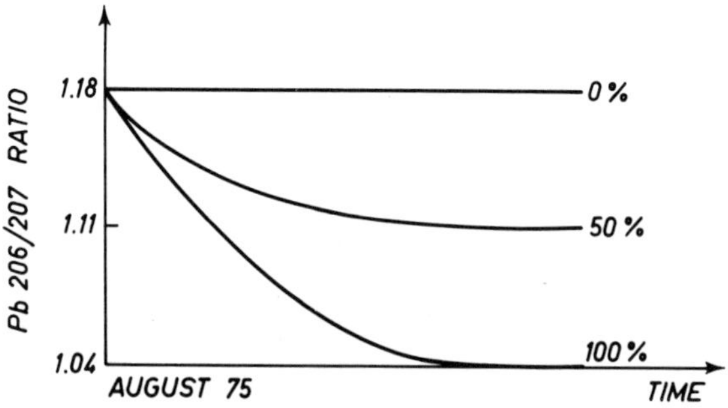

Fig. 5.

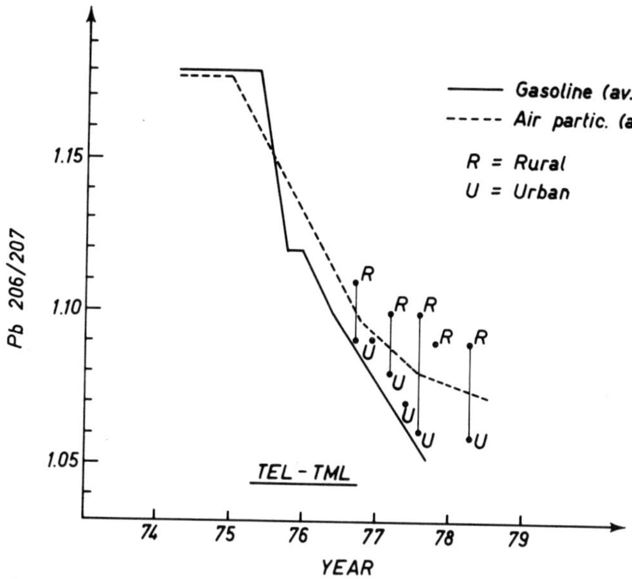

Fig. 6.

It may also be interesting to gain some knowledge of the total lead content of the atmospheric airborne particulate of Turin and surrounding rural areas, where subjects undergoing periodic blood sampling usually reside. On the basis of 1976 results, expressed in terms of annual averages, the following data can be given:

Turin, city area	- 4.5 μg/m^3	24 hr sampling
Countryside	- 0.6 μg/m^3	24 hr sampling
Turin, city area	- 6.3 μg/m^3	4 hr sampling

LEAD IN CHILDREN

From December 1975 until May 1978, five groups of children aged 2 to 12 years old have been examined, the entire group amounting to a total of 206 subjects (Table 6). The first three groups are made up of children hospitalized because of illnesses of various origins. The fourth and fifth groups (147 children) have been formed with pupils from three primary schools located in the city of Turin. Characteristics of the various groups examined, namely age, sex, and place of residence, have been reported in Tables 7, 8, and 9. While the program with adults provides that the same subjects are periodically checked, the initial program for children

TABLE 6. Groups of Children Sampled in Piedmont Area

Group	No.	Origin
A	17	Hospital
B	21	Hospital
C	21	Hospital
D	21	School
E	126	School
Total	206	

TABLE 7. Age of Children Groups

Group	<5		5 - 10		>10	
	No.	%	No.	%	No.	%
A	–	–	10	59	7	41
B	5	24	9	43	7	33
C	2	10	11	52	8	38
D	–	–	21	100	–	–
E	–	–	126	100	–	–
Total	7	3	177	86	22	11

TABLE 8. Sex of Children Groups

Group	Male No.	%	Female No.	%
A	9	53	8	47
B	14	67	7	33
C	13	62	8	38
D	12	57	9	43
E	62	49	64	51
Total	110	53	96	47

TABLE 9. Domicile of Children Groups

Group	No.	%	No.	%
A	13	76	4	24
B	18	86	3	14
C	14	67	7	33
D	21	100	–	–
E	126	100	–	–
Total	192	93	14	7

took into account lead isotope composition measurements in blood through a succession of subject groups, each one sampled twice.

The original program scheme is shown in Figure 7. In this way, it might have been possible to create a continuation of groups in order to observe possible variations of lead content and isotope composition in blood as a function of time. Unfortunately, it was not possible to fulfill such a sampling program because of the practical difficulties encountered in sampling children. As a consequence, it was possible to repeat the second sampling for one group of children only, eight months after the first one was completed (Fig. 8). However, it is hoped that the results obtained in the future with the present modified program will contribute positively to the solution of the problem of lead pollution.

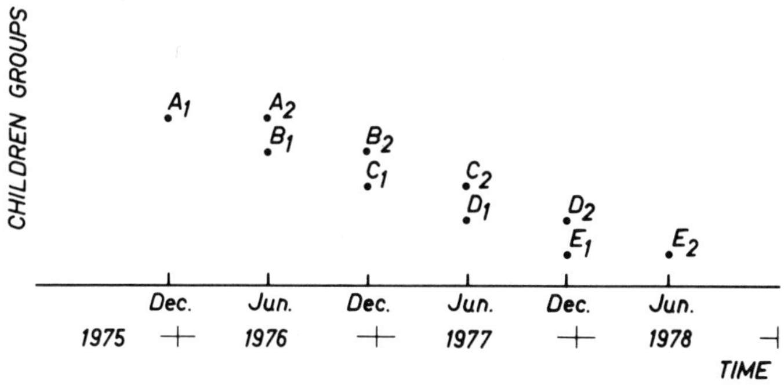

Fig. 7.

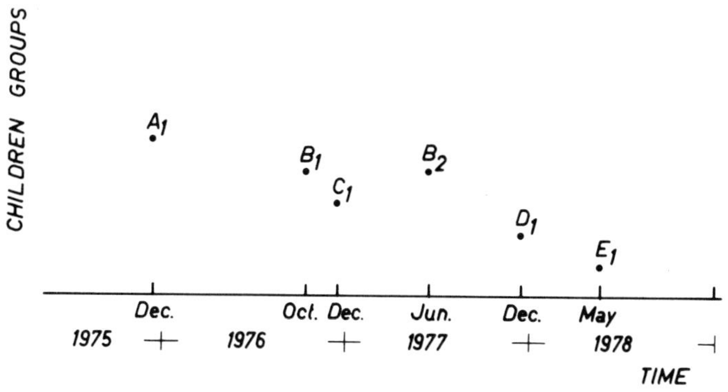

Fig. 8.

DATA AVAILABLE TO DATE

All blood is analyzed in terms of lead content immediately after sampling. An aliquot of the sample is stored in a freezer and its isotope composition determined as soon as the instrumentation is available.

A. Lead Concentration in Blood

Although the main interest of this research work lies in finding the trend for the isotope composition of lead contained in blood, it is believed that it may also be of interest to show some concentration data which may be helpful in interpreting the isotope composition results. Table 10 illustrates the distribution of lead levels in the blood of the five children's groups examined, while the graph contained in Figure 9 shows the distribution of lead levels when the 207 subjects examined are considered together. Fifty percent (50%) of the examined subjects show lead concentration values lower than 16 μg/100 ml and 80% show values lower than 20 μg/100 ml.

B. Isotopic Composition

Only a part of the samples collected could be analyzed to determine isotope composition. A preliminary evaluation of available data is given in Table 11. A slight decrease from 1.157 to 1.141 was the only notable variation of the average value of 206/207 isotope ratios of the various groups which had been sampled at different times between December 1975 and June 1977. However, no definite conclusions can be drawn at this moment, because the examined samples were taken during the transition phase and because

TABLE 10. Distribution of Lead Concentration in Blood[a]

Lead Conc. g/100 ml	*Group A 17 Subj. Dec. 75 %*	*Group B 21 Subj. Oct. 76/ Jun. 77 %*	*Group C 21 Subj. Dec. 76 %*	*Group D 21 Subj. Dec. 77 %*	*Group E 126 Subj. May 78 %*
<10	*23*	*5*	*14*	*43*	*1*
11-15	*23*	*38*	*52*	*33*	*17*
16-20	*18*	*29*	*19*	*24*	*56*
21-25	*23*	*19*	*10*	-	*18*
>26	*12*	*9*	*5*	-	*8*

[a]*In the case of the B group, for which blood sampling was made twice for each subject, the average value is given.*

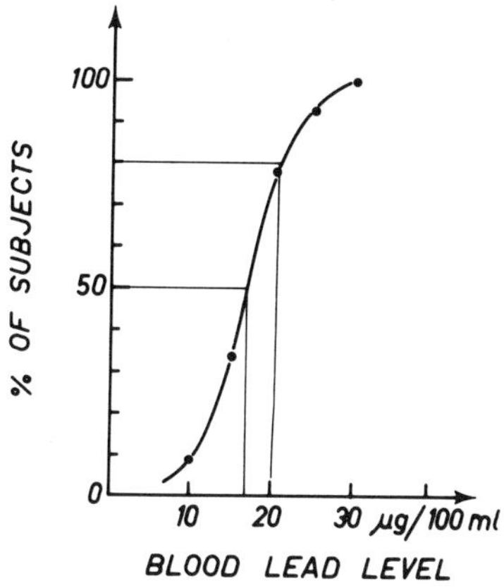

Fig. 9.

not all of the collected samples have been analyzed.

Some preliminary analyses made on samples coming from the D1 group (December 1977) appear to show a further slight decrease of the 206/207 ratio. It should be kept in mind, however, that in the course of the analytical work it has been found by the use of an internal standard and the control of the blanks that a certain contamination level did actually affect the entire treatment of the samples. This, in turn, caused the undertaking of a large unfinished effort to completely eliminate such undesired occurrences.

It is deemed preferable, therefore, to wait until the whole series of isotope composition data, particularly from those last sampled, becomes available before any conclusions are drawn.

TABLE 11. Isotopic Composition Data Available up to Date

Group	*A 1*	*B 1*	*C 1*	*B 2*	*D 1*	*E 1*
Date	*Dec. 75*	*Oct. 76*	*Dec. 76*	*Jun. 77*	*Dec. 77*	*May 78*
Samples	*17*	*21*	*21*	*15*	*21*	*126*
Analyses	*4*	*18*	*8*	*13*	-	-
Average	*1.157*	*1.145*	*1.136*	*1.141*	-	-
SD	*0.0095*	*0.0104*	*0.0074*	*0.0210*	-	-

DUST LEAD CONTRIBUTION TO LEAD IN CHILDREN

J. Sayre, M.D.

University of Rochester
Rochester, New York

Epidemiologic studies in the last ten years have suggested that there may be more factors involved in childhood lead exposures than the formerly accepted concept of eating paint chips. The possibility that lead-containing dust may result in significant child and adult exposure has been highlighted by studies in El Paso, Texas, Kellogg, Idaho, and Toronto, on people living adjacent to smelters, where lead-containing particles are emitted from stacks and deposited in the environs. Three separate episodes of childhood exposure are reported wherein the occupational lead contact of parents resulted in lead dust being carried on the clothing of the parents into their own homes. The children have then apparently picked up the lead dust from their clothes and other surfaces identified inside the houses. These industrial sources have added perspective on the mechanisms of exposure, but are not common enough to account for widespread exposures such as those we encounter in our cities.

The idea that small particulate lead might be of importance in childhood exposure is not new, however. In studies in Brisbane, Australia, of children with lead poisoning at the turn of the last century, this idea was suggested by Gibson, an ophthalmic surgeon, who wrote:

> "Dust, of course, is capable of being both swallowed and inhaled. A greater danger, however, is adhesion of the paint either by nature of its stickiness or by nature of its powdery character,on fingers and nails by which it is carried to the mouths of children, especially in the case of those who bite their nails, suck their fingers, or eat with unwashed hands ...
>
> "... To ascertain whether the powdery substance so easily detachable from painted surfaces which have been exposed to the weather was or was not composed

ISBN 0-12-460520-6

> largely of lead, I supplied the Government Analyst with two pieces of calico.... The total lead on one of the pieces of calico was 0.3 grammes ..."

It is an interesting twist that those Australian children were generally from well-to-do families wealthy enough to be able to paint the railings of their verandahs with lead-containing paint. During the long wet seasons, children spent hours on those verandahs.

In Rochester, New York, we began in 1973 to explore the possibility that children might derive exposure to lead-containing dust from the interior of their houses through contamination of their hands. We were looking for a source of exposure of a lower magnitude than the large one of lead paint chips. In those studies, we found that there was a good correlation between hand lead and household surface lead. These values were found to be high among 50 children known to have modestly elevated lead levels and low among a control population of children from homes in suburban Rochester. Since there were many children who lived in the inner-city who apparently are <u>not</u> lead exposed, we felt it important to do similar and more complete studies on an exclusively inner-city population.

The purpose of this study, then, was to test the concept that interior household dust constitutes the principal source of lead exposure responsible for the augmented body lead burden found in inner-city children. To this end, we proposed the following hypothesis: Children, ages 1½ to 6 years, with blood lead (PbB) levels between 40 and 79 µg%, will have significantly more lead on surfaces in their household environments than will a matched group of children whose blood levels are below 30 µg%. The (unproved) implication of this hypothesis is that the PbB elevation is due to the lead contaminating the children's hands and then being ingested by the normal hand-to-mouth activity of these children.

Children for the study were collected from the same general geographic area of central Rochester. Home visits were made, location of the home was noted, and in-the-home interviews were conducted with the child's mother. Proximity of the house to heavily trafficked streets was noted. Water and food sources were not studied. Prior studies on the water supplies of Rochester had not suggested that water contamination was a significant exposure source.

During home interviews, families were asked about pica and mouthing activities of the children, where the children played (both inside and outside), and in what other locations they spent their time. An assessment of socio-economic status was made. Specimens of dust were collected from specified areas. Paint samples were obtained where loose paint was noted. Soil specimens were collected if appropriate.

SELECTION AND STUDY OF SUBJECTS

Rochester has a number of on-going programs of blood lead level surveillance in health centers and clinics. The children of this study were selected from the records of the Anthony L. Jordan Health Center located in the center of Rochester, comprising approximately three-fourths of the patients, and the pediatric clinic of the Strong Memorial Hospital, comprising one-fourth of the subjects.

From family records, the home address was obtained to ascertain that the patient lived in the study area. Children were selected whose ages ranged between 18 and 71 months. Blood lead levels were reviewed for those children who had been recently tested. It is the general policy of these facilities to test children in these age groups on a yearly basis. If levels are above 30 μg/dl, testing is done more frequently.

Two groups were defined for study. Those in the "low lead" group had blood lead values of less than 29 μg/dl and free erythrocyte protoporphyrin (FEP) less than 59 μg/dl. There were 50 in this group. The "high lead" group consisted of children with levels between 40 and 79 μg/dl and FEP values over 59 μg/dl. There were 49 children in this group. To qualify for inclusion in the study, patients in the "high" group had to have had at least two PbB levels in this range over a period of three to four months. In reviewing the charts of the children in both groups, prospective study subjects were excluded if at any time preceeding or within several months following the interview time the blood lead determinations strayed from the chosen ranges. It was our specific intention to exclude those children with blood lead levels between 30 and 39 μg/dl in order to define more sharply the two study groups.

Children were taken into the groups over the same interval of time. As closely as possible, children in the high and low groups were matched for age, and, as previously stated, area of residence. Socio-economic status and race, two other controlled factors, were determined at the time of the home visit. The family was then approached by a postcard announcement, which was followed up by a home visit by one of the two research associates. Written informed consent was obtained in all instances.

During the home visit, questions, information, and specimens were obtained in the following areas:

(1) The identification of all locations in addition to the home where the child under study has spent "significant" amounts of time over the preceeding four months. We defined "significant" as at least eight waking hours per week for at least four weeks. A four month interval was selected for a retrospective review because it appears likely that in human ingestion, at least in adults, a steady state of exposure may be reflected in blood lead levels in approximately that length of time. A longer time of re-

view, while perhaps desirable, was thought to be impractical, and probably unreliable.

(2) The parent or guardian was questioned about mouthing activities of the child, using specific examples such as finger sucking, mouthing of toys, coins, pencils, or articles of clothing. The parent was asked in similar fashion about specific objects the child had eaten, such as newspaper, cigarettes, or dirt, and whether the child had ever been seen with paint chips in his/her mouth. The child was considered to have pica for paint even if the answer to this question was vague or remote in time. All answers were scored yes or no, and a "yes" answer if the interviewer, herself, observed these behaviors during the visit. Information was obtained about the employment of the mother, the number of years of her education, and whether she was on Welfare. The Medicaid status was usually learned from the medical chart.

(3) While in the house, the interviewer obtained specimens of loose or peeling paint from any area where loose paint could be seen, regardless of its accessibility to the child.

(4) If the child was said to play outdoors, specimens of soil were taken from that area reported to be used most frequently.

(5) Towel wipe specimens were taken in a method we have used previously (Sayre *et al.*, 1974; Vostal *et al.*, 1974). The interviewer proceeded in a specific order. First, she washed her own hands with soap and water and dried them on a specimen towel. This was saved for analysis. The child's hands were then sampled, rubbing both front and back surfaces of both hands. Next, a windowsill sample was taken from any area the family identified as being commonly used by a child in daily play. From such areas, samples of one square foot of floor area were taken. Lastly, prior to departure from the house, the examiner once again sampled her own hands, without previous washing.

LABORATORY METHODS

Paint analyses were prepared by homogenization of the specimens. They were sent to one of two laboratories, depending on the size of the sample. Those of more than approximately 10 mg were sent to the Health Research Laboratories of the State of New York. The smaller specimens were analyzed by the Environmental Sciences Associates of Burlington, Massachusetts. Soil analyses were done in the laboratory of the Environmental Health Sciences Center at the University of Rochester by a single technician, using nitric acid digestion and measurement of lead (as μg/gm) by atomic absorption spectrophotometry. The analysis of towels measured for total lead on the towel. It was done by an overnight elution with 0.1 N hydrochloric acid. The eluate is examined for lead by atomic absorption spectrophotometry.

For purposes of calculation of surface lead exposure of the child in the study, the percentage of time at a given site was approximated based on a maximum of 90 waking hours/week. From each site, the highest towel lead value was taken to represent surface lead level at that site. This somewhat arbitrary decision was made on the rationale that the highest available lead source would provide the closest approximation of the child's maximal exposure. The alternative of averaging surface lead values from multiple sites in one home would tend to dilute the effect of one large exposure source.

In instances in which a child had spent a significant amount of time at more than one site, the percentage time at each site, expressed as a decimal, was multiplied by the highest surface lead value at that site. All such derived values of multiple sites were summed to yield a single lead dust source.

RESULTS

The 49 children and households studied in the high group and the 50 of the low group were, for the most part, well-matched. The residences of all subjects studied was in the same general area of town and had the same composition of census tracts. Thirteen (13) of the sites were located on or near streets that might be considered moderately heavily travelled, that is, over 10,000 average daily transits; seven were in the low group and six were in the high group. The age of the children was close in the high and low groups (Table 1).

The racial balance between black and white was not as close, unfortunately; there was a somewhat higher number of blacks in the high group (Table 1). Blacks numbered 45 in the high group and 36 in the low group. Socio-economic status, as measured by years of education of the mother, employment, and Medicaid status, was fairly close, as illustrated in Table 1. There were slightly more subjects in the high group on Medicaid support (41 compared with 31).

The two groups showed some significant differences in mouthing and pica activity of the children (Table 2). All of these activities appeared more commonly in the high lead group. Differences particularly notable were: eating newspaper, pencils, and reported to have eaten paint chips. More high lead children were said to be dirt eaters. As might be expected, these characteristics appear age-related (Table 3). Paint chip pica, reported in 12 of the 19 subjects of the high lead group under 3 years of age, was reported in seven of the 28 children 3-6 years old. However, since the paint chip question was asked as "Has he ever put paint chips in his mouth," we would not know for certain in how many children ingestion was current.

Among the 18-35 month old children, newspaper eating, finger

TABLE 1. Comparison of the High and Low Groups[a]

Factor	High PbB Group	Low PbB Group	
Age of the child (months)	N	N	
18-35	19	19	
36-47	15	14	p=.93
48-72	15	17	
Ethnic group			
White	1	8	
Black	45	36	p=.02
Puerto Rican	3	6	
Mother's educati-n (grade completed)			
2- 8	9	8	
9-11	27	24	p=.50
12-14	13	18	
Head of household employed?			
Yes	15	21	p=.25
No	34	29	
Medicaid support			
Yes	41	31	p=.02
No	8	19	

[a]p values by chi square analysis.

TABLE 2. Mechanisms of Lead Ingestion[a]

Mechanism of Ingestion (from history)	High PbB Group (n=49)	Low PbB Group (n=50)	p Value[a]
	(percent affected)		
Habitual finger sucking	73.5	50	.04
Mouthing toys	82	58	.007
Chews pencils	50	24	.001
Eats paint chips	43	4	.001
Eats outside dirt	48	20	.001

[a]p values by chi square analysis.

TABLE 3. Paint Chip Pica by Age Group

Age (mos.)	High PbB Group	Low PbB Group
18-35	12 (n=19)	2 (n=19)
36-47	5 (n=15)	0 (n=14)
48-71	2 (n=13)	0 (n=17)

sucking, and dirt eating were not unusual in the low lead group and were seen in about half of the high lead group (Table 4). Among the older children, eating newspaper and dirt tended to persist in the high lead group and were reported less often in the low lead group. Finger sucking, however, was reported in half of the low lead group older than three years.

The household dust values in the high lead group were higher than those of the low lead group when compared by mean values as well as by median values (Table 5). When the house dust scores in the two groups were compared, looking at the percentage of values above and below the combined median of 93 μg, there were approximately twice as many above the combined median in the high lead group. Lead on the hands of the children was also higher in the high lead group than the low by mean values and by median values (Table 6). Twice as many of the high group fell above the combined median as the low.

There was a correlation between hand and household lead levels, with an r value of .25, but a linear relationship could not be seen. Low household levels under 50 μg were usually associated

TABLE 4. Pica and Mouthing by Age Group and Material

Age	Newspaper		Suck Fingers		Eat Dirt	
(mos.)	High PbB	Low PbB	High PbB	Low PbB	High PbB	Low PbB
18-35	11 (n=19)	4 (n=19)	11 (n=19)	10 (n=19)	10 (n=18)	5 (n=19)
36-47	6 (n=15)	1 (n=14)	13 (n=15)	7 (n=14)	8 (n=15)	2 (n=14)
48-71	6 (n=12)	0 (n=17)	12 (n=15)	8 (n=17)	3 (n-11)	3 (n=17)

TABLE 5. Household Dust Lead Levels

Household Dust (μg/sample)	*High PbB Group*	*Low PbB Group*	
Mean	*265 (±288)*	*123 (±160)*	$p \leq .01$
Median	*149*	*55*	
Combined Group Median (93 μg)			
Above median	*32*	*17*	$X^2=9.7$
Below median	*17*	*33*	$p=.005$

TABLE 6. Hand Lead Levels

Hand Pb Level (μg/sample)	*High PbB Group*	*Low PbB Group*	
Mean	*49 ± 69*	*21 ± 28*	
Median	*30*		
Combined Group Median (20 μg)			
Above median	*31*	*16*	$X^2=9.70$
Below median	*18*	*34*	$p=.005$

with low hand values below 20 μg, but the converse was not uniformly true; high dust values did not always correlate with high hand values. This is probably understandable in light of the possibility that children might have been playing elsewhere or possibly washed or wiped their hands shortly before testing. One interesting corollary was that lead found on the hands of the interviewer contained significantly more lead at the termination of the home study in the high group than in the low lead group (10.4 versus 6.2 μg). The values were approximately equal in the specimens taken at the beginning of each house study.

Peeling paint was found in approximately two-thirds of the houses of both groups. We can make no assertion that all of the loose paint found was readily or commonly available to the children, since the decision had been made that samples of any loose paint in any of the homes would be taken.

Paint and soil values are shown in Table 7. On the lower line, the percentage of paint specimens with lead values over 1% by weight are compared. There are almost twice as many specimens

TABLE 7. Significant Sources of Lead[a]

Source of Lead	*High PbB Group*	*Low PbB Group*	*p Values*
Soil (mean, μg/g)	*1563 (±1325) (n=36 samples)*	*1008 (±889) (n=27 samples)*	*.04*
Paint (% of homes with lead ≥1% in peeling household paint samples)	*46*	*26*	*.05*

[a]*Soil Pb values compared by t test, paint values by chi square analysis.*

over this value in the high lead group. Soil lead values, recorded as μg/g of soil, are on the upper line. Soil taken adjacent to the homes of the high lead children contained somewhat more lead than in the low group, that is, 1563 versus 1008 μg/g.

Correlation coefficients calculated using blood lead as the dependent variable show some degree of correlation in most of the factors discussed, but with an r value no greater than .45, indicating that no single variable can explain more than 20% of the variance in values (Table 8).

Finally, a stepwise multiple regression analysis was performed using 23 different variables in the analysis, including several combined variables for a given subject, such as history or persistent finger sucking and/or household dust levels. The multiple regression studies are recorded in two ways, analysis of all children as a whole group (Table 9), and subdivided by the three separate age groups (Table 10). Because of somewhat different representations of black children in the high lead and low lead groups, the analysis has been separately recorded in the figures on the righthand column looking at blacks only. Multiple regression analysis is a useful means of teasing out those factors of many which may account for the largest part of the variance in

TABLE 8. Correlation Coefficients Versus Blood Lead Level

	r
Pica-for-paint	*.45*
Hand lead	*.42*
Household dust lead	*.41*
Eats outside dirt	*.33*
Lead in soil	*.22*
Sucks fingers	*.22*

TABLE 9. *Stepwise Multiple Regression for All Ages. Variables in Order of Significance ($p < .0001$).*

Total Sample $n = 99$ $r^2 = .44$ $F = 12.6$	*Blacks Only* $n = 81$ $r^2 = .40$ $F = 10.8$
1. Hand lead	1. Finger sucking x hand lead
2. Pica-for-paint	2. Pica in general
3. Eats outside dirt	3. Household dust lead
4. Household dust lead	4. Pica-for-paint

TABLE 10. *Stepwise Multiple Regression by Age Group*

Age (mos.)	*Total Sample*	*Blacks Only*
	$r^2 = .77$ $F = 11.1$	$r^2 = .67$ $F = 12.1$
	1. Soil lead	1. Household dust lead
	2. Household dust lead	2. Pica in general
	3. Pica in general	
	4. Eats outside dirt	
	$r^2 = .49$ $F = 10.1$	$r^2 = .40$ $F = 8.8$
	1. Finger sucking × hand lead	1. Finger sucking × hand lead
	2. Pica in general	2. Pica in general
	$r^2 = .77$ $F = 14.1$	$r^2 = .91$ $F = 16.0$
	1. Hand lead	1. Pica-for-paint
	2. Pica-for-paint	2. Hand lead
	3. Finger sucking × household dust lead	3. Finger sucking × household dust lead
	4. Mouthing in general	4. Mouthing in general
		5. Soil lead

blood lead levels. The figure r^2 gives the clue as to the percentage (actually a decimal here) of total variance explained by the factors listed.

From this analysis, it appears that five or fewer factors added together could greatly increase the amount of variance (ranging from 40% for a-1 blacks, to 91% for black children ages 48-71 months). Hand lead level, household dust level, history of pica, and lead content of outside dirt appear to be additive factors. Lead in paint was not selected out, however, as a strong independent factor in any of the multiple regression analyses.

DISCUSSION

The analysis of these data suggests that hand and household lead do correlate with blood lead and appear to be important sources of lead exposure in inner-city children. It further appears that other factors are important, as well. Soil lead, surprisingly, was higher in samples taken from the environs of the high lead group and appears high on the list of factors in our multiple regression analysis of the younger children. The standard deviations of the soil values of both groups were large, however. Lead in interior household paint, though differing in high and low groups, did not come out as an important factor. Mouthing activity was greater in the high lead group, especially among older children studied, which suggests, of course, that lead in the household environment is more likely to be ingested if the child has such a propensity.

How these various sources and mechanisms interplay with one another can be viewed in Figure 1. Lead-containing paint, airborne lead, and soil lead may all contribute directly to the child's exposure through ingestion or respiratory intake. They may all three contribute to dust formation and accumulations on interior household surfaces, being pulverized, or chalked in the case of paint; leaking into the house, in the case of airborne lead; or tracked in, in the case of soil. Surface lead, then, contributes the potential source of hand contamination which the mouthing-prone child can ingest, causing significant additional burden to his other known sources (as food and water) and resulting in mild to moderate increases in his body burden of lead.

IMPLICATIONS OF THE STUDY

I believe we must now look at ways we can reduce household dust lead accumulation. This means, of course, that we continue to monitor the condition of painted surfaces for chipping and flaking areas, correcting such conditions where appropriate. I do not favor continuation of campaigns to strip all the lead from the insides of our houses; sufficient risk of doing so is amply demonstrated. Surely we are as safe in homes with intact leaded paint as we are when we have our x-ray taken in a room with half a ton of lead in its walls. Parents can and should be warned of soil lead risks. We can, if necessary, do soil surveys in our older cities. Our city has planted grass and has sodded around a number of city-owned housing developments, which helps to reduce some of this risk.

Housecleaning and houskeeping touch on some rather sensitive strings, but they are also important factors here. We have some good evidence that tap water, a brush, and paper towels reduce these surface lead values to almost zero values. Studies now in progress suggest to us that surface lead is higher on windowsills

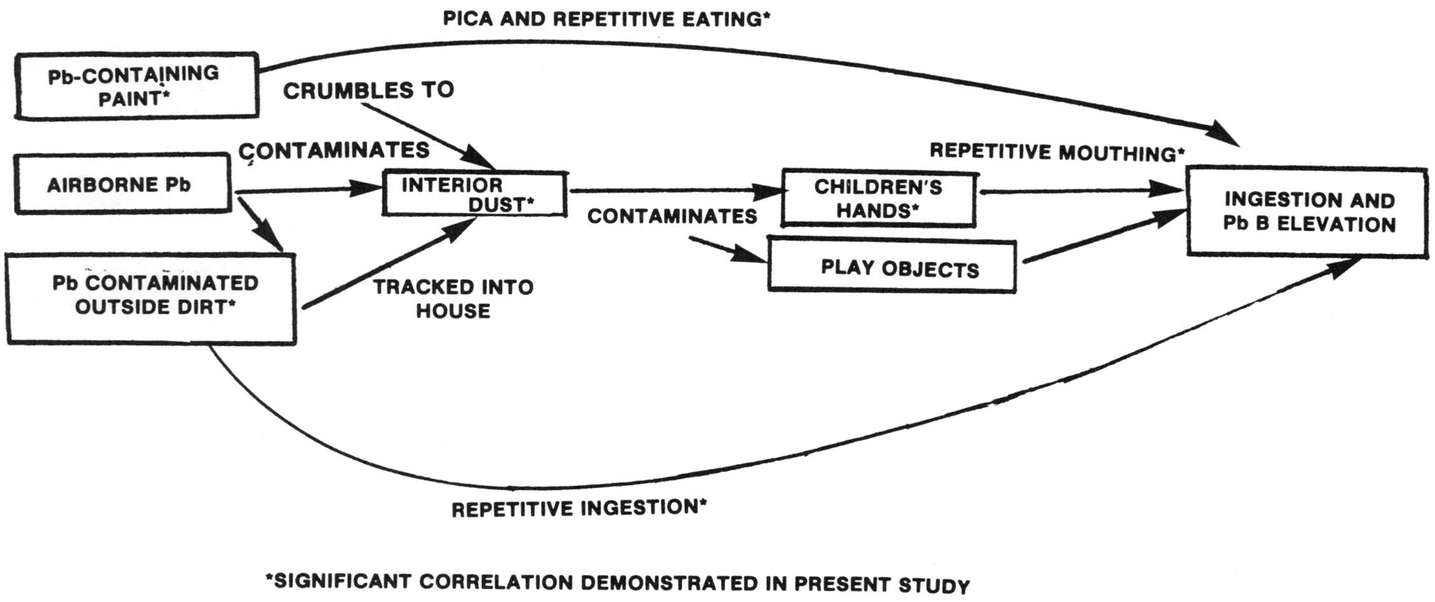

Fig. 1. Possible mechanisms for blood Pb elevation of inner city children.

and adjacent windows and is much lower in homes that have more tightly fitted windows. My personal bias is that the lead exposure of inner-city children may be reduced by means short of condemning large blocks of our cities and putting up more expensive, perhaps less durable, housing.

REFERENCES

Sayre, J., Charney, E., Vostal, J., *et al.* (1974). House and hand dust as a potential source of childhood lead exposure. *Am. J. Dis. Child 127*, 167-170.

Sayre, W. and Katzel, M.D. (1979). Household surface lead distribution: its accumulation in vacant homes. *Environ. Health Perspect. 29*, 179-182.

Vostal, J. J., Taves, E., Sayre, J. W., *et al.* (1974). Lead analysis of house dust: a method for the detection of another source of lead exposure in inner city children. *Environ. Health Perspect. 7*, 91-97.

DISCUSSION OF PAPER BY DR. JAMES SAYRE

Mahaffey: As I'm sure you know, there are a number of reports in the literature indicating higher incidence of lead poisoning cases in warmer months. Have you, or can you, analyze your data to see if there is a seasonal variation in lead content or total dust level?

Sayre: That's a very good point. I must confess I hadn't even thought about that. I think we could do it, but what I'm afraid is that we would find that the variation from one site to the next which we presently observe anyway, would almost overpower the effect of seasonal sampling. Of course during the summer time the windows are open a bit more than they are in the winter time, and I don't know what effect that would have. It might change the value some

Kagey: When the mothers were interviewed about the pica experiences of their children, were they aware of the blood lead levels of their children?

Sayre: The answer is yes. I'm very glad you asked that question. I think that it plays into the reporting that we got from the mother about eating paint. It's a common phenomenon that when you take histories from people and when you ask them repeatedly, as one does with a child who has an elevated blood lead, you tend to get the right answer if you ask the question a number of times. I think that is what we saw with the larger number of children with pica in the high lead group.

Kagey: You'd also mentioned that the interviewer was observing the child during the interview. How did you handle the discrepancy between the mother's answer and the interviewer's answer?

Sayre: We felt that in some cases the interviewer's impressions, the interviewer's observations, right there on the scene, were more valid than what the mother said.

Kagey: Once you gained entry into the home, did you ever think of taking water samples?

Sayre: We did not do that.

Hammond: First, I would like to begin with a comment concerning a study which we completed a year ago. We found that in no instance could we isolate from the feces of children with high blood lead levels, and I'm talking about hundreds of fecal samples, in no instance could we find any paint chips. This occurred in spite of the fact that some of the homogenized fecal samples were grossly elevated as to lead concentrations, not to the point that you would attribute to a paint chip. I would say that a high value was approximately 300 micrograms per gram of stool, which is not really very much. In any event, this also leads to an impression we had from this study that particulate lead is not as important perhaps as lead in some finely divided state.

My question is in regard to the source of the dust and soil lead, that is, apart from chips. How did you sort this out? For example, was there a correlation between interior or exterior paint and lead in dust, interior or exterior? Why not? In an earlier study you did report, I believe, that there was a higher level of lead in dust in pre-World War II houses than in post-World War II houses.

Sayre: In this study we took samples from paint outside the house as well as inside the house. We felt that as long as we were in the home, we might as well get samples of outside paint as well as inside paint samples. As posed in our hypothesis, however, we concentrated on interior sources. However, the data are computerized so that we can examine the exterior paint sources, particularly from railings of porches and from floors of porches, which, incidentally, were enormously high in almost all of the samples that we took. What you see here is entirely interior household sources. There might have been two children, but only one that I can recall clearly, who were eating soil from the flowerpots inside the house. All of the samples reported in these data were from exterior sources, and were collected from the area closest to the place where the parent let the child play. Does that answer your question?

Hammond: I'm sorry, I don't think it does. Maybe I didn't catch the key word. But in any event, was there a correlation between the paint concentration inside the house or even the outer surface of the house and the concentration of lead in the dust inside the house? I ask this because of an earlier paper in which you indicated, from my reading, that for housing which was post-

World War II in the city of Rochester, the dust lead concentration was considerably lower than in the houses in which the paint predated World War II, and by implication was high in lead. So I'm asking whether the dust came from paint or whether the dust came from automobile exhaust.

Sayre: Those comparisons were not made with the present data. They clearly could be. As to the material about the lead content of the older houses as opposed to the newer houses, that was not reported. We do have that material on some housing developments that we studied. However, that has not been published.

Worth: I'm making some comments today that are based on a paper we will present tomorrow. I'm making it for completeness of discussion of Dr. Sayre's work for people who may not be here tomorrow. I would prefer not to go into too much detail about what I'll present tomorrow. First, I'd like to compliment Dr. Sayre on the thoroughness of his analysis of the contribution of mouthing to blood lead in relationship to dust. This has been a deficiency in our study. We recognized, as we began to analyze our data, that we had not asked specific enough questions about pica. However, in the course of a study of the possible contribution of water lead to blood lead, in designing the study, we were aware of the potential confounding possibility for the dust lead in houses. We did study the level of lead in dust in houses where we took water samples for the purpose of analyzing the relationship with water. We did not have such a beautifully refined method for collecting dust samples, and our methods also may be in question. We collected dust samples from vacuum cleaners of the houses of those persons who had vacuum cleaners. In addition, our population was a completely white population in three working class neighborhoods in the Boston area and so represent a different socioeconomic group of people. In these houses we found a positive correlation with dust lead collected from the vacuum cleaners and the blood lead of all residents of all ages. We also found a correlation between a "yes" reply to "does the child have pica?" for children under the age of six and blood lead levels. We were unable to go further with our analysis of the sources of either lead in dust or the way in which the child ingested it. We did have paint measures on all surfaces in the house which the child would ever be in contact, and we found no correlation between maximum paint lead level and dust lead levels. However, in only two of our houses was flaking paint observed, which, again, shows that we were dealing with a different housing population than the population that Dr. Sayre has reported.

We measured distance from major highways, and we were able to find a correlation by the measures that we used between dust lead and proximity to traffic. I don't wish to make further statements other than to say that I do not see this as inconsistent with what Dr. Sayre has reported. In fact, I think that it supports the idea that dust lead is an important household contributor to blood lead and that the role of paint has, perhaps, not been sufficiently explicated and that further studies are necessary. I do believe that it's very important to recognize that a study done in one geographic area in the United States is not necessarily representative of what one would find in other geographic areas. There will have to be additional studies, because of differences in the housing, the traffic, and other potential contributions, including the obvious differences between smelters, rural areas without smelters, and urban areas with and without different traffic. This research really indicates a need for additional research, I believe.

Sayre: Thank you very much.

Hine: Dr., I would like to ask if you have arrived at what you consider to be an acceptable soil lead level. We had an area in California, in the Bay Region, where a housing development was built on a former industrial site. There was no lead paint used in the housing, and the survey of children indicated elevated blood lead levels in about ten percent, that is over 30 μg/dl. The mean was raised considerably. Some children were as high as 50 μg/dl and were treated. The only source was the soil. The lead content in the soil around the area ranged from about 800 to 3,000 ppm, and the health department took quite drastic action about cementing over areas. Do you have an acceptable soil level that you think would be reasonable? Our parks analyzed about 300 to 500, I think.

Sayre: I'm afraid I really don't know a good answer to that because, obviously, its quantity in the soil times amount ingested. I'm sure that in the group that we studied there were children who received, perhaps, 90 percent of their lead body load from soil. As a matter of fact, as the study progressed, we began to tell people at the health center, watch out for the kinds who are eating soil, because, case by case, it was possible to see that the household dust surfaces were low and we so appraised the doctors taking care of the kids to find out if the children are eating a lot of soil. I don't know the answer to your question. I do know that I am comfortable that our values are not as high as those reported for Boston in September at the Washington meeting. We're a little better off than Boston, but I don't know how well off we ought to be.

Dr. Henrietta Sachs: I want to mention an additional source of lead that affects black children in particular, something that we're likely to see when warm weather comes. Many black children, especially the very little ones, do not go out to play. When the warm weather comes, in late April or in May, they stand at an open window and look out at other children playing, and then they reach out their little hands and eat paint from the window ledge, outside the window. At one time, I think during the month of May, we tested about 100 children who had, all of them, positive x-rays of the abdomen to see whether or not there was freshly ingested paint, which might explain, to some extent, this additional source.

Pb γ/towel

	High Pb	*Low Pb*
Windowsills		
Mean	*267.74*	*87.9*
S.D.	*32.8*	*116.*
S.E.	*50.*	*16.2*
Floors		
Mean	*105.4*	*39.*
S.D.	*110.*	*49.*
S.E.	*16.*	*7.*

Slide 23 (left)
Slide 24 (right)

Household Surface Lead

Towel Pb vs. XRF Reading

AREA	*TOWEL Pb*	*XRF*
Dresser top	*8*	*0.35*
Wall	*64*	*0.0*
Ceiling	*4*	*0.0*
Molding of mirror	*32*	*18.2*
Bathroom door	*94*	*18.4*
Sill, inside	*1072*	*1.9*
Sill, outside	*1408*	*20.5*
Bath, room divider	*16*	*17.5*

Towel levels: γPb/towel
XRF readings: mg. Pb/CM² mean of 2 readings

Study 3/77

Sayre: No, we did not do any clinical studies on patients. We drew no blood. We only examined the hands and the households. That is again a very important source and it's a hard one to control. These houses, many of them continually break down, and they shed quite a lot of paint, in addition to some of the old putty that contains also sizable quantities.

Sidney Lerne, University of Cincinnati Medical Center: You indicated a rather small concern for paint that was well-fixed onto the wall, and I wonder if we might be dealing with some paint which is shedding off, wearing off, if you will, and then actually getting into the dust and forming lead in dust from paint. I think this is what Dr. Hammond was talking about. We haven't really clearly identified, at least I don't think so, where this lead in dust is coming from.

Sayre: Well, I think that certainly is a potential source. The studies that our people have done in the health department using the XRF machine suggests that the walls, at least of the houses in Rochester, don't have that much lead in them, at least not compared with some of the other sites. I'd like to show one thing that came out of this. Would the projectionist turn to slide number 23? This slide gives a little insight into this matter of where it comes from. This is a comparison between the lead values found on the windowsills and the lead values found on the floors. The windowsill areas vary a lot, but we found them to be approximately 70-85 percent of a square foot. Now look at the difference between the windowsill values and the floor values. As we began to see this we got very much more suspicious about those windowsills. We then went on to do a couple of more things. The next slide (number 24) is a comparison of the towel lead values and the reading on an XRF machine. You can see that there is no correlation between the amount of lead in the paint by XRF and the surface towel lead on dresser tops, ceilings, or molding

TABLE III. Towel Lead by Distance from Window (number represent micrograms per towel at each location)

Location	Windowsill	Floor Adjacent Window	2 Meters into Room	3 Meters into Room
1	301	384	125	106
2	792	880	37	72
3	933	968	406	347
4	2640	3819	418	354
5	933	669	171	171
6	1267	413	130	200
Median	933	774	150	185
Mean	1144	1189	214	208

Slide 25

of mirrors. Now look at the two values on the windowsills, one taken inside, and one taken outside. Towel levels close, XRF readings greatly differing. Now I'd like to show you another slide (number 25). Here are some studies that we've done on some empty houses, houses not occupied, owned by HUD. In fact, the study has been done with the support of HUD. We measured the lead levels on the windowsills and then the area immediately under the windowsill and then step-wise two meters and then three meters into the room from there. I think you can see there is a considerable drop-off in lead values, and there is also a cer-

and the level immediately underneath the windowsill. We then took our water and apil and did some cleaning. We found that if we washed these places we could reduce by a considerable magnitude, the lead values obtained on these surfaces. Now this lends, I think, some credence to the fact that there is something that's deposited on the surface of the windows, on the windowsill. That's why I think that the question has to remain very much open as to whether the lead we find on the windowsills, doesn't, perhaps, come from exterior sources, very possibly airborne, rather than being specifically related to paint.

Cole: I would like to ask a question regarding that point. Wouldn't it also be possible that since you saw the high levels, relatively high levels, of lead on the outside of the sills, I think the XRF reading was 20 something, that this paint being on the outside is more prone to oxidation, and, perhaps, dusting and could not that in itself be a source. Paint on the inside, not subject to that kind of oxidation might not be the same kind of source when you wipe, just as speculation.

Sayre: Quite so. Quite possible, but still very much unproven.

PROGNOSIS FOR CHILDREN WITH CHRONIC LEAD EXPOSURE

H. K. Sachs

Chicago, Illinois

During the years in which I was associated with the Lead Clinic in Chicago, I was impressed by the absence of neurological sequelae following symptomatic lead poisoning (LP). The opportunity to reexamine these patients after an interval of several years was provided by a grant of funds from ILZRO beginning in 1974.

Since the task of locating highly mobile families is difficult, we channeled our efforts toward finding the 175 patients I had treated between 1966 and 1972 whose maximum blood lead values (PbB) were over 99 µg. To date, we have evaluated 107 of the 175, and located another 25 who have left Illinois or who have failed appointments. We have also evaluated 108 children whose blood lead was 40 to 99 µg, many of whom are siblings of the patients above 100 µg. We examined an additional 49 siblings whose blood lead was below 40 µg and who provide a control group.

During the evaluation, we obtained interval and developmental histories from the mother. In the interval since treatment, the usual febrile illnesses and accidents occurred, but no illness was accompanied by recurrence of symptoms of LP, and no child was reported to have developed neurological impairment, two conditions which have been cited as sequelae to LP.

Each child was given medical, neurological, and psychological examinations and the following laboratory tests: EEG, NCV, urinalysis, hematocrit, hemoglobin electrophoresis, and blood lead determination.

The medical examination revealed an essentially healthy study population. No hypertension was observed, and urinalysis did not indicate any renal problems. Blood lead of the patient groups was higher than that of the controls, averaging 36 against 23 µg/dl.

The neurological examinations were negative. Soft signs, such as mirror movements, adiodochokinesis, or mixed right and left dominance, were present in the same frequency as in the general population and were not correlated to blood lead concentrations.

Nerve conduction velocities of 133 patients and ten controls were all within the normal adult range (Table 1). There was no

ISBN 0-12-460520-6

TABLE 1. Nerve Conduction Velocity Following LP; 133 Patients and 10 Sibling Controls

PbB in		Right Peroneal		Left Peroneal		Right Ulnar		Left Ulnar	
μg/dl	No.	Mean	Range	Mean	Range	Mean	Range	Mean	Range
≥200	8	52.2	48-60	52.7	44-60	60.6	53-68	60.9	56-68
150-199	18	52.1	46-69	53.4	45-64	59.6	50-74	60.7	49-74
100-149	42	51.6	44-61	52.1	45-60	58.4	48-70	57.6	47-66
80- 99	27	52.2	47-57	52.7	47-58	59.1	50-70	57.5	47-67
60- 79	20	51.8	44-62	53.6	48-62	60.1	53-70	58.5	46-70
40- 59	18	53.8	46-65	53.7	43-58	57.7	50-66	58.2	47-64
<40	10	54.3	49-61	54.1	46-66	57.5	50-64	60.3	56-68

TABLE 2. EEG Tracings Following Lead Poisoning; 167 Patients and 33 Siblings

PbB μg/dl	Number	Abnormal	%
≥100	89	16	18
40 to 99	78	13	17
<40	33	4	12

correlation between NCV and previous blood lead.

EEGs were obtained from 167 patients and 33 controls (Table 2). Twenty-nine of the patients and four of the controls had abnormal tracings, but these included about 10 positive spike dysrhythmias, which are not regarded as abnormal by all neurologists. We observed no EEG pattern which we could identify as characteristic of lead poisoning.

Patients and controls were given several psychological tests: the Wechsler IQ, or Stanford-Binet if under 4 years of age, the Beery-Buktenica Visual-Motor Integration, a figure drawing test, and the Ravens progressive matrices, a non-verbal test of perception and abstract reasoning. For comparison of the results, patients were separated into six groups by blood lead concentration. This was a clinical division, based on severity of illness. All patients above 200 μg had had symptoms and had been hospitalized; between 150 and 199 μg, 13 of 29 had required hospitalization; between 100 and 149 μg, only 6 of 60; and below 100 μg, none.

There was no significant difference between the six groups in the IQ means (Table 3). High individual scores were scattered across all lead groups, although the group of 13 children above 200 μg scored lowest. The highest scores were made by the group of 43 patients between 80 and 99 μg, 49% of whom had IQs above 90. Mean IQ scores of the patients were Verbal 84, Performance 88, and Full Scale 85; the means of the controls were 88, 87 and 85.

In the subtests of the Wechsler, we again found no significant difference between lead groups and no pattern of failure characteristic of LP, that is, comprehension, abstract reasoning, or association did not seem to be specifically affected (Table 4). In the tests of visual motor integration, figure drawing, and the Ravens, only the Ravens showed a significant difference, attributed to the low scores of the group above 200 μg.

The effect of symptomatic lead poisoning on IQ was examined in 29 patients who presented with vomiting, irritability, drowsiness, or ataxia. As Table 5 shows, there was little or no difference between symptomatic and non-symptomatic patients, or between patients and unaffected siblings, although the siblings have a slightly higher verbal score.

We also examined the effect of lead lines on IQ. Wide metaphyseal bands are presumptive evidence of lead poisoning. A wide radiopaque band indicates current or recent toxicity; series of bands or lines extending into the diaphysis, that is, the body of the long bone, indicate previous or recurrent episodes of LP. Nineteen of the patients in this study had series of lines when admitted to the lead clinic. Although their verbal scores were at the mean, performance and full scale scores were three points lower. These children may have sustained unrecognized LP for some time before they were detected and treated. Some now attend EMH classes, others are average students, but a few are in the upper percentiles. Thus, even serial lines have not proved to be reliable predictors of academic failure or success.

TABLE 3. Psychological Test Results Following Lead Poisoning; 212 Patients Compared to 47 Unaffected Siblings

PbB in		Wechsler IQ						Stanford-
		Verbal		Performance		Full Scale[a]		
μg/dl	No.	Mean	Range	Mean	Range	Mean	Range	Binet
≥200	13	83	57-100	82	58- 96	81	53- 98	-
150-199	29	83	54-108	88	57-110	84	52-113	92, 90
100-149	60	84	60-117	89	54-113	85	54-114	93, 53
80-99	43	87	62-119	90	64-124	88	64-115	105, 92
60-79	41	85	57-113	87	58-118	85	54-115	-
40-59	26	83	64-109	87	63-108	84	59-106	-
<40	47	88	69-128	87	67-112	85	67-123	107, 62 54, 39

[a]Full scale includes Stanford-Binet scores.

TABLE 4. Psychological Test Results Following Lead Poisoning; 212 Patients Compared to 47 Unaffected Siblings

PbB in		Visual Motor Integration		Figure Drawing[a]			Raven's Percentiles	
μg/dl	No.	Mean	Range	Mean	Range	No.	Mean	Range
≥200	13	76	43- 97	82	61- 97	11	10	4-25
150-199	29	83	58-111	84	50-108	21	29	4-90
100-149	60	80	55-134	86	58-129	48	36	1-99
80-99	43	79	60-122	85	62-107	32	44	4-96
60-79	41	84	55-120	86	61-134	29	41	4-96
40-59	26	86	52-115	85	59-121	22	29	4-96
<40	47	75	43-110	81	50-144	38	30	1-99

[a]Male and female figure drawing scores combined.

TABLE 5. IQ Scores of Patients With and Without Symptoms Compared to Siblings

Symptoms	No.	Wechsler Test of IQ V	P	FS[a]
Present	29	84	87	85
Absent	183	84	88	85
Siblings	47	88	87	85

[a]Combined with Stanford-Binet.

We examined the age when LP was recognized to determine whether the youngest children were more susceptible to intellectual impairment. The 38 who were between 14 and 18 months at diagnosis had the highest mean lead concentrations. Nevertheless, their IQ scores were one point above the mean (Table 6).

The groups between 31 and 48 months, whose mean scores were lowest, may also have sustained longer periods of unrecognized LP. However, the low scores of this group are probably due to a chance preponderance of children who had both pica and genetic retardation, since several had PbBs well below 80 µg.

I should add that culturally deprived children are quite numerous in the population at high risk for LP. Therefore, we must be on guard not to carelessly attribute their low IQ scores solely to increased blood lead concentration. The low IQ scores of inner-city and particularly of rural black children of low socio-economic levels have been adequately documented over the past 50 years.

The effect of a higher socio-economic level is illustrated by 28 patients from privately-owned homes. Although their mean PbB was as high as the children's from rented homes, about 110 µg, their IQ means were significantly higher, in the lower 90s.

During 1977 and 1978, we recalled 100 children, initially above 100 µg, and 28 control sibs for neuropsychological examina-

TABLE 6. Mean Psychological Scores: Age at Diagnosis

Age at Diagnosis (mos.)	No.	No. ≥100 µg/dl	Maximum PbB µg/dl Mean	Range	FS Wechsler & S-B Mean	Range
14-18	38	25/66%	115	52-230	86	53-114
19-24	65	29/45%	114	41-471	85	53-115
25-30	46	23/50%	116	40-365	86	52-113
31-36	28	12/43%	106	48-328	82	54-102
37-48	22	8/36%	100	55-208	82	53-104
49-72	14	4/29%	92	54-168	90	73-115

tion using the Halstead-Reitan battery. These are essentially tests of motor skills and of non-verbal intelligence. Differences between patients and controls were minimal. For example, name-writing speed of the group above 200 μg was the same as that of the controls, 10 seconds.

We also tested finger-tapping rapidity, to which some investigators have attributed considerable importance. Again, the fastest rates were achieved by the group above 200 μg. The controls were slowest, although there were only four points difference between high and low scores. Children with lower IQs tended to score below the mean, but we observed no correlation to PbB.

In a separate study, to examine genetic and cultural determinants of IQ, we compared each of 47 patients to an unaffected sibling. Twenty-three of the sib pairs, at all levels of blood lead concentration, varied in IQ from 0 to 10 points (Table 7). Among the remaining pairs, cohorts in all five blood lead groups were found to have similar IQ means (Table 8).

We have just completed another study in which about 90 children were examined by a pediatric psychiatrist. We are now in the process of examining our data to learn why one particular child in the family developed pica and why his habit of eating paint was ignored until he developed lead poisoning. At this point, depression and emotional withdrawal in the maternal-child relationship are emerging as important factors.

RLB
1

TABLE 7. Analysis of IQs in Patient-Control Sibships by Wilcoxon Matched Pairs Signed-Ranks Test

Test	Patient (P) or control (C)	No. of pairs	Mean scores and range	2-tailed P value
Visual-motor	P	46	83(43-135)	
integration	C		75(43-110)	0.05
Figure drawing - male	P	45	86(61-131)	
	C		82(60-127)	0.06
Figure drawing - female	P	45	84(66-129)	
	C		80(50-144)	0.11
Ravens Percentiles	P	38	36 (4-99)	
	C		30 (1-99)	0.23
WISC-WIPPSI-Stanford Binet*				
Verbal	P	43	87(69-113)	
	C		88(69-128)	0.46
Performance	P	43	90(54-118)	
	C		87(67-112)	0.20
Full scale[a]	P	47	87(53-115)	
	C		85(39-123)	0.50

[a]Stanford Binet score included with full scale.

TABLE 8. Psychologic Test Scores of Sibling-Patient Pairs Grouped by Blood Lead Concentration

PbB in μg/dl	No. of pairs	WISC, WPPSI, and SB IQ	Visual-motor integration	Figure drawings[a]	Ravens percentiles
≥200	6	75	71	81	11
C		72(39, 54)	77	81	29
150-199	9	90(92)*	84	86	40
C		84	72	76	28
100-149	14	88	88	87	48
C		90(62)*	80	86	46
80-90	8	88	86	87	37
C		86	73	80	22
50-79	10	86	79	85	34
C		86	75	83	27
Total	47				

*SB scores.

[a]Male and female drawing combined.

DOES LEAD PRODUCE HYPERACTIVITY IN RATS OR MICE?

R. L. Bornschein
L. S. Rafales
L. Hastings
R. K. Loch
I. A. Michaelson

As many of you are probably aware, several investigators within the Department of Environmental Health including Drs. Michaelson, Rafales, Cooper, and Hastings, have been interested in and involved in the use of animal models for the investigation of possible neurobehavioral toxicity of lead. I would like to summarize some of our recent efforts for you.

Since some may not be fully aware of this literature, I would like to begin with a brief review. In the period between 1972 and 1974, several papers appeared in the literature which purported to establish a link between lead exposure and hyperactivity. The first of these was a clinical study wherein hyperactive children, subjected to diagnostic chelation challenge, were shown to have higher lead burdens than their controls. I am referring, of course, to the work of Dr. Oliver David (1,2).

Following almost immediately after were two reports in animals which indicated that neonatal lead exposure in mice and rats resulted in hyperactivity. Here I am referring to the work of Drs. Silbergeld and Goldberg (3,4) at Johns Hopkins, and Dr. Michaelson (6,7) here at the University of Cincinnati. This afternoon, I would like to review these animal studies and indicate what has happened in this area in the ensuing years. I will begin with the general exposure protocol used in these studies.

The basic experimental paradigm was first developed by Pentschew and Garro (8) about 1965. They showed that when lead carbonate was added to the food supply of lactating dams, lead was transmitted via the dams' milk to the offspring. Little was known at that time about the external exposure level of the pups, or the resultant blood lead levels during the exposure interval. We first attempted to establish what this external exposure was in the various animal models.

Figure 1 depicts the external exposure level of the neonates

ISBN 0-12-460520-6

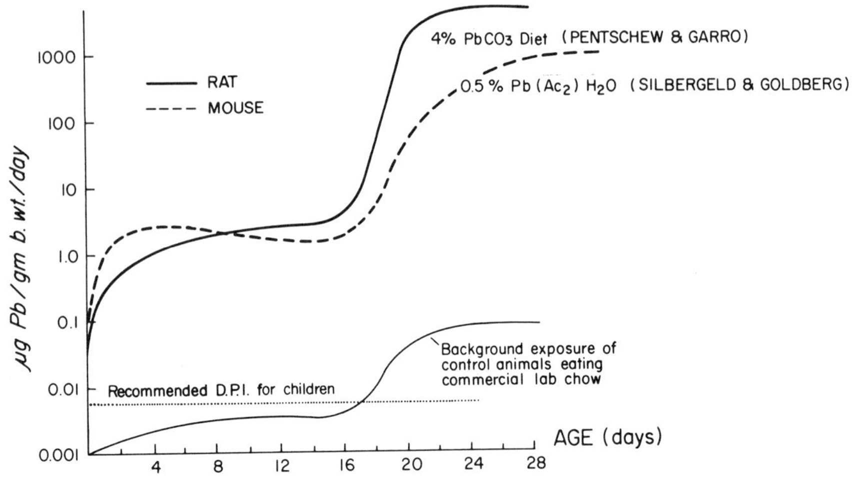

Fig. 1. Daily lead exposure in rat and mouse: pre- and post-weaning.

during the lactation period (cf. ref. 9). On the vertical axis we show the external exposure in μg Pb/gm body wt/day, or mg/kg/day, and on the x-axis we see the age of the animals during the lactation period. The upper solid line indicates the estimated exposure level of these animals during the lactation period, which is approximately 1 mg/kg/day, up through about Day 16. At that age the animals begin to gain access to the food supply and water supply of the dam, which contains lead. We see a marked rise in the external exposure to these pups, such that by weaning, at Day 20-22, we see a very elevated external exposure, in the range of 600-1000 mg/kg/day. Shortly after weaning onto the dams' food supply, these animals became paraplegic, and a large percentage of them died. During the lactation period, when the estimated external exposure is approximately 1 mg/kg/day, the blood lead level is approximately 200 μg lead/dl whole blood.

We've also shown the background exposure level in our control animals nursed by, non-exposed dams. Again, at about Day 16, the pups get into the lab chow. Since there is approximately 0.5-4 ppm lead in the lab chow, we get measurable lead exposure, even in control animals.

A frame of reference relative to childhood exposures is also provided. The maximum daily permissible intake (DPI) has been recommended to be no more than .01 μg/g/day. Our controls are well below what we would expect to see in non-lead exposed children while our lead exposed animals are about two orders of magnitude above this level.

I should also point out that with this exposure model we see a marked reduction of body weight at the time of weaning. This is most likely because of an aversion on the part of the lactating dam for the lead-containing food or water (cf. ref. 10). We see a 30-50% reduction in body weights relative to controls, at weaning.

I would like now to compare, or, perhaps, contrast the testing conditions in these early reports of hyperactivity. Although both authors used the term hyperactivity, a closer scrutiny of the reports shows that they were measuring quite different aspects of behavior.

Figure 2 shows the original data from the Michaelson and Sauerhoff (7) report of hyperactivity. I might note that in this model the lead exposure was 4% in the diet up through Day 16. At that point, the diet was replaced with a diet containing 40 ppm lead. So, the external exposure of the pups remained about 1 mg/kg/day throughout the period. On Days 26-27, and 28, the locomotor activity of rats was measured. An increase in activity of lead exposed rats above controls of approximately 50-100% was observed. The animals were tested in groups of four littermates at a time in their home cage. The testing occurred over a 24-hour period using two separate devices. The apparatus employed was a Selective Activity Meter.® The instrument is sensitive to almost all movement, i.e., to preening behavior, feeding, drinking, as well as ambulation and rearing. We know now that this is not a good apparatus to use in this kind of testing. Each instrument is unique unto itself and it is very difficult to calibrate the two instruments to produce equivalent results.

I'd like to turn now to the reports of hyperactivity in lead exposed mice. In this case, mice were tested between 40-60 days of age. They were continuously exposed via the dams' milk during the first 21 days, and via direct access to lead-containing water following weaning. Testing occurred on three consecutive days, three hours per day. The lead-exposed mice, apparently hyperactive, show a 2-3 fold increase in activity above the controls at

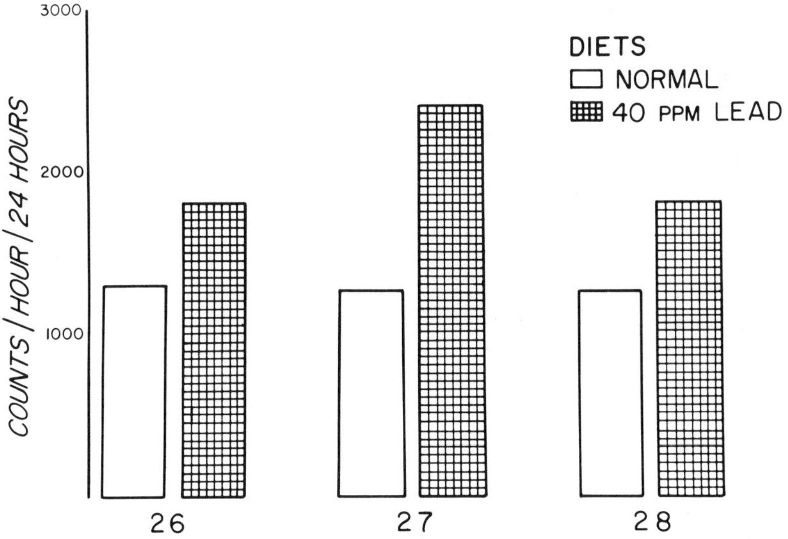

Fig. 2. Total activity counts per 24 hours for lead exposed weanling rats tested as a group of four littermates per test cage.

Figure 3. Motor activity[a] in lead exposed mice.

Hour	Lead Exposed	Controls
1	1280 ± 300	575 ± 187
2	942 ± 80	290 ± 127
3	860 ± 227	197 ± 65

[a]Mean ± SEM N = 21/gp
From ref. 4

any point during testing (see Fig. 3). This magnitude of increase in activity is quite large, being comparable to that obtained in normal animals given a maximum effective dose of amphetamine.

Mice were tested individually. The apparatus, four cages constructed from mouse cages, was sensitive to locomotion only. As the animals stepped from one plate to the next in the floor of the cage, a count was registered. In this case, short-term reactivity to the stress of being removed from a home cage environment to this strange test environment was being measured. It was a quite different situation from the one reported in the Michaelson and Sauerhoff paper with rats.

These two animal reports of hyperactivity suggested to some that lead might be an etiological factor in the production of hyperactivity in children. The appeance of these two reports less than a year after the report by David (2) of elevated body burdens of lead in hyperactive children tended to strengthen this association between lead and hyperactivity.

The following year, Drs. Silbergeld and Goldberg (5) reported the results of pharmacological testing in these same mice. Their hypothesis was that if this was a valid model of childhood hyperkinesis, then the lead-exposed mice should respond in a similar manner as hyperkinetic children. That is, in particular, they would expect a paradoxical response to amphetamine, a calming effect with amphetamine, and an exaggerated response to phenobarbital or excitation following phenobarbital. These latter data appeared to demonstrate a firm link between lead and hyperactivity.

Figure 4 shows the data from both the control and lead-exposed animals subjected to various pharmacological challenges. Notice that in the pre-drug data, the controls have an activity level around 250 counts/hour and the lead-exposed animals some three times more active than that. Following amphetamine challenge in the controls, we see the expected increase in activity in the control mice and the expected decrease in activity following phenobarbital administration. Methylphenidate, also a stimulant, increased activity in the controls.

When we look at the lead-exposed animals, we see that d-amphetamine and l-amphetamine challenge produced what appears to be a paradoxical decrease in activity compared to the control animals and a marked increase in the activity following phenobarbital ad-

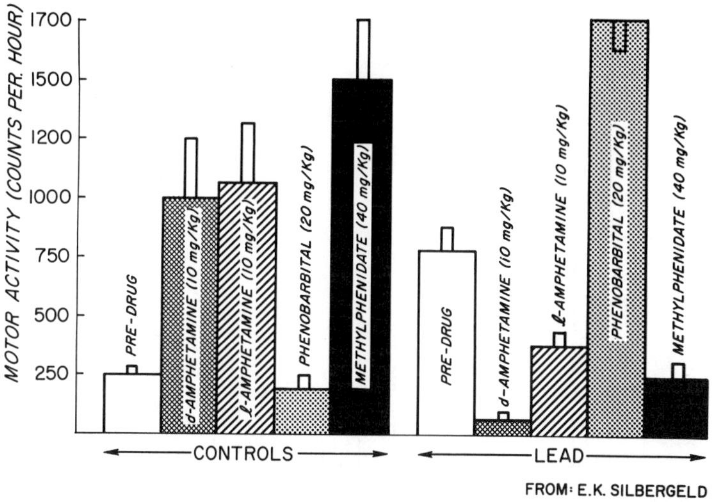

Fig. 4. Effects of drug on motor activity in control and PbAC treated mice. (cf. ref. 5)

ministration. Thus the animal models, particularly the mouse model, appeared to be a valid model for further study of lead-induced hyperactivity.

In the years that followed these reports, many neurochemists, neurophysiologists, toxicologists accepted the validity of the animal models and launched into various mechanistic studies. Many of the investigators refer to their animals as hyperactive, not because they measured the activity themselves, but rather because they had put lead in the drinking water or the diet of the dam and assumed that the offspring would be hyperactive, based on these early reports.

Since 1975, our laboratory has been involved in a series of experiments which have attempted to replicate these early reports, to account for some of the discrepancies and to document more accurately the conditions under which they occur.

Figure 5 illustrates one of our early attempts to replicate the report of hyperactivity in mice. Mice were exposed to 0.5% lead acetate (2780 ppm lead) in the drinking water. The lead concentration in the dams' milk during lactation was about 40 ppm; at weaning the blood lead of the pups was about 200 μg% and they weaned onto this high lead-containing water and continued to consume it. The mice at 35 days of age were placed in test chambers and activity was recorded for a three-hour period. We've employed the same strain, testing paradigm, and essentially the same equipment as was used in the initial reports. We obtained very little evidence of hyperactivity in the lead-exposed mice. If anything, the mice were slightly hypoactive as compared to the controls.

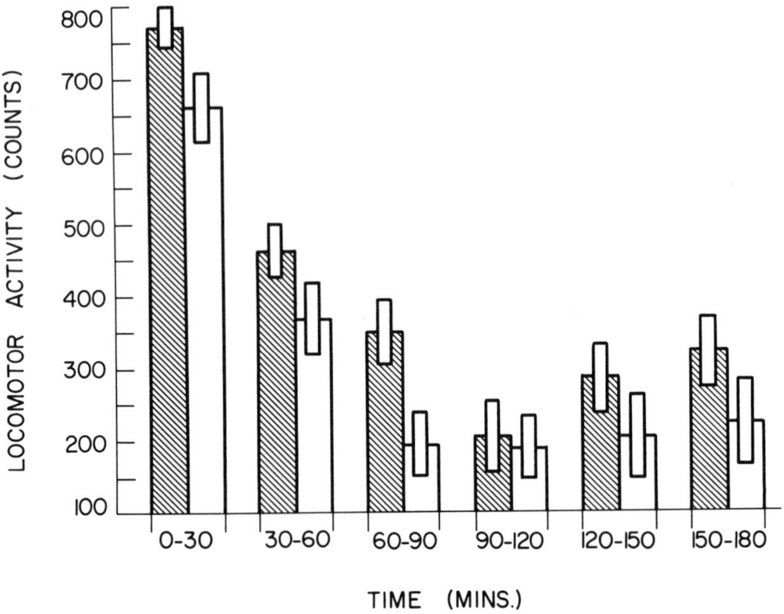

Fig. 5. Locomotor activity in 35 day old Pb-exposed mice: lead in dam's water, control (<.05 ppm ▧), exposed (2780 ppm □).

Another more recent attempt, carried out by Dr. Rafales used larger number of litters per treatment group and tested for a larger number of days, again three hours per day, but for five consecutive days. We've also explored the effects of terminating lead exposure at weaning, as well as continuing lead exposure. Again, we saw no marked hyperactivity, or difference between the two lead exposure groups and the control group.

This was very disconcerting, and a lot of effort went into identifying possible differences between our studies and the early reports of hyperactivity. At the same time we were carrying out this work, Dr. Reiter at EPA-North Carolina was also attempting to replicate the early reports. He was also having little success. After a meeting between Drs. Goldberg and Reiter, and discussing these problems, Dr. Goldberg offered to prepare animals and ship them for subsequent testing in Dr. Reiter's lab.

Figure 6 summarizes the results of several efforts at replication (13). The far left shows the initial report of hyperactivity in mice. The second set of data is from one of our attempts at replication in our laboratory, using essentially the same testing paradigm. We see, if anything, a slight hypoactivity. The third set of data was one of the early attempts by Drs. Reiter and Gray at EPA to replicate hyperactivity in mice. They saw, again, no difference between the experimental and control animals. Finally,

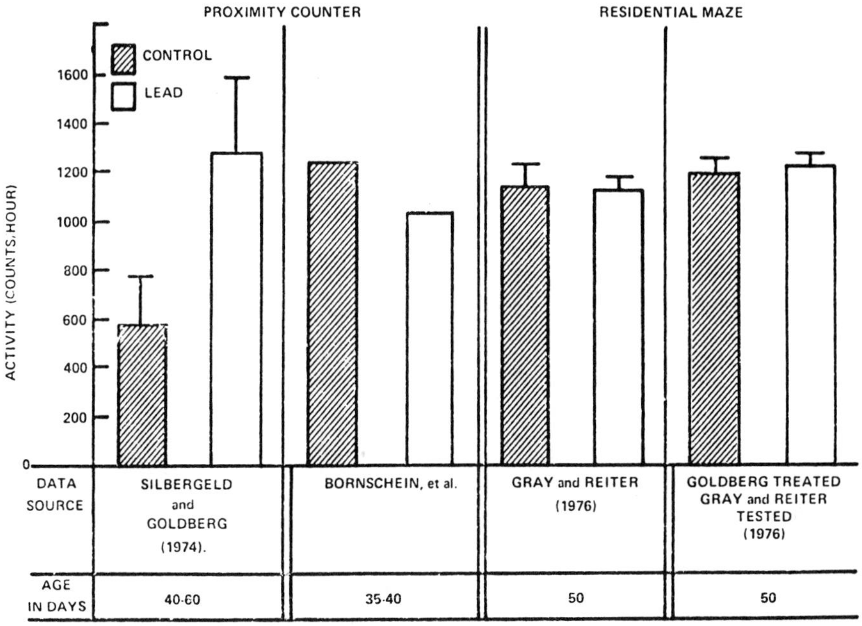

Fig. 6. The effect of lead administration (5 mg/ml in the drinking water) on the first hour of exploratory activity of male (CD-1). Data obtained from different laboratories from L. Reiter.

at the far right, we see animals that were prepared by Dr. Goldberg and sent to Dr. Reiter for testing. Again, they saw no indication of hyperactivity in these animals. Thus, at the present time, there does not appear to be support for the contention that lead induces a state of persistent hyperactivity under the exposure and testing conditions which I have described to you.

What about the reports of the pharmacological responses in these animals? Here, the findings are mixed. Dr. Silbergeld reported a paradoxical response to amphetamine, that is, after amphetamine treatment, the activity was less than that seen prior to amphetamine treatment, an apparent calming effect by a stimulant. We have been unable to replicate that finding, and to my knowledge, no one else has observed a paradoxical response in lead-exposed mice.

However, there have been numerous reports, both in rats and mice, of an attenuated response following lead exposure. Figure 7 shows the activity response in lead-exposed animals prepared and and tested by Dr. Reiter. They were challenged with two doses of amphetamine - 5 and 10 mg/kg. You can see that the lead-exposed animals showed an attenuated response. Their activity increased following amphetamine, but it was less than what was seen in nor-

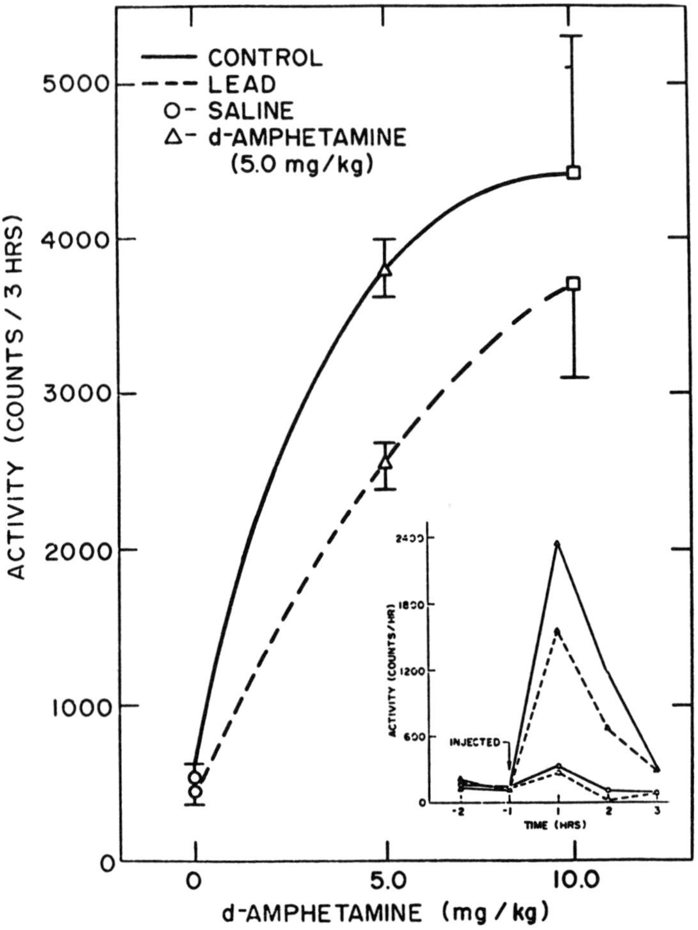

Fig. 7. Locomotor response of lead-exposed and control animals following the administration of saline or varying doses of amphetamine. **From L. Reiter.**

mal animals. The lower right hand corner shows the time course for the 5 mg/kg dose.

We have carried these studies further by examining the amphetamine response of mice subjected to lead exposure only during the first three weeks of life. We are also examining lower exposure levels, that is, mice with blood lead levels in the range of 80-150 μg%. Figure 8 shows the response to amphetamine challenge in controls and two groups of mice exposed to 0.1% and 0.02% lead acetate in the drinking water. The exposure occurred only during the first three weeks of life. The amphetamine challenge and testing occurred when the animals were approximately 150 days of age. At that point, the blood leads were within the control

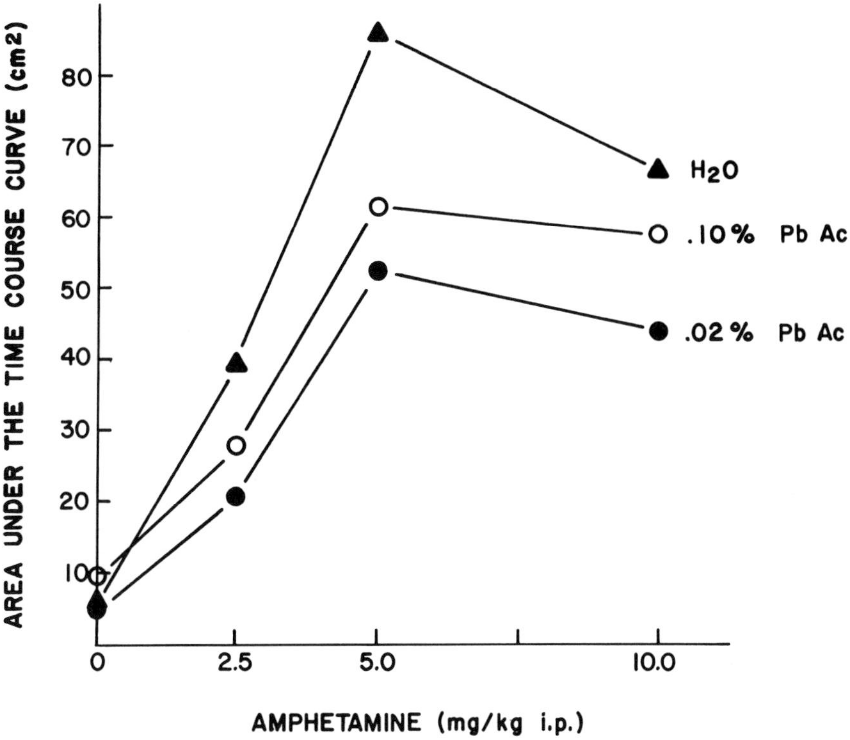

Fig. 8. Locomotor activity response of 110 day old lead-exposed mice tested individually following an infection of amphetamine.

range. We see an attenuated response that persists well beyond the termination of exposure. This effect is evident as early as 21 days of age but does not become statistically significant until about 40 days of age. We feel that this attenuated amphetamine response is a reliable finding and deserves further study.

Dr. Rafales of our laboratory has just completed a study of other pharmacological responses in mice (11). I won't present the data but will merely summarize the findings by stating that the previously reported aberrant responses to methylphenidate, apomorphine and phenobarbital were not confirmed.

We made several observations during the course of the pharmacological studies, which resulted in the study of the role of growth retardation and how it might interact as a confounding variable in these lead exposure studies. You will recall that exposure to 0.5% lead acetate in drinking water produced an aversion to the water by the lactating mouse. This most likely results in a reduction in the amount of milk produced and resultant growth retardation. I would also add that this retardation is a highly variable response, both across litters and across consecutive studies.

We noted that the smallest of the lead-exposed animals appeared to become active and ataxic after phenobarbital administration, whereas the larger of the lead-exposed mice appeared to become sedated, as might be expected. This suggested to us that the undernutrition and subsequent growth retardation might be one of the factors in the seemingly aberrant pharmacologic responses in these animals. In order to further explore this issue, we decided to raise mice in litters of varying size. As a result of competition in large litters, a spectrum of growth rate resulted in the offspring.

In the studies I'm going to report on now, no lead exposure occurred, but the offspring were subjected to the same testing paradigms as had been used in the previous lead hyperactivity studies. Figure 9 shows the growth curves for animals raised in intermediate size litters, that is, approximately eight pups per litter [and in large size litters, of 16 pups per litter], wherein we obtained growth retardation. The points show the mean litter weights, although recall that we do have a whole spectrum of body weights within our large litter groups, the more competitive individuals being equal in body weight to the controls, and the less successful being markedly retarded in body weights.

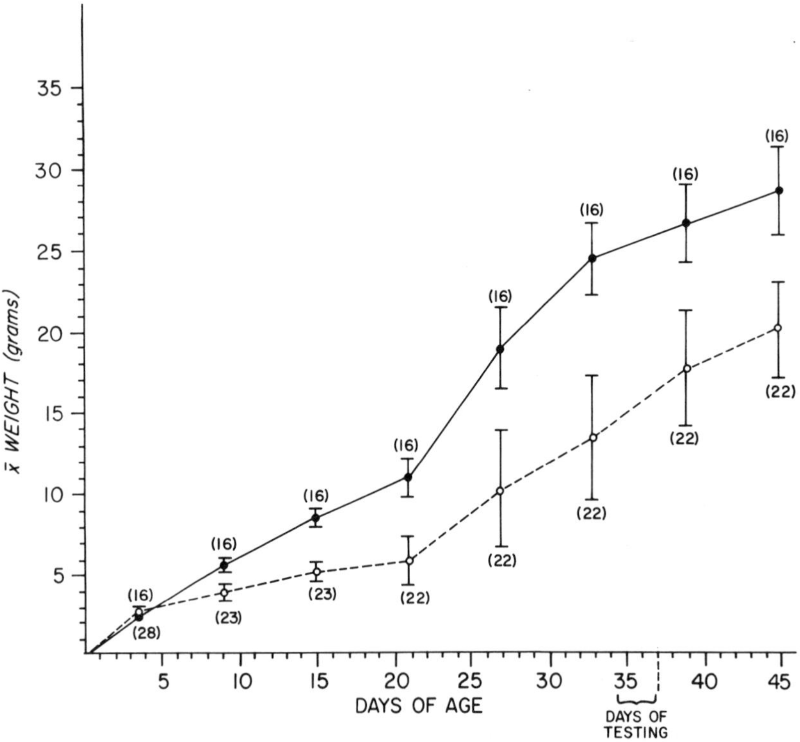

Fig. 9. Growth ($\bar{x}$ ± SD) of CD-1 mice raised in small (●—●, n = 8 pups) and large (o---0, n = 16 pups) litters.

The mice were tested in the interval between 35-37 days of age, comparable to that in previous lead-exposure studies. Figure 10 shows the results of that study. During the first two hours, the animals are habituating or adjusting to the testing environment. You'll notice that during this pre-drug stage, the growth retarded, small animals are showing a slightly reduced activity during the first hour. During the second hour, however, we see an increase relative to the controls. Following drug administration, we see that the group mean activity for the small animals, those raised in large litters, is reduced and is even more markedly reduced during the second hour.

If we examine the data on an individual animal basis, we find an interesting correlation between body weight, activity, and drug response. Figure 11 plots the change in activity against body

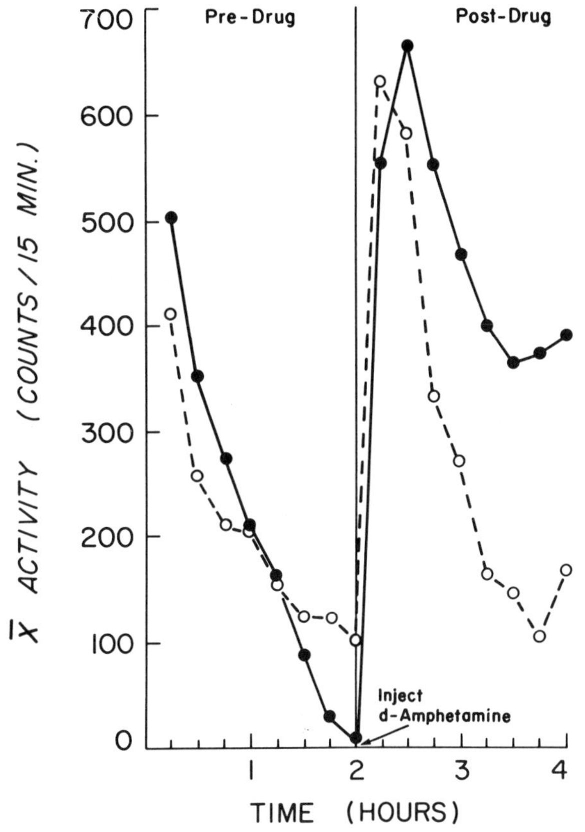

Fig. 10. Pre- and post-drug (d-amphetamine, 10 mg/kg, i.p.). Activity in CD-1 mice raised in large (o---o, n = 16) and small (●---●, n = 8) litters.

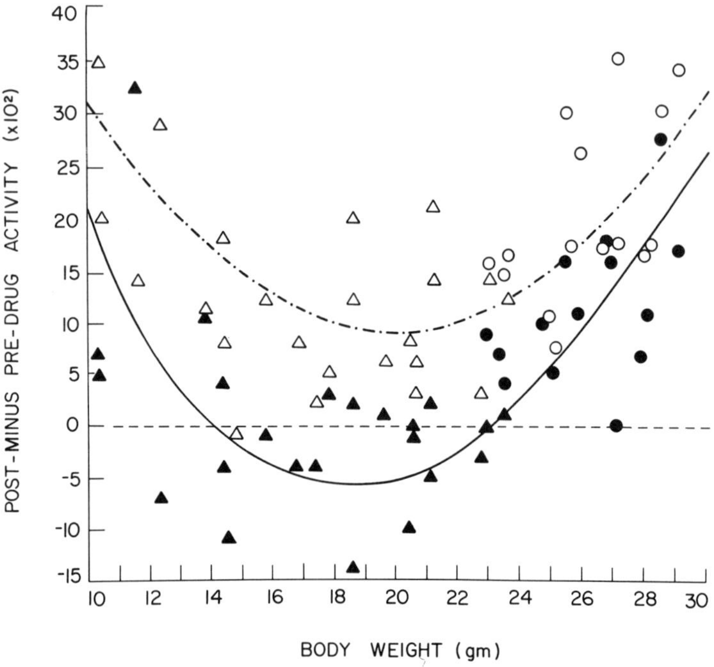

Fig. 11. Relative locomotor activity in mice one hour (-•-•-) and second hour (——) post-d-amphetamine (10 mg/kg, i.p.) as a function of body weight. Large (▲, △, n = 16) and small (●, ○, n = 8).

weight. We've subtracted the post-drug administration activity from the pre-injection, baseline activity, and plotted that against the body weight of the animal. You'll notice that the animals raised in normal litters, of approximately eight pups per litter, show the expected increase in activity; these are the animals indicated on the right hand portion of the graph as circles. We see an increase in activity following drug administration.

If there was no drug effect, the points would fall approximately on the dashed "zero effect" line. Even with normal decreases, we see mice with low-normal body weights show a smaller drug response. When we examine animals whose body weights are approximately 30-50% below controls, that is, equivalent to lead-exposed mice, we find that not only do we see an attenuated amphetamine response, but a fair percentage of the animals actually show a reduction in absolute activity, that is, their activity is now less than it was prior to the injection of amphetamine - an apparent "paradoxical response" to the amphetamine injection.

A few mice with very low body weights appear to show a normal drug response. These animals were markedly growth retarded. We lost a few of them as a result of the injection. It is quite likely that the activity response that we saw was a toxic response occurring prior to death. The 10 mg/kg dose is very toxic to severely growth retarded animals.

Figure 12 summarizes the change in activity following drug challenge. If we make a comparison between this present study and the early reports in lead-exposed animals, we see the following picture. We're plotting the ratio of the drug response over the pre-injection baseline. On the left hand portion of Figure 11, we see the data from the initial reports of lead-induced hyperactivity. We see that the lead-exposed animals during the first hour following amphetamine injection show an attenuated response. During the second hour we can see something that might be interpreted as a paradoxical response.

On the right hand side, we have the data that we've generated in growth-retarded animals, merely by raising them in large litters and retarding the growth rate. We see essentially the same profile, that is, animals raised in large litters with reduced growth rates

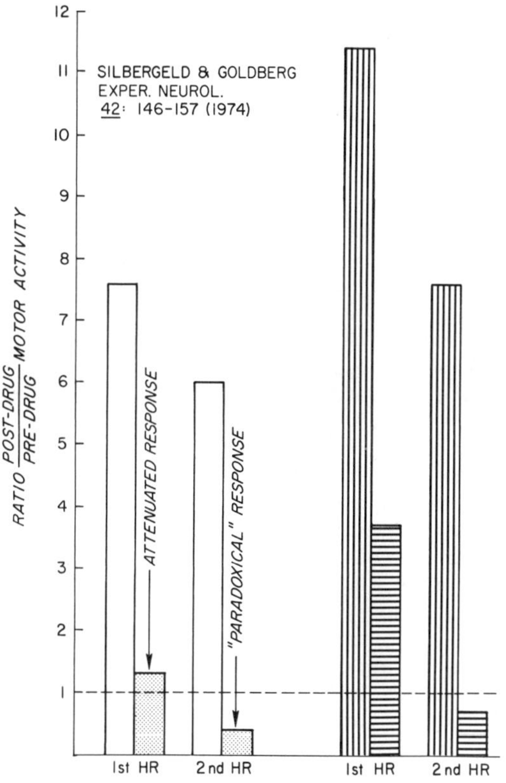

Fig. 12. Effect of methylphenidate (40 mg/kg) on control ☐ and lead exposed ▩ mice (left panel) and d-amphetamine (10 mg/kg) on normal ▥ and undernourished ▤ mice (right panel) methylphenidate data replotted from ref. 4.

show an attenuated response during the first hour after injection and a paradoxical response during the second hour following injection.

These data do not absolutely prove that growth retardation was the sole factor responsible for the behavior of the lead-exposed mice. However, they do indicate that growth-retardation is a confounding factor in most of the previous studies. Age-matched controls are not sufficient in these studies. We are also reminded that a wide diversity of insults, for example, undernutrition, various forms of brain lesions, carbon-monoxide exposure, x-radiation, can result in altered activity levels, if the insult occurs early in development.

I would like to turn attention now to another aspect of these early reports, that is, the report that the animals were irritable or aggressive. Drs. Silbergeld and Goldberg (3,4) have reported that mice exposed to lead have a heightened frequency of fighting, as determined by the incidence of bites observed on litter mate males that were housed together. Sauerhoff and Michaelson have reported an increased aggressiveness in lead-exposed rats during the fourth week of development (7).

In neither case, however, was there a reported attempt to quantitate these observations of increased aggresion. They were, at that point, merely observations. There have been, since that time, several attempts to quantitate various aspects of aggression. Drs. Gray and Reiter have examined the aggressive behavior of mice exposed to lead. The aggressive behavior was measured by introducing adult mice intruders into the home cage of individual experimental male mice. This procedure produces a high level of aggressive behavior in mice. Control males wounded the intruder about 85 times during a 14-hour test period, whereas lead-exposed or pair-fed animals wounded the intruder approximately 32-35 times during a 14-hour test session. If anything, they observed a reduced incidence of inter-male aggression in mice, not an increase (cf. ref. 13).

However, the reduced aggression seen in these experiments cannot be accounted for by lead-exposure alone, since similar reductions were seen in pair-fed controls. So, again, we see an instance where growth retardation during early development may be at least partially responsible for the early reports of altered behavior, although in this case, when an attempt was made to quantify the behavior, what was seen was decreased aggression, not increased.

Dr. Hastings, in our laboratory, has examined irritability-induced aggression. He has exposed lactating rats to lead, 0.02% and 0.1%, during the first three weeks of life, resulting in blood leads below 40 µg%. The lead treatment did not produce a change in growth or development of the offspring. Individual pairs of male offspring derived from the same treatment groups were tested at 60 days of age for susceptibility to shock-elicited aggression.

The lead-exposed groups showed significantly less aggressive behavior than did the controls. It's possible that if the animals had different pain thresholds, they might react differently. However, prior testing indicated the animals had the same pain thresholds.

On the left of Figure 13 we see a set of trials in which the animals adopted a "boxing" stance. The controls adopted an aggressive "boxing" stance approximately 40-42% of the time, whereas the two lead-exposed groups exhibited a decrease incidence of this irritable, shock-elicited aggression. The right hand side shows non-aggressive behavior; these are other body positions the animals might adopt in response to each other following shock. Again, we see that the non-aggressive stances were adopted a higher percentage of the time in lead-exposed animals. Thus, in both the rat and the mouse, a quantitative examination of aggressive behavior suggests that the lead exposure, if anything, causes a decrease rather than an increase in aggression.

I don't want to leave this audience with the impression that neonatal lead exposure in rodents is without effect. Certainly, some of the recent studies wherein lead-exposed rodents have been evaluated in learning/performance paradigms, as well as the

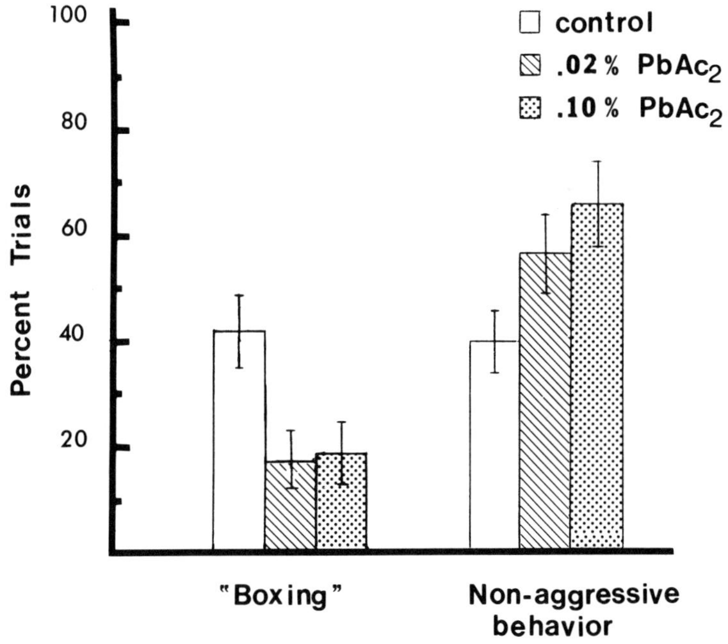

Fig. 13. Frequency of shock-elicited aggression in adult 17 week rats following neonatal lead exposure (n = 16).

reports I've shown here of an apparent attenuated amphetamine response, would indicate that lead is having some effect on behavior. However, these attempts to replicate the early and often-cited studies do not support the notion that neonatal lead exposure in rodents produces an animal model of childhood hyperkinesis (cf. ref. 14). Nor should investigators assume that their exposed animals have altered activity levels without direct substantiation.

REFERENCES

1. David, O. J. "Association between lower level lead concentrations and hyperactivity in children" (1974). *Env. Health Perspect.*, May, 17-25.
2. David, O., Clark, J., and Voeller, K. "Lead and hyperactivity" (1972). *Lancet 1*, 900-903.
3. Silbergeld, E. K., and Goldberg, A. M. "Hyperactivity: A lead-induced behavior disorder" (1974). *Environ. Health Perspect. 7*, 227-232.
4. Silbergeld, E. K., and Goldberg, A. M. "Lead-induced behavioral dysfunction: An animal model of hyperactivity" (1974). *Exp. Neurol. 42*, 146-157.
5. Silbergeld, E. K., and Goldberg, A. M. "Pharmacological and neurochemical investigations of lead-induced hyperactivity" (1975). *Neuropharmacol. 14*, 431-444.
6. Michaelson, I. A., and Sauerhoff, M. W. "An improved model of lead-induced brain dysfunction in the suckling rat" (1974). *Toxicol. Appl. Pharmacol. 28*, 88-96.
7. Sauerhoff, M. W., and Michaelson, I. A. "Hyperactivity and brain catecholamines in lead-exposed developing rats" (1974). *Science 182*, 1022-1024.
8. Pentschew, A., and Garrow, F. "Lead encephalomyelopathy of the suckling rat and its implications on the porphyrinopathic nervous diseases with special reference to the permeability disorders of the nervous system's capillaries" (1966). *Acta Neuropath. 6*, 266-278.
9. Bornschein, R. L., and Michaelson, I. A. "Methodological problems associated with the exposure of neonatal rodents to lead" *In* Behavioral Toxicology: An Emerging Discipline, H. Zenick and L. Reiter, eds., U.S. Govt. Printing Office Washington, D.C., 1978.
10. Morrison, J., Olton, D., Goldberg, A., and Silbergeld, E. "Alterations in consummatory behavior of mice produced by dietary exposure to inorganic lead" (1975). *Devel. Psychobiol. 8*, 389-396.
11. Rafales, L. S., Bornschein, R. L., Michaelson, I. A., Loch, R. K., and Barker, G. F. "Drug induced activity in lead-exposed mice" (1979). *Pharmacol. Biochem. Behav.*, in press.

12. Reiter, L. W., Anderson, G. E., Ash, M. E., and Gray, L. E. "Locomotor measurements in behavioral toxicology: effects of lead administration on residential maze behavior" *In* Behavioral Toxicology: An Emerging Discipline, H. Zenick and L. W. Reiter, eds., U.S. Govt. Printing Office, 1978.
13. Gray, L. E., and Reiter, L. W. "Lead-induced developmental and behavioral changes in the mouse. Presented at the 16th Annual Meeting of the Society of Toxicology, Toronto, Ontario, Canada, March, 1977.
14. Krehbiel, D., Davis, G. A., Leroy, L. M., and Bowman, R. E. "Absence of hyperactivity in lead-exposed developing rats" (1976). *Environ. Health Perspect. 18*, 147-157.

DISCUSSION OF PAPER BY DR. ROBERT L. BORNSCHEIN

Ter Haar: We spent several years arguing about the subject you just talked about here, and now we're talking about learning deficits. Are we faced with the same problem that it has not been adequately evaluated as well?

Dr. Robert L. Bornschein, University of Cincinnati Medical Center: I think that the problem will arise from an inappropriate use of the word learning. A learning paradigm, much as an activity paradigm, measures a large number of behaviors. It is a screening test. There is no reason to expect that a deficit in performance in a learning paradigm should imply an impairment of learning ability in children. We should not try to make that type of extrapolation.

Ter Haar: Could you explain why?

Bornschein: Because, as I indicated, the learning test, as usually run, is a screening test. The decrement in performance could arise as a result of a number of factors--motivational differences, sensory differences, or memory differences. We're obviously picking up some kind of treatment effect when we see a deficit in a "learning task." But that does not mean that we are looking at the ability of an animal to form a pure association, which is the real definition. When you get right down to what learning is, it's the ability to form an association between two stimuli. We're not looking at a unitary behavior in a learning paradigm, anymore than in those tasks where we're recording activity. You saw the diversity of testing situations in what was recorded in those two tasks that were called activity or hyperactivity assessments. In one case, we're talking about individual animals. We're really looking at a reactivity for a short period of time. In another case, we were looking at long-term chronic activity of grouped animals in their home cage. Those are quite different situations.

Ter Haar: Are you satisfied that the investigators who are studying the animals and, what I call a learning deficit, if that's the wrong word for it forgive me, are in universal agreement with you on this?

Bornschein: As to whether or not a learning deficit in an animal would lead to a prediction of a learning deficit in the human?

Ter Haar: Right.

Bornschein: I think that most of them would be very hesitant to go out on a limb like that and predict a learning deficit in a human as a result of that kind of data. That is, the learning studies are initial screens. When a deficit is located, then a whole series of studies have to be undertaken to determine whether the deficit is on the sensory side: Can the animal see the stimuli and see the problem or gather information from the environment? Then we have to look at such things such as memory and motivation.

Ter Haar: Related to lead, do you feel those studies have yet been done and completely defined?

Bornschein: No.

Ter Haar: Thank you.

Lawrence: I assume from your studies with the mouse that you have been using an outbred strain of mouse. Is that true?

Bornschein: These mice are CD-1's, which is an outbred strain. That's correct. The reason for using them was essentially the fact that they were used in the initial reports.

Lawrence: Are they being commercially purchased or bred within the local environment?

Bornschein: The animals in any one study are being commercially purchased.

Lawrence: Have you considered the possibility for the difference between past studies and the present studies to be the possibility of genetic drift?

Bornschein: I think that is a possibility. I think there is reason to use outbred mice in this situation. I think that the use of inbred strains, which would possibly alleviate some of the problems of genetic drift, are more appropriate for studies of mechanisms of action. Once we have an effect, and we want to look at the mechanism, then we want to hold the genetics as tightly as possible. In this type of study, which is essentially a screening study, I think we would want to have the maximum possible genetic representation available in the pool. I might point out that in both the activity studies and in the "learning studies" we do see a fraction of the animals being affected by the lead exposure. In other words, they do not appear to be equally affected. We have some lead exposed animals which perform equally well, if not better, than our control animals in spite of the fact that they have a large lead body burden. On the other hand, we have other animals, again with the same lead exposure, the same lead body burden, that are performing at levels three to five standard deviations outside the control mean. So, it would appear that in these random outbred strains there may be a portion of the animals which are selectively sensitive to the lead exposure.

Lawrence: If you are trying to reproduce earlier studies, though, the genetic drift can play a very definite role. Lately, we've become involved in lead research. Obviously, we've used inbred strains of mice and outbred strains of mice and fortuitously, or whatever the case may be, came across the fact that different inbred strains of mice prefer drinking different levels of lead. At about 15 millimolar, or between 10 and 15 millimolar, you can see different preferences in drinking of the inbred strains of mice as opposed to the outbred strains of mice.

It could also be related to the aggressive behavior of the mice. Some mice have natural genetic predeliction for aggressive behavior as opposed to other strains of mice. For example, we've looked at the CBA/J, which is a naturally passive mouse, as opposed to the C57Bl/6, which is a naturally aggressive mouse. And lead aggraviates those natural tendencies. In other words, the C57Bl/6 mouse becomes more aggressive, and the CBA mouse maintains its passive behavior. There seems to be a slight correlation between the drinking preference and the quantity of lead in the water for those particular strains as well, which might be of interest in your studies.

Bornschein: I think those observations are quite interesting. I think the genetic factor is something that we should be looking at in the animal models. It would be of interest to determine whether or not there is any biochemical marker that we could use which would allow us to predict which animals are going to be affected, regardless of what our behavioral task might be. I think that this is one of the areas that should be pursued in the animal behavioral studies as opposed to concentrating and focusing on behavior, per se. Use behavior as an overt readily quantifiable index of lead effects and then start looking at dose response relationships or interactions of lead with various stress situations to determine those factors which might render one animal more susceptible than another.

Donald R. Lynam, International Lead Zinc Research Organization, Inc.: While I realize that your work has involved animal models, would you care to comment on the earlier studies on humans and hyperactivity?

Bornschein: For the most part, the early reports used rating scales. It is not altogether clear what the behavior was, exactly what was meant by hyperactivity no more so than in the animal studies. So, it makes it difficult to determine what is meant when they say that a group of children are diagnosed as hyperactive. What symptoms are they really exhibiting? I noticed in the Oliver David papers that three different rating scales were used and those rating scales quantitate different aspects of behavior. So we have three different rating instruments used by physicians, by teachers, and by parents, to classify children as hyperactive. There's a real mixture of behaviors included in those groups. So, it's difficult to interpret the data on that basis.

NERVE CONDUCTION AND NERVE BIOPSY IN MEN EXPOSED TO LEAD[1]

F. Buchthal
F. Behse

University Hospital (Rigshospital)
Copenhagen, Denmark

The toxic hazards in our environment are fashionable topics. Lead is one of the most dangerous and insidious metallic poisons known to man. Lead poisoning most commonly occurs after the inhalation of lead fumes or finely divided lead dust. Lead poisoning also follows the ingestion of food or drink contaminated with lead.

An early detectable adverse effect is interference with the production of hemoglobin; another early sign is interference with kidney function. Lead also has effects on the nervous system. Children are most endangered by acute encephalopathy, which is the end result of excessive accumulation of lead in the body. The most common cause is the ingestion of paint with a high lead content in the slums of American cities. A tiny flake of paint from painted plaster, wood, or wall paper may contain as much as 100 mg of lead.

The present study deals with the possible effect of lead on peripheral nerve in adults. Though chronic lead poisoning from industrial exposure is still an occupational hazard, its incidence has steadily decreased. Clinical symptoms and signs referable to peripheral nerve do not occur with blood levels of less than 90 to 100 μg/100 ml.

Our study was motivated by the fact that some investigators found normal maximum conduction velocity at blood lead levels exceeding 80 μg/100 ml (Catton *et al.*, 1970), while others found slight slowing in conduction when blood lead levels were 50 to 60 μg/100 ml (Seppäläinen and Hernberg, 1972; Seppäläinen *et al.*, 1975; Araki and Homma, 1976). In these studies, the incidence of abnormalities or the degree of slowing increased with increasing

[1]*A discussion of this study also appears in the British Journal of Industrial Medicine 36:135-147, 1979.*

ISBN 0-12-460520-6

concentration of lead in the blood. The slowing in conduction is believed to be the first step towards a classical lead neuropathy and to have bearing on behavior. It led to the suggestion adopted at the Second International Workshop on Occupational Lead Exposure in Amsterdam (Zielhuis, 1976), that the permissible blood lead level be decreased from 70 μg/100 ml to less than 60 μg/100 ml.

Since presumably not everyone in this audience is acquainted with the structural and electrophysiological background of a study of this type, let me just give you a brief account of it. A 3-cm long portion of the sural nerve is removed at the ankle for histological study (Fig. 1). This does not harm the subject, since the nerve regenerates and its function is taken over by adjacent nerves. The sural nerve conducts sensory impulses to the central nervous system. In normal subjects, the nerve contains 5,000 to 9,000 myelinated fibers (Fig. 2); they vary in a characteristic manner in diameter, in that there are large fibers 7-14 μm in diameter, responsible for the conduction of tactile and positional information, and small fibers 2-7 μm in diameter that transmit information about painful stimuli and temperature.

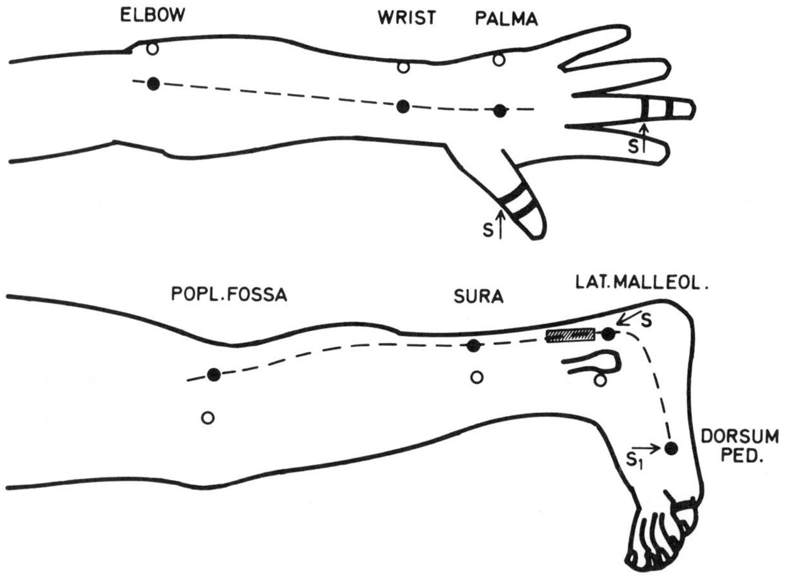

Fig. 1. Placement of the near nerve (●) and remote (o) needle electrodes to record from different segments of the median (above) and the sural nerves (below). "Sura" denotes the site of recording 12 cm procimally to the lateral malleolus. The shaded area shows the 3 to 5 cm long segment of the sural nerve taken in toto as biopsy. (From Buchthal et al., 1975).

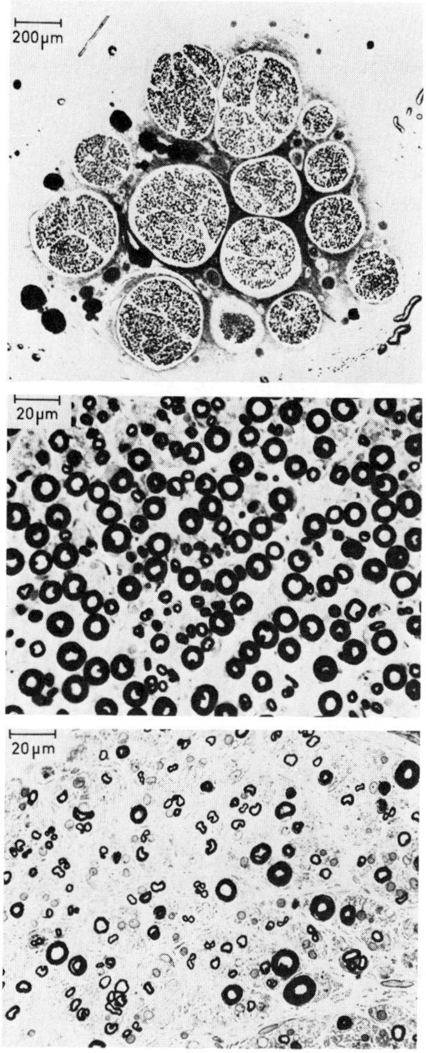

Fig. 2. Above: Cross-section of a normal sural nerve removed in toto just proximally to the lateral malleolus. The endoneurial area was 0.99 mm^2. Middle: Representative sample of the normal nerve (above) at 10 times higher magnification. Below: Representative cross-section of the sural nerve from the patient with lead neuropathy (of Fig. 13). Note the reduced number of myelinated nerve fibers and the clusters of three or more small myelinated fibers indicating regeneration. The method is described in Behse and Buchthal (1977).

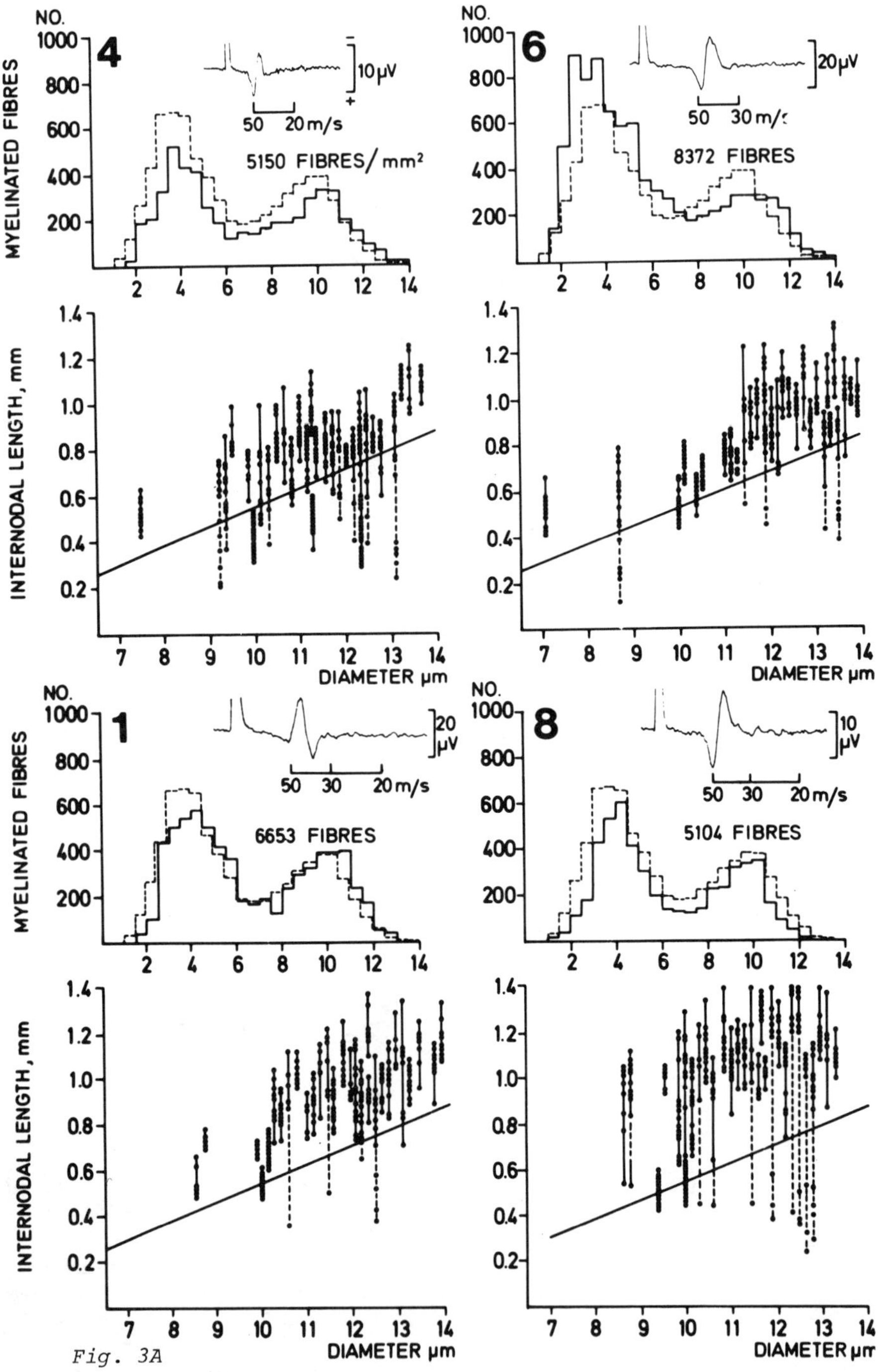
NO.
1000
800
600
400
200
4
10µV
50 20m/s
5150 FIBRES/ mm²
MYELINATED FIBRES
2 4 6 8 10 12 14
INTERNODAL LENGTH, mm
1.4
1.2
1.0
0.8
0.6
0.4
0.2
7 8 9 10 11 12 13 14
DIAMETER µm
6
20µV
50 30 m/s
8372 FIBRES
1
20 µV
50 30 20 m/s
6653 FIBRES
8
10 µV
50 30 20 m/s
5104 FIBRES
Fig. 3A

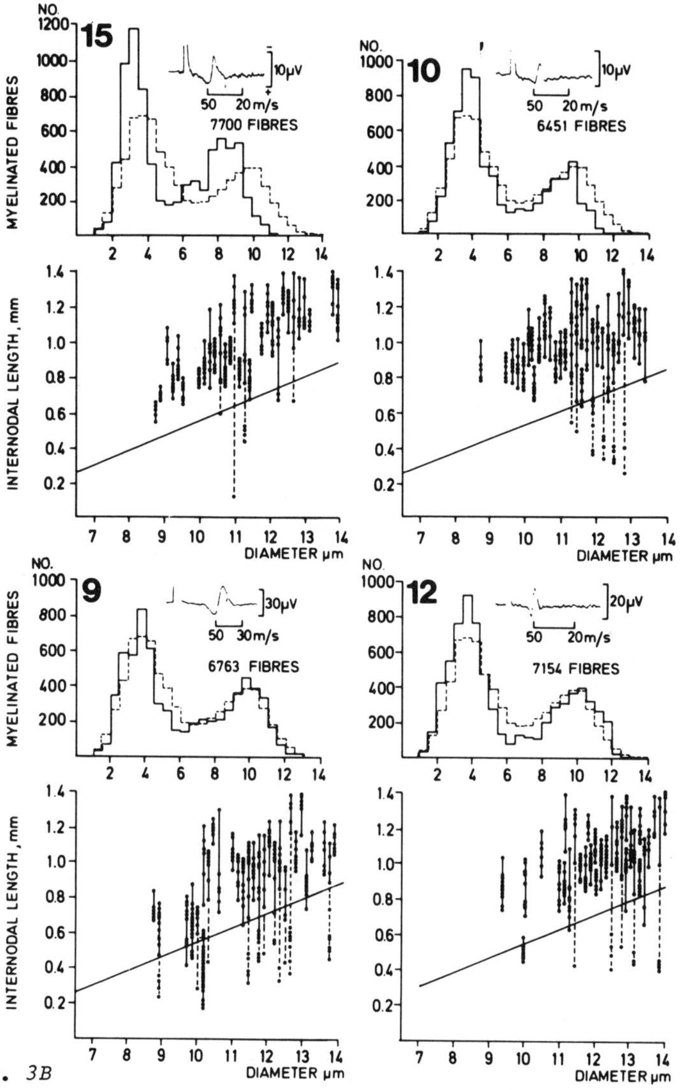

Fig. 3B

Fig. 3A and 3B. Distribution of diameter of myelinated nerve fibers, the sensory potential evoked at the lateral malleolus and recorded at midcalf (inset) and internodal length as a function of maximum diameter of teased fibers. Data for each of 8 sural nerves from men exposed to lead. The plots of internodal length are shown for 30 representative teased fibers. Normal segments are connected by full vertical lines, remyelinated segments by dashed lines. The oblique line is the lower 95 percent confidence limit of normal nerve. The histograms with dashed lines are from controls.

In addition to myelinated fibers, the nerve contains 20,000 to 50,000 unmyelinated fibers. They are much smaller in diameter than the myelinated fibers - 0.2 to 3.0 μm - and in the sural nerve, they transmit information on painful stimuli (Sjö *et al.*, 1976).

The myelinated fiber contains an axon, which connects to a cell in the dorsal root of the spinal cord. The axon is surrounded by a series of contiguous Schwann cells, each about 0.5-1.5 mm long. The region where two contiguous Schwann cells meet along the nerve fiber contains a stretch of 1.5 μm in which the axon is devoid of meylin, the Ranvier node. The myelin sheath is a specialized part of the Schwann cell membrane. It consists of lipids of low conductivity and insulates the nerve fiber electrically. The nerve impulse jumps from Ranvier node to Ranvier node, the exchange of sodium and potassium at the Ranvier node insuring the transmission of the nerve impulse.

There are two types of structural changes in myelinated fibers that occur most commonly in diseased nerve:

1. Axons degenerate and their myelin sheaths disintegrate. This results in fewer nerve fibers (Fig. 2), often followed by regeneration of new nerve fibers. In some disorders, the large fibers are more apt to be affected than small fibers, thereby transforming the bimodal distribution of diameters to a unimodal one. If there is regeneration, the number of small fibers increases, and, though small fibers are also lost, they may appear to be of a normal number because the regenerating fibers are small. Whether there is loss of axons is estimated by counting the number of myelinated nerve fibers in a cross section of the nerve. To determine whether large or small fibers are differently involved, the distribution of fiber diameters is determined (Figs. 3a and 3b). Degeneration of myelinated fibers is associated with a diminution in the amplitude of the action potential recorded from the nerve, because there are fewer generators to produce electricity (Fig. 4). Moreover, since the conduction velocity increases proportionally with the diameter of the fibers (Gasser and Erlanger, 1927), loss of the large fibers results in a slowing in impulse conduction.

2. The other most common structural change which may occur in diseased nerve is a pathological process which is quite different from axonal degeneration. It involves the Schwann cell, which surrounds the nerve fiber, more specifically it concerns breakdown of the myelin sheath (segmental demyelination) without interrupting the continuity of the axon. The loss of myelin usually starts near a node of Ranvier (paranodal demyelination) and may then spread to involve the entire internodal segment (Fig. 5). The axon survives and the nerve is normal on either side of the demyelinated segment or segments.

Whether there is damage to the myelin sheath and how often it occurs cannot be studied on cross sections of the nerve. Abnormalities in the myelin sheath are evaluated in teased fibers, that

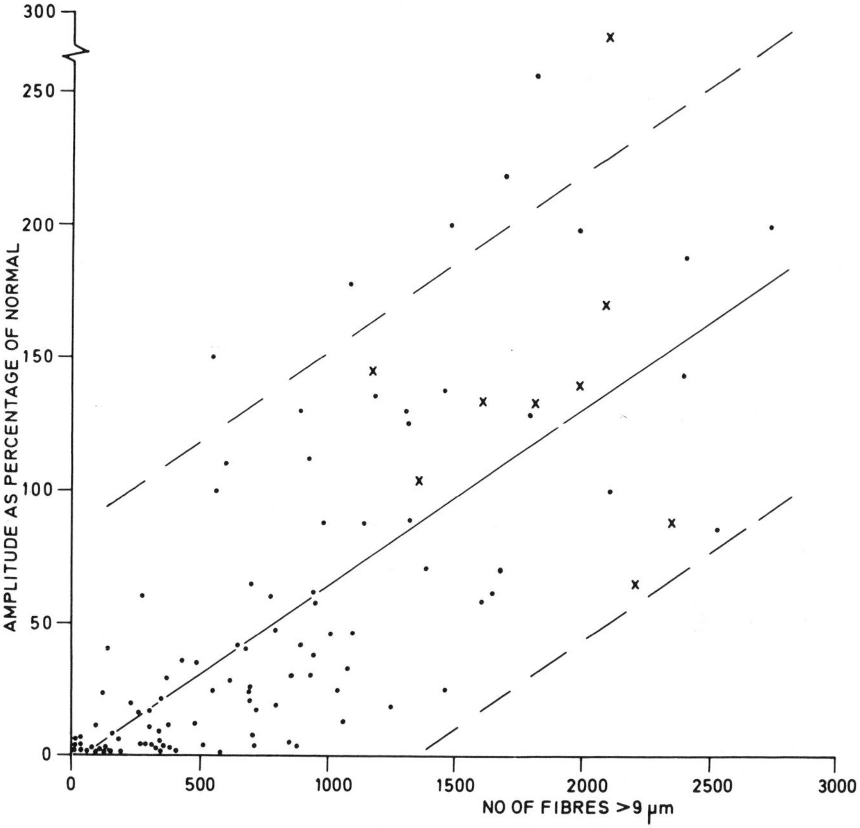

Fig. 4. Amplitude of the sensory potentials in sural nerves, given as percentage of the normal average matched for age, as a function of the number of myelinated nerve fibers more than 9 μm in diameter in patients with acquired and heredodegenerative polyneuropathy (•), in men exposed to lead (•) and in controls (×). The full line is the regression line (P < 0.001); the dashed lines indicate the 95 percent confidence limits. Nerves along which conduction was slowed by 20 percent or more than to be expected from the diameter of the largest fibers are not included. (From Behse and Buchthal, 1978).

is, by isolating single nerve fibers and by counting the number of fibers with segmental demyelination or other abnormalities in the myelin sheath. Also, demyelination manifests itself by a diminution in conduction velocity. A nerve fiber deprived of its insulating myelin conducts the impulse much slower than when the myelin is intact.

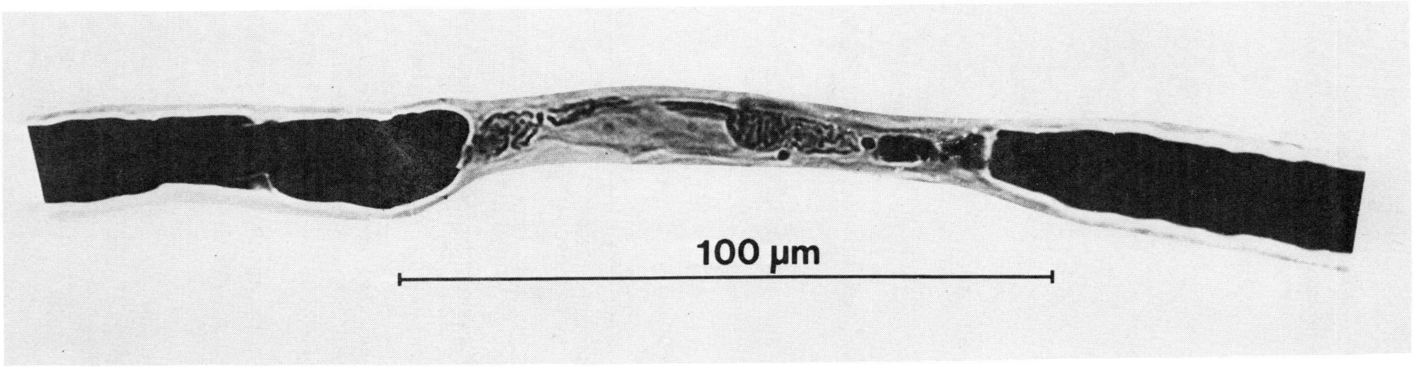

Fig. 5. Nerve fiber teased from the sural nerve to show paranodal demyelination. The nerve biopsy was obtained from a 30-year-old patient one month after the onset of a post-infectious neuropathy. Thirteen percent of 53 teased fibers showed signs of segmental or paranodal demyelination and 4 percent showed evidence of remyelination. The number of myelinated nerve fibers (6430) and the distribution of diameters were normal; the cross sections showed 50 nerve fibers with an abnormally thin myelin sheath also indicating remyelination.

The maximum conduction velocity along the sural nerve was normal, the amplitude of the sensory potential was diminished by 80 percent. Four months later the maximum conduction velocity and the amplitude were normal in the contralateral sural nerve.

One of the first disorders of the myelin sheath studied in teased fibers was experimental lead neuropathy in guinea pigs. Gombault (1880) found segmental demyelination to be the characteristic lesion produced by lead poisoning. This was confirmed in a later study (Fullerton, 1966). Ever since, this finding has been assumed to be valid for lead neuropathy in humans as well. In the context of this paper, there is no need to discuss myelin damage other than demyelination. I wish to mention only that certain mild changes in the myelin sheath are concomitant with early axonal damage (Dyck *et al.*, 1971).

The identification of the type of the primary lesion is complicated by the fact that demyelination is nearly always associated with axonal loss (Behse and Buchthal, 1978). Hence, a diminished number of nerve fibers found in the cross section of a nerve does not exclude demyelination as the primary lesion in a neuropathy. This problem can be solved in two ways, either by determining the incidence of demyelination in teased fibers or by combining the determination of conduction velocity with that of the distribution of fiber diameters in the cross section of the nerve

As mentioned, the maximum conduction velocity along the nerve (Fig. 6) depends on the fibers of largest diameter present in the biopsy. By means of a conversion factor, one can calculate the maximum velocity from the diameter of the largest fibers (Gasser and Erlanger, 1927). If the slowed conduction velocity corresponds to the velocity calculated from the fiber diameter, the slowing is due to fiber loss and the primary lesion is axonal loss. If the slowing is more pronounced than to be expected from the diameter, it is usually due to demyelination.

Finally, there may be mild slowing in conduction velocity that is associated neither with fiber loss nor with demyelination. The cause of this slowing is probably some slight change in the membrane at the node of Ranvier (Behse and Buchthal, 1978).

The subjects of this study are 20 men, 22 to 60 years old, employed in a lead smelting and refining factory for four months to 33 years. Seventeen had been exposed for more than one year and all but one had maximum blood levels of more than 70 μg/100 ml (50-144 μg/ml; average levels: 41-90 μg/100 ml). The blood levels were monitored every two to three months and the lead levels in each sample of blood were determined by: (1) Environmental Sciences Associates, Inc., Bedford, Massachusetts (R. M. Griffin), (2) National Occupational Hygiene Service Ltd., Manchester, England (E. King), and (3) Medi-Lab, Inc., Copenhagen, Denmark (P. Persson) (NIOSH 1977). The lead content in the blood determined by (1) and (2), and later also by (3), showed satisfactory agreement within the variability of lead determinations.

None of the subjects had a disease that predisposed to peripheral neuropathy or had anemia; the hemoglobin was within the range of normal (7.9-10.5 mmol/l). None had a lead line on the gums. A careful history was taken from the subjects and their fellow workers to exclude men who were or had been alcoholic. None complained of any of the effects attributable to chronic ex-

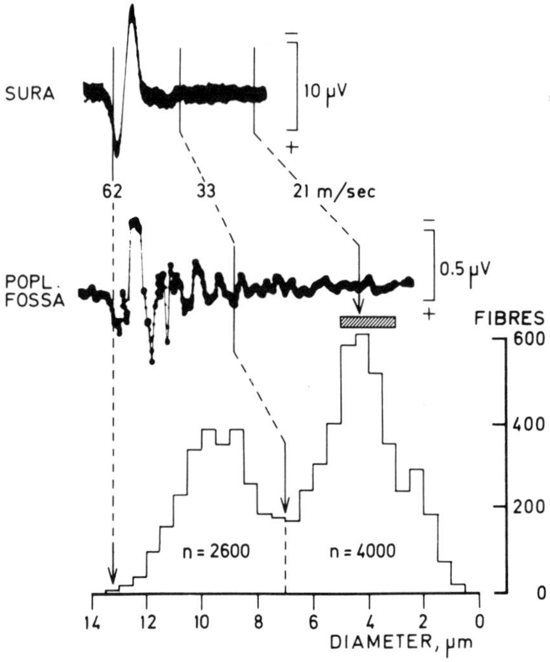

Fig. 6. Components of the sensory potential and distribution of myelinated fibers in a normal sural nerve. Above: The nerve was stimulated maximally at the lateral malleolus and the potentials were recorded 15 cm proximally to it (sura) and in the popliteal fossa by electronic averaging of 200 responses. The temperature near the nerve was 36° to 37°C. Below: Distribution of diameters of all 6,600 myelinated fibers obtained from a biopsy of the same nerve taken 2 cm proximally to the lateral malleolus. Note that the diameter of the fibers is plotted from right to left. The dashed lines point to the corresponding fiber diameter. The conversion factor was determined from the components conducted at 62 m/s and the fibers of 13 to 13.5 μm in diameter (from Behse et al., 1975).

posure to large amounts of lead: fatigue, loss of appetite, headache, tremor, loss of memory, or constipation. There was no weakness and no wasting in the muscles of the extremities and flexion and extension of the neck were performed with normal force.

Discrete neurological abnormalities were found in seven men. One complained of intermittent paresthesia and restless legs, present before he was exposed to lead. Three men had weak or absent tendon jerks, confined to one leg in 2 and to both legs in 1 subject. Four men had absent perception of vibration in the big toes, probably due to compression by the safety shoes; vibratory stimuli were perceived for a normal length of time on the medial malleolus; one also had diminished perception of vibration in the digits.

These signs and symptoms can hardly be considered evidence of neuropathy; they were neither related to the level of lead in the blood nor to a history of colic present in four men. One of these men had a high level of lead (144 μg/100 ml) at the time of the colic. Another man complained of frequent gastrointestinal cramps and diarrhea, probably not related to the exposure to lead (maximum level of lead in the blood 81 μg/100 ml).

Since the slowing in conduction velocity reported by Finnish (Seppäläinen and Hernberg, 1972; Seppäläinen *et al.*, 1975) and Japanese (Araki and Homma, 1976) investigators was significant only when the subjects were taken as a group, we have determined motor and sensory conduction along 12 nerves or different portions of the same nerve. Moreover, since the morphometric study of the nerve biopsy is a more sensitive indicator of neuropathy than nerve conduction, biopsies were obtained from sural nerves of eight of the lead-exposed men.

In the individual man, motor and sensory conduction and the amplitude of sensory and motor responses were normal or borderline (Figs. 7 and 8). The incidence of borderline abnormalities was not related to the maximum or average levels of lead in the blood, to the levels of lead in the blood at the time of the electrophysiological study, or to the time of exposure. An exception is the marked prolongation in distal motor latency (Fig. 7) in the deep peroneal nerve. Sensory and motor conduction in the different nerves of each individual lead-exposed man are shown in Figure 9.

In the group as a whole, the motor latency was longer and the average maximum sensory and motor conduction was slightly slower than in controls (Table 1). Notwithstanding the slowing in conduction, the average amplitude of the sensory potentials was or tended to be higher in lead-exposed men than in controls (Table 2). When all nerves are pooled, the increase in average amplitude is highly significant ("omnibus test," Fisher, 1944), suggesting that the higher amplitude is a general property of the nerves of lead-exposed men. There was no trace of a dose-effect relationship of the abnormalities. Average values of conduction velocities and amplitudes in 10 men with maximum lead concentrations in blood of less than 85 μg/100 ml were the same as in men with concentrations of more than 85 μg/100 ml. Similarly, the average conduction velocity and amplitude did not differ significantly in 10 men with average blood levels of 70 μg/100 ml or less from the average conduction velocity in 10 men with average blood levels of lead of more than 70 μg/100 ml.

The number and size distribution of myelinated fibers obtained by light microscopy of the cross section of the nerve *in toto* and of unmyelinated fibers was within the range of controls (Fig. 10). An increased incidence of clusters of regenerating fibers was present in one nerve. The only abnormality on electron microscopy of unmyelinated fibers was a 3% incidence of fibers undergoing degeneration compared with 1.8% in normal nerve.

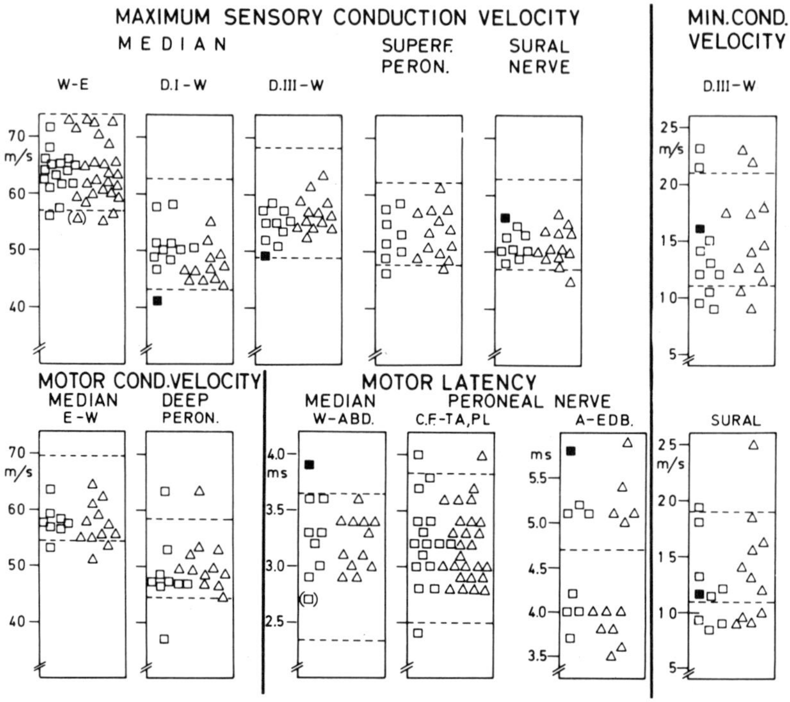

Fig. 7. Maximum sensory and motor conduction velocity (left) and minimum sensory conduction velocity (right) in meter per second and motor latency in milliseconds along the nerves indicated on the Figure in twenty lead-exposed men and in the patient with lead neuropathy (■); □ men in whom a sural biopsy was obtained, △ no sural biopsy; (□) man with carpal tunnel syndrome. The dashed lines denote the lower and the upper 95 percent confidence limits from controls matched for age. *<u>Abbreviations</u>: W, wrist; E, elbow; DI, digit I; DIII, digit III; ABD., m. abductor pollicis brevis; C.F., capitulum fibulae; TA, m. tibialis anterior; PL, m. peroneus longus; A, ankle; EDB, m. extensor digitorum brevis.*

he maximum conduction velocity calculated by means of a conversion factor from the diameter of the largest fibers (Fig. 6) corresponds to that determined from the sensory potential (Fig. 11), indicating that the conduction velocity, whether normal or borderline, is determined by the diameter of the largest myelinated fibers (Behse *et al.*, 1975).

Of special interest are findings in teased fibers, since demyelination has been assumed to be the cause of slowing in man as in the guinea pig (Seppäläinen *et al.*, 1975). In 400 fibers teased from eight nerves, segmental and paranodal demyelination was virtually absent (Fig. 12). Also, other abnormalities in teased fibers were unimpressive. In seven nerves, the incidence

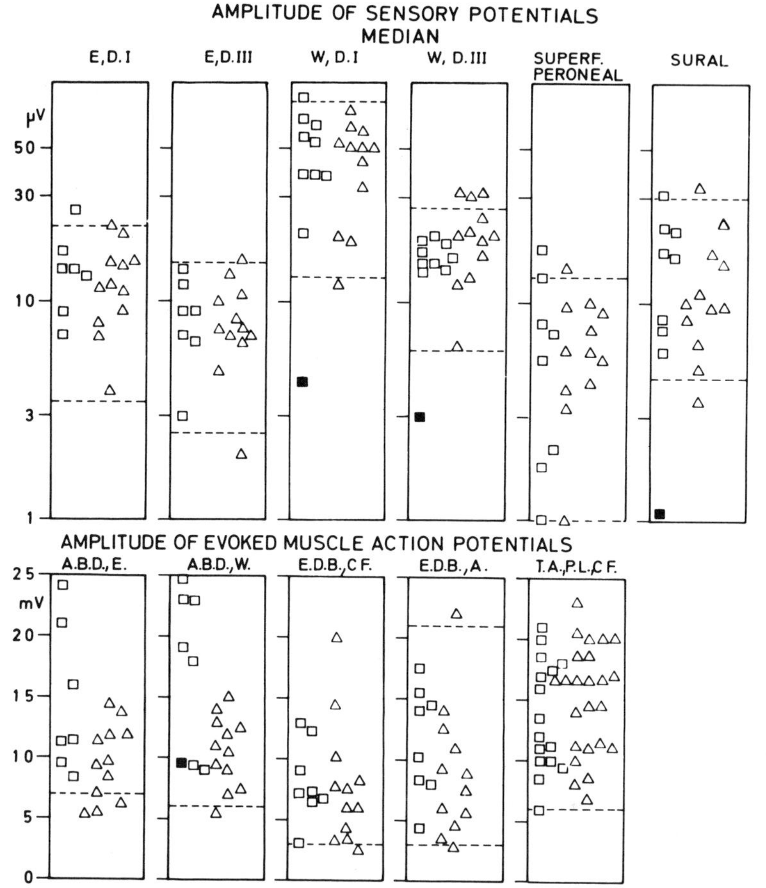

Fig. 8. Amplitude of sensory potentials and of evoked muscle action potentials in 20 lead-exposed men and in the patient with lead neuropathy (■). The dashed lines denote the lower and the upper 95 percent confidence limits from controls matched for age. Abbreviations and symbols as in Figure 7.

of fibers with solitary intercalated segments (for definition see legend of Fig. 12) was slightly in excess of that in controls, and one nerve had an increased incidence of fibers with segmental remyelination at multiple sites. Finally, six of the nerves contained up to 12% of fibers with multiple internodes of normal length but inappropriately small diameter, as described in diabetic neuropathy (Thomas and Lascelles, 1966).

In several respects, the electrophysiological and the histological changes in nerves of lead-exposed men differed from those

TABLE 1. *Mean Nerve Conduction in Lead-Exposed Men and Controls, Matched for Age*

Nerve	Lead-Exposed Men		Controls[a]	Difference		Significance[b]
	No.	Mean	Mean	Mean	S.D.	P
Motor						
Median nerve						
Wrist-m.abductor pollicis brevis[c]	19	3.2 ms	2.8 ms	0.39 ms	0.23 ms	<0.001
Elbow-wrist	19	58.1 m/s	63.6 m/s	-5.50 m/s	4.03 m/s	<0.001
Peroneal nerve						
Ankle-m.extensor digitorum brevis[d]	20	4.5 ms	3.9 ms	0.67 ms	0.77 ms	<0.001
Capitulum fibulae-ankle	20	50.1 m/s	51.0 m/s	-0.90 m/s	5.15 m/s	NS
Capitulum fibulae-m.tibialis anterior[e] and-m.peroneus longus[e]	40	3.2 ms	3.0 ms	0.20 ms	0.32 ms	<0.001
Sensory						
Median nerve						
Digit I-wrist	19	48.5 m/s	54.5 m/s	-5.96 m/s	3.33 m/s	<0.001
Digit III-wrist	19	56.2 m/s	61.2 m/s	-5.05 m/s	3.36 m/s	<0.001
Digit III-wrist, minimum velocity	19	14.9 m/s	16.0 m/s	-1.09 m/s	4.40 m/s	NS
Wrist-elbow (Digit I)	20	63.7 m/s	67.1 m/s	-3.40 m/s	5.09 m/s	<0.01
Wrist-elbow (Digit III)	20	63.9 m/s	67.2 m/s	-3.36 m/s	4.87 m/s	<0.01
Superficial peroneal nerve						
Retinaculum superior-capitulum fibulae	20	53.0 m/s	55.5 m/s	-2.51 m/s	4.50 m/s	<0.025
Sural nerve						
Malleolus lateralis-midcalf	20	50.7 m/s	54.6 m/s	-3.83 m/s	2.98 m/s	<0.001
Malleolus lateralis-midcalf, minimum velocity	19	13.3 m/s	15.0 m/s	-1.65 m/s	4.53 m/s	NS

[a]Obtained from the regressions of nerve conduction as a function of age. [b]NS: not significant. [c]Corrected to a standard distance of 6.5 cm. [d]Corrected to a standard distance of 9.0 cm. Corrected to a standard distance of 10.0 cm (Wagner and Buchthal, 1972). (From Buchthal and Behse, 1979,

[illegible] Evoked Potentials in Lead-Exposed Men and Controls, Matched for Age

Nerve	Lead-Exposed Men		Controls[a]	Log. Difference		Significance[b]
	No.	Mean	Mean	Mean	S.D.	P
<u>Motor</u>						
Median nerve						
m.abductor pollicis brevis, stimulus at wrist	19	13.4 mV	14.9 mV	-0.09 mV	0.19 mV	NS
m.abductor pollicis brevis, stimulus at elbow	19	11.6 mV	11.0 mV	-0.02 mV	0.18 mV	NS
Peroneal nerve						
m.extensor digitorum brevis, stimulus at ankle	20	10.0 mV	10.0 mV	-0.01 mV	0.37 mV	NS
m.extensor digitorum brevis, stimulus at capitulum fibulae	20	8.0 mV	8.0 mV	-0.06 mV	0.24 mV	NS
m.tibialis anterior and m.peroneus longus, stimulus at capitulum fibulae	40	14.5 mV	15.0 mV	-0.04 mV	0.15 mV	NS
<u>Sensory</u>						
Median nerve						
Wrist, stimulus to digit I	19	39.8 μV	35.7 μV	0.05 μV	0.20 μV	NS
Wrist, stimulus to digit III	19	18.1 μV	13.8 μV	0.12 μV	0.14 μV	<0.005
Elbow, stimulus to digit I	19	12.0 μV	9.8 μV	0.09 μV	0.18 μV	<0.05
Elbow, stimulus to digit III	19	7.8 μV	6.4 μV	0.08 μV	0.23 μV	NS
Superficial peroneal nerve						
Capitulum fibulae, stimulus at retinaculum superior	20	5.5 μV	3.6 μV	0.22 μV	0.33 μV	<0.01
Sural nerve						
Midcalf, stimulus at malleolus lateralis	20	11.3 μV	10.7 μV	0.02 μV	0.28 μV	NS

[a] Obtained from the regressions of log. amplitude of the evoked potential as a function of age.
[b] NS: not significant. (From Buchthal and Behse, 1979).

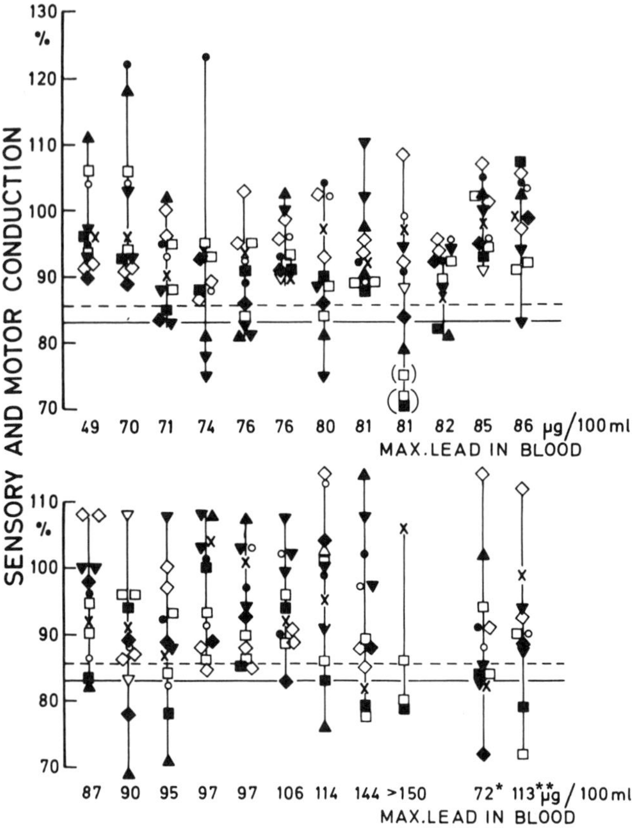

Fig. 9. Sensory and motor conduction in the different nerves of each individual lead-exposed man.

*Each vertical line connects findings in one man. The maximum conduction velocity is given as a percentage of the normal average matched for age; motor latencies are corrected to a standard distance and given as reciprocal latencies (Wagner and Buchthal, 1972). The figures below each vertical line indicate the maximum level of lead in the blood monitored within the past year. The value in the patient with lead-neuropathy, indicated by >150 was extrapolated from urinary lead and coproporphyrin. Below, right: * excluded because of multiple joint swellings; ** excluded because of chronic alcoholism. ( ) data from the median nerve in the patient with the carpal tunnel syndrome. The full horizontal line represents the lower 95 percent confidence limit for motor and the dashed line for sensory conduction. Symbols: Motor nerves: median, elbow-wrist (◆), wrist-m. abductor pollicis brevis (■); peroneal, capitulum fibulae-ankle (●); capitulum fibulae m. tibialis anterior and m. peroneus longus (▼); ankle-m. extensor digitorum brevis (▲). Sensory nerves: median, wrist-elbow (stimuli to digit I and digit III) (◇), digit I and digit III-wrist (□); superficial peroneal nerve, retinaculum superior-capitulum fibulae (○); sural nerve, lateral malleolus-midcalf (×). (From Buchthal and Behse, 1979).*

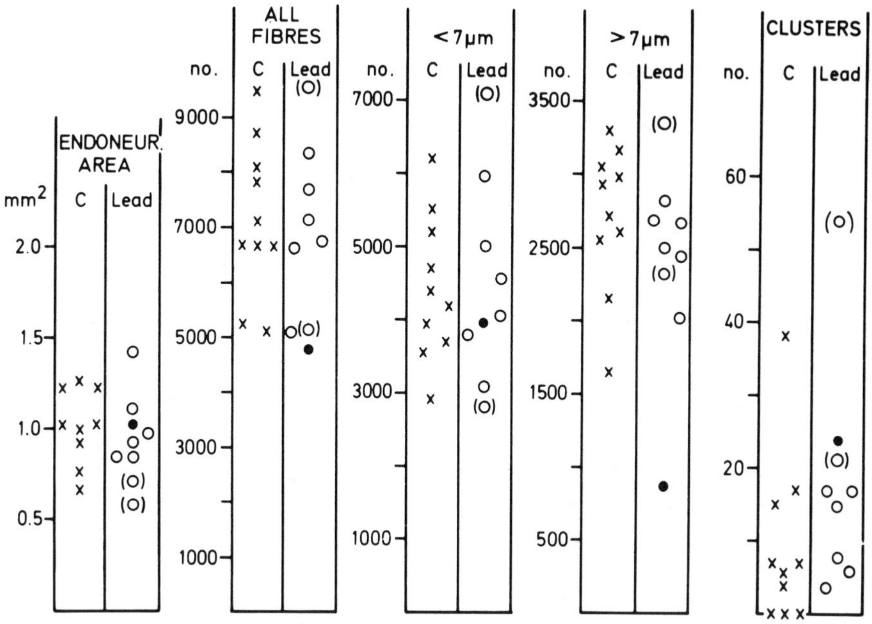

Fig. 10. Endoneurial area, number of all and of small (<7 µm) and large (<7 µm) myelinated fibers, and number of clusters of 3 or more regenerating myelinated fibers in sural nerves of 8 lead-exposed men (o) of a patient with lead neuropathy (•) and in nerves of 8 control persons (×); (o) partial biopsy, numbers are given per mm^2. (From Buchthal and Behse, 1979).

in lead neuropathy. In a patient with excessive blood lead levels and typical signs and symptoms of lead neuropathy, electrophysiological and histological abnormalities are shown in Figure 13. The number of large myelinated nerve fibers was markedly diminished (Fig. 2); there were no abnormalities in teased fibers, particularly no demyelination. This is similar to findings in the baboon (Hopkins, 1970) and differs from those in the guinea pig (Fullerton, 1966). The amplitude but not the conduction velocity of the sensory potential was diminished by 90%, and electromyography showed signs of chronic denervation with loss of motor units and fibrillation potentials.

Unlike the study of Seppäläinen *et al.* (1975), 40 muscles examined by electromyography in lead-exposed men showed neither fibrillation potentials nor signs of loss of motor units. The duration and amplitude of 30 motor unit potentials sampled in each muscle were normal. The only relevant abnormality that could be demonstrated in eight of 20 lead-exposed men was a prolonged latency from the ankle to the extensor digitorum brevis muscle, probably due to local compression by the metal-lined safety shoes (Fig. 14).

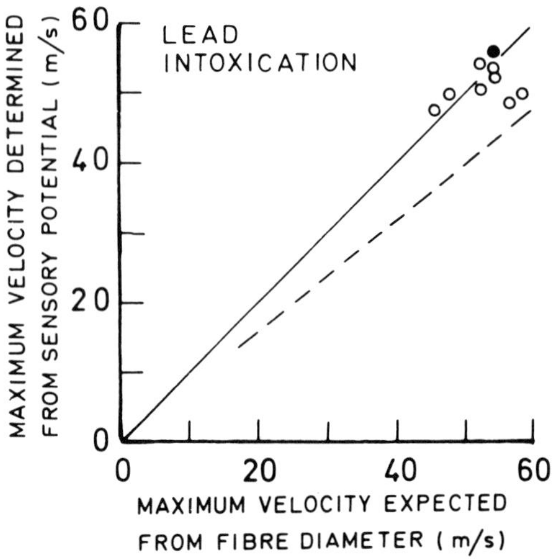

Fig. 11. Maximum conduction velocity along the sural nerve, determined from the sensory potential as a function of the velocity predicted from the diameter of the largest myelinated fibers. In the lead-exposed men (o) and in the patient with lead neuropathy (•). Velocities grouped about the solid line with slope 1, intercept 0 or were within 20 percent of the expected velocity (dashed line, scatter in normal nerves). (From Buchthal and Behse, 1979).

In summary, in 20 lead-exposed men with blood lead levels of 70 to 144 μg/100 ml, motor and sensory conduction in most of 12 different nerves or different portions of the same nerves examined in each man showed a slight slowing when considering the groups of men as a whole. The amplitude of the sensory potential was increased and electromyographic abnormalities were absent. The slight slowing in conduction was without clinical significance and without relevance to the health and well-being of the lead workers.

Since there was neither axonal degeneration nor abnormalities in the myelin sheath to account for the slowing, it seemed to be due to a slight change in the properties of the nodal membrane that might prolong the rise time of the transmembrane potential.

The incidence and degree of the electrophysiological and of the histological change did not show any sign of a dose relationship, nor were they dependent on the time of exposure to lead. Some of the changes in the lead-exposed men were basically different from those in manifest lead neuropathy. It seems doubtful, therefore, that the slight change in conduction should be considered to constitute an early stage in the development of clinically manifest neuropathy.

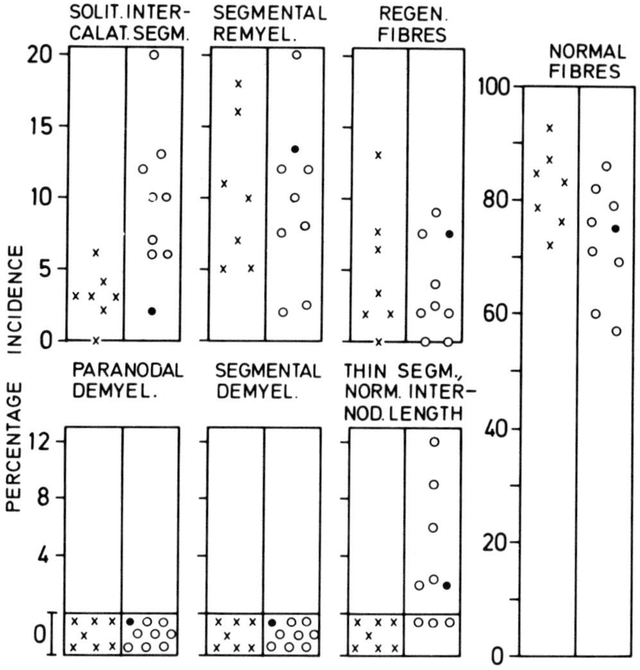

Fig. 12. Incidence of teased fibers with abnormalities from sural nerves of 8 lead-exposed men (○), of a patient with lead neuropathy (●) and of 7 controls (×). Abbreviations: Solitary intercalated segments, i.e., paranodal remyelination. Segmental remyelination. Regenerated fibers. Paranodal demyelination. Segmental demyelination. Thin segm., norm. internod. length: internodes with inappropriately small diameter. (From Buchthal and Behse, 1979).

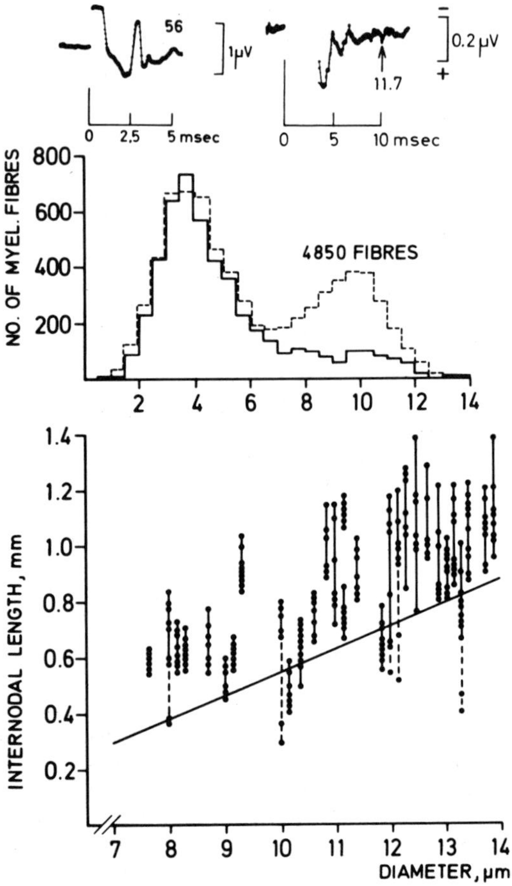

Fig. 13. Patient with lead neuropathy and with more than 150 μg/100 ml lead in the blood. Above: Sensory potential of the sural nerve, evoked by a maximal stimulus at the lateral melleolus and recorded at midcalf; left, shows the 90 percent diminished amplitude and the normal maximum conduction velocity (figures above the trace); right, minimum conduction velocity (arrow and figures below the trace). Middle: Histogram of diameter of myelinated fibers from the nerve of the patient (full line) and from controls (dashed line). Note the diminution in number of large fibers in the patient. Below: Internodal length as a function of maximum diameter of teased fibers, shown for 30 representative fibers. Normal segments are connected by full vertical lines, segments with paranodal remyelination (intercalated segments) by dashed lines. The oblique line is the lower 95 percent confidence limit of normal nerve. (From Buchthal and Behse, 1979).

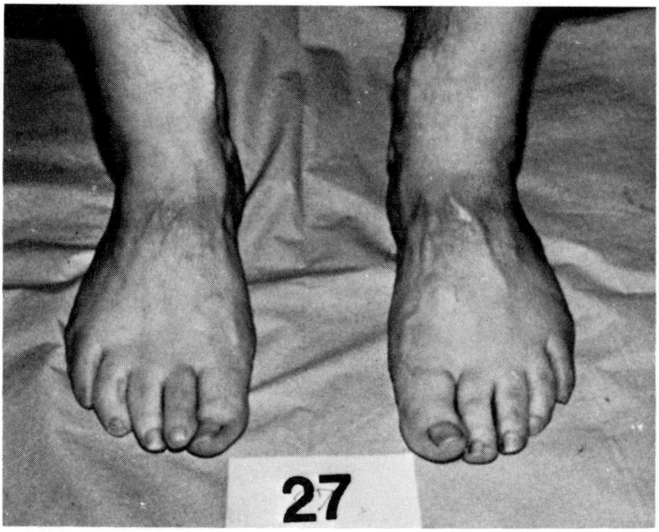

Fig. 14. Example of the pressure bands caused by the metal-lined safety shoes. (From Buchthal and Behse, 1979.)

Acknowledgments

This work was supported by grants from the Michaelsen Foundation, Copenhagen, Denmark, and the Muscular Dystrophy Association, New York; we acknowledge the assistance of the International Lead Zinc Research Organization, Inc., New York, for compensating volunteers for lost income during their participation in the study.

REFERENCES

Araki, S., and Homa, T. (1976). Relationships between lead absorption and peripheral nerve conduction velocities in lead workers. *Scand. J. Environ. Health 4*, 225-231.

Behse, F., and Buchthal, F. (1977). Alcoholic neuropathy: Clinical, electrophysiological and biopsy findings. *Ann. Neurol. 2*, 95-110.

Behse, F., and Buchthal, F. (1978). Sensory action potentials and biopsy of the sural nerve in neuropathy. *Brain 101*, 473-493.

Behse, F., Buchthal, F., and Rosenfalck, A. (1975). Sensory conduction and quantitation of biopsy findings. *In* Studies in Neuromuscular Diseases (K. Kunze and J. E. Desmedt, eds.), S. Karger, Basel, pp. 229-231.

Buchthal, F., and Behse, F. (1979). Electrophysiology and nerve biopsy in men exposed to lead. *Brit. J. Ind. Med. 36*, 135-147.

Buchthal, F., Rosenfalck, A., and Behse, F. (1975). Sensory potentials of normal and diseased nerves. *In* Peripheral Neuropathy (P. J. Dyck, P. K. Thomas, and E. H. Lambert, eds.), Vol. 1, W. B. Saunders, Philadelphia, pp. 442-464.

Catton, M. J., Harrison, M. J. G., Fullerton, P. M., and Kazantzis G. (1970). Subclinical neuropathy in lead workers. *Brit. Med. J. 2*, 80-82.

Dyck, P. J., Johnson, E. W., Lambert, E. H., and O'Brien, P. C. (1971). Segmental demyelination secondary to axonal degeneration in uremic neuropathy. *Mayo Clin. Proc. 46,* 400-431.

Fisher, R. A. (1944). "Statistical Methods for Research Workers," 9th ed., Oliver and Boyd, Edinburgh and London, pp. 1-350, §21, 1.

Fullerton, P. M. (1966). Chronic peripheral neuropathy produced by lead poisoning in guinea pigs. *J. Neuropathol. Exper. Neurol. 25,* 214-236.

Gasser, H. S., and Erlanger, J. (1927). The role played by the size of the constituent fiber of a nerve trunk in determining the form of its action potential wave. *Amer. J. Physiol. 80,* 522-547.

Gombault, M. (1880-1881). Contribution à l'étude anatomique de la névrite parenchymateuse subaiguë et chronique: Névrite segmentaire péri-axile. *Arch. Neurol.* (Paris) *1,* 11-38 and 177-190.

Hopkins, A. (1970). Experimental lead poisoning in the baboon. *Brit. J. Ind. Med. 27,* 130-140.

Seppäläinen, A. M., and Hernberg, S. (1972). Sensitive technique for detecting subclinical lead neuropathy. *Brit. J. Ind. Med. 29,* 443-449.

Seppäläinen, A. M., Tola, S., Hernberg, S., and Kock, B. (1975). Subclinical neuropathy at "safe" levels of lead exposure. *Arch. Environ. Health 30,* 180-183.

Sjö, O., Buchthal, F., Henriksen, O., and Sejrsen, P. (1976). The absence of fibers in the sural nerve mediating vasoconstriction. *Acta. Physiol. Scand. 98,* 379-380.

Thomas, P. K., and Lascelles, R. G. (1966). The pathology of diabetic neuropathy. *Quart. J. Med. 35,* 489-509.

U.S. Department of Health, Education and Welfare (1977). Lead in blood: Physical and chemical analytical method NO. 195. *In* Manual of Analytical Methods, 2nd ed., Vol. 1. National Institute for Occupational Safety and Health, Cincinnati.

Wagner, A. L., and Buchthal, F. (1972). Motor and sensory conduction in infancy and childhood: Reappraisal. *Dev. Med. Child Neurol. 14,* 189-216.

Zielhuis, R. L. (1977). Second International Workshop on Permissible Levels for Occupational Exposure to Inorganic Lead. *Int. Arch. Occup. Environ. Health 39,* 59-72.

DISCUSSION OF PAPER BY DR. FRITZ BUCHTHAL

Dennis Malcolm, Chloride Group: I would like to congratulate Professor Buchthal on this very fine paper. I was responsible for Chairing the editorial committee which reported the findings of the Second Amsterdam Conference. In the First Amsterdam Conference, there was, in fact, a general agreement. However, in the

Second Conference, in one of the drafts, the words "concensus opinion" crept in and I didn't recognize this. It should have been a compromise because there was considerable difference of opinion as to the meaning of nerve conduction time and the views varied from recommending a blood lead standard of 40 micrograms up to 80 micrograms/100 ml. I also queried this question of nerve damage and progressive neuropathy which the Finish work has suggested. On advice I put forward the hypothesis that this might be interference with a biochemical mechanism, namely the membrane transfer. There was no evidence of that, nor was there any evidence of progressive neuropathy. Therefore, I think your work is very timely. The question I would like to ask is this, are these changes reversible? If somebody has a slowing of nerve conduction velocity and is taken off lead work or even chelated would the slowings be reversible? So far as clinical meaning of this change is concerned, I don't believe that they affect the general health or well being of the lead worker. Thank you.

Dr. Fritz Buchthal, Rigshospitalet: I have no follow-up on these people. Most are still working at that factory and they still have episodes of high lead levels- The person with lead neuropathy I've followed over 15 years. There the changes were not reversed, and, of course, he had clear cut clinical signs and symptoms of neuropathy.

Jaroslav J. Vostal, General Motors Research Laboratories: I wonder if you could tell us also some possibilities on compilation of your data on lead neuropathy with some other neuropathies observed in some other diseases. For example, we know very well the patients with diabetes mellitus could suffer from the neuropathy which is explained by deteriorated blood perfusion of the nerves. You have mentioned, and this was brilliant documentation, that it is mainly the diameter of the nerve not of the myelin cover which plays the role. Now what role do you attribute to the blood transfusion?

Buchthal: In old times the lesion in diabetic neuropathy was considered to be axonal loss. Whether this axonal loss is due to vascular deficiency, I don't know. It is very well possible because you have vascular changes in diabetic neuropathy, though I don't believe that this is the sole explanation, because while the vessels are impaired, the blood flow is not impaired to a degree that you can associate with a true hypoxia. Then, after this period of axonal loss, came the period of demyelination. People had seen demyelination in teased fibers and jumped to the conclusion that the demyelination could explain the slowing in conduction velocity in diabetes. Until a few years ago, it has been used as the example par excellence for a demyelinating neuropathy. But this is incorrect. The demyelination concerns, at most, 10 percent of the fibers. In diabetes, there is both a demyelinating lesion and axonal loss. But the slowing in conduction velocity is beyond doubt explicable on the basis of axonal loss. The 10 percent demyelination has no bearing on conduction velocity at all.

Charles H. Hine, University of California: I should like to ask whether you could comment on the validity of the observations which have been made regarding amyotrophiclateral sclerosis and lead exposures. Does the increased amount of lead in the nervous tissue, if it is true that this occurs, does it suggest there's a relationship? Do you have any figures about lead concentrations in axons and nerves, in general?

Buchthal: Kinnier Wilson (*Neurology (London) 1*, 729, 1940) drew attention to certain symptoms and signs of lead poisoning which were very like those of amyotrophic lateral sclerosis. The question was taken up anew by Simpson *et al.* (*J. Neurol. Neurosurg. Psychiat. 27*, 536-541, 1964) and by Campbell *et al.* (*J. Neurol. Neurosurg. Psychiat. 33*, 877-885, 1970). To my knowledge lead has not been determined in the nerves in these studies nor in ours. That lead can be an etiological factor was indicated by the improvement of the otherwise progressing disease by treatment with chelating agents.

Philippe Grandjean, Mount Sinai Hospital: I enjoyed, of course, immensely to hear my former teacher in neurophysiology give this exciting paper. Before I left the country I had the opportunity to study the workers at the, I guess, the same secondary lead smelter, where the patients Dr. Buchthal studied came from. The blood lead levels that the labor inspection and I found were below 150. I believe the highest one was 110, or so. I would, therefore, like to ask where you obtained the blood lead results. Was it from the medical laboratory?

Buchthal: The lead determinations were made in three laboratories, in England, in the United States and in Denmark. At the beginning of our study there was quite a bit of disagreement between the Danish and the English and American data. The English and American data were in agreement. Every person for whom I have given you values had determinations from each of these three laboratories. Where the Danish Laboratory, which initially had difficulties in methodology, deviates from the other two laboratories, I have taken the values from the American and English laboratories.

Grandjean: I think there's no doubt that these 20 workers that you studied had a severely increased lead absorption. But I have some doubt that the blood lead is the best indicator of the exposure. I think that the ZPP measurements that I did on that same plant have been made available to you. If they haven't I'll be glad to make them available to you. I wonder if it was possible for you to correlate your nerve conduction velocities with the measurements of zinc protoporphyrin which were done by my laboratory in that same group of workers? Did you actually have access to these data?

Buchthal: No, I didn't have access to these data. In fact, I thought and I felt that the best measure of the lead exposure really was the lead values in the blood. What evidence do you have that the other values are better?

Grandjean: I'd be happy to make them available to you, too. Another thing I wonder, who were your controls? Were the control subjects patients from the hospital or were they students?

Buchthal: No, they were a group of about 300 people collected over many years.

Grandjean: Matched for age and so on?

Buchthal: The controls are from the same age group; they are the standard controls which we use, which are from normal subjects, from volunteers, and from paraneurological patients who have no neurologic disease, but it is by no means exclusively a hospital population.

Grandjean: The last question I would like to raise is the following. Recently there have been three reports on central nervous system dysfunctioning in lead workers. One was done by Dr. Valciukas and published first in *Science* and later in the *International Archives of Occupational and Environmental Health.* One was done by the Finish group and was published at the Congress in Yugoslavia and also published in the *Journal of Occupational Medicine*. And the third study, I'm happy to say, is a Danish study published in the *Scandinavian Journal of Work Environment and Health*. If you look at these three studies you'll see that there are indications that there is severe CNS involvement at levels below the lead levels that you studied. I wonder if you, in your study of the peripheral nervous system, can see any clue as to the origin of such central nervous system dysfunction?

Buchthal: I have not seen obvious clinical evidence of central nervous system damage in this group of 20 lead workers. We haven't done any more detailed investigation of the central nervous system. The electrophysiological investigation of this type takes about six hours. It was impossible to get these people to submit to other tests or any other investigations, and I don't think we would have got much out of it. What are the tests you mentioned?

Grandjean: Dr. Valciukas used a battery of four tests. Two of them were from Wechsler Intelligence Scale.

Buchthal: They are psychological tests, in other words.

Grandjean: Right, and the Santa Anna dexterity test. He also used the inbedded figures Dr. Bender has described. The Finish people used the same tests and some others from the Wechsler Battery. The Danish study used a complete Wechsler Scale and Ruth Anderson's tests, which were developed at the University of Copenhagen.

Buchthal: I don't know how much confidence I would have in these tests unless these people had been seen before they were exposed to lead.

Vostal: With your permission, Mr. Chairman, I would like to ask a question of Dr. Grandjean from Denmark. I think he has raised a very important viewpoint, and I wonder if he could tell us what is the reason he feels that zinc protoporphyrin levels would be a better indicator of the exposure than the blood lead levels. We don't expect, this is what I assume, that zinc protoporphyrin levels have anything to do with any effect on the peripheral nervous system. Could he explain to us why?

Grandjean: If the Chairman allows me, I'll be happy to answer that question. We are here talking about chronic effects and it's well known that the blood lead level varies from day to day. Lead has a short biological halflife in the blood. It's in the area of about 20 days, the halflife of lead.

Vostal: What about the studies of Dr. Kehoe? We are in Cincinnati.

Grandjean: If I can conclude my remarks, I would add that studies of zinc protoporphyrin have shown that zinc protoporphyrin is relatively stable over long periods of time. For instance, Dr. Valciukas' study showed that zinc protoporphyrin was a better indicator, I'm not saying cause, I'm saying indicator, a risk estimator of the central nervous system dysfunctioning, than was blood lead. It has also been found in other studies that zinc protoporphyrin is a better risk indicator for anemia so it shouldn't be surprising to you, Dr. Vostal.

HUMAN PERFORMANCE IN RELATION TO OCCUPATIONAL LEAD EXPOSURE

G. W. Crockford

London School of Hygiene and Tropical Medicine
London, England

E. Mitran

Institute of Hygiene
Bucharest, Romania

INTRODUCTION

In this paper I am reporting on some of the findings of a project which involved investigators from the factories involved in the study, The National Occupational Hygiene Service Ltd., and the Department of Neurology at the Middlesex Hospital. A large part of the work has already been presented in an unpublished report (Mitran, 1978). However, the views expressed in this paper are my own and do not necessarily represent those of my colleagues.

My own interest in the effects of lead on man were aroused when it was drawn to my attention that nerve conduction velocity (NCV) was said to decrease and then stabilize when workers were exposed to lead. This response looked to me typical of a homeostatic mechanism responding to an environmental stress, i.e., a period of change and then stability at a new level (Schwarz, 1975; Repko, 1976). If this were so, the terms which, at the time, were being used to describe the changes in NVC, such as "damage" and "injury," were inappropriate. A test of whether the nerves were being damaged might be to determine if their functioning was affected in such a way as to influence the performance of workers. If it was not, the results would support the hypothesis that the NCV changes were indicative of one or more homeostatic mechanisms associated with NCV which were operating within their range.

Archibald (1976) in the UK has studied a group of lead workers and shown that their motor nerve conduction velocity (MCV) was reduced (Table 1). At about the same time, we heard Repko (1976) had found that lead workers were showing signs of impairment in a

ISBN 0-12-460520-6

TABLE 1. Motor Nerve Conduction Velocity (m/s), Blood Lead (μg/100 ml) and Age (years) of Control and Lead-Exposed Men in Factory I (Archibald, 1976).

Parameter	*N*	*Exposed*	*Controls*	*P*
Age	*94*	*39±9*	*39±9*	*
Blood lead	*85*	*60±15*	*24±9*	*<0.0005***
Ulnar MCV	*94*	*53.4±4.1*	*55.6±4.3*	*<0.0005***
Median MCV	*94*	*55.9±3.9*	*57.3±3.9*	*<0.01***

*Not significant
**Significant

number of performance tests, the results supporting the damage hypothesis.

In designing an experimental program to measure differences in performance between groups of people or within a group, it is most important to be sure that the differences one is looking for were not there before the exposure. In the world in which we live, pressures are acting on the population which grade people in terms of many performance parameters. It cannot be taken for granted, therefore, that some form of selection for ability is not operating within the factory or between factories.

As it happened, the factory that Archibald had used for her study, and in which she had demonstrated the reduced MCV, was unusual in that on entering the factory, new employees do the job that is vacant, they have no choice of job, and stay there at least three years. The conditions of employment are good and all workers are paid the same, so that there appears to be no monetary inducement to move to other jobs within the factory. The work force is also very stable due to wages and conditions of work being good compared with neighboring factories.

It appeared, therefore, that there were sound reasons for believing that those pressures that might lead to stratification of the work force were absent and that we might be dealing with a randomized sample vis-a-vis ability and lead exposure. We had, therefore, almost the ideal population for a performance test study. Farther south there was another factory belonging to the same company, which we were able to investigate, also.

METHODS

Six performance tests were done by 64 exposed workers, with a mean blood lead (PbB) value of 52 μg/100 ml, at factory I and 73 control workers, with a mean PbB of 29 μg/100 ml, drawn from factory I and nearby power stations (Table 2a). The control group

TABLE 2a. Age (years), Weight (kg), and Height (cm) of the Lead-Exposed Group (EXP), Office Controls (CO), and Manual Controls (CM) at Factory I and Factory II

	Factory I			Factory II	
	EXP	CO	CM	EXP	CM
N	64	45	28	29	29
Age	40	38	40	43	43
SD	9	9	9	7	10
Weight	77	75	74	76	74
SD	9	8	10	10	8
Height	173	176	173	173	173
SD	6	6	7	6	8

TABLE 2b. Blood Lead, Urinary Lead, and ALA in Urine of the Lead-Exposed Group (EXP), Office Controls (CO), and Manual Controls (CM) at Factory I and Factory II

		Factory I			Factory II	
		EXP	CO	CM	EXP	CM
Blood lead	mean	51.9	25.5	28.8	50	31.5
(μg/100 ml)	SD	11.2	9.2	8.6	10.8	8.4
Urinary lead	mean	70.1	27.5	21.2	78.4	36.2
(μg/liter)	SD	35.8	25.6	11.4	50.2	15.2
ALA in urine	mean	4.4	2.5	2.6	3.6	2.8
(mg/liter)	SD	3.6	1.7	1.5	3.2	1.7

contained office staff and manual workers. At factory II, 29 exposed (mean PbB 50 μg/100 ml) and 29 manual workers (mean PbB 30 μg/100 ml) were tested (Table 2a), the tests in this factory including nerve function tests. The blood analysis for both factories (Table 2b) was undertaken by the National Occupational Hygiene Service Ltd.

The performance tests were: adding test, tapping single and double plates, grip strength and endurance, one-hole pin test, reaction time to a visual stimulus and to a touch stimulus on the arm and leg. At factory I, physiological tremor was also measured. The administration of the tests and instructions to the subjects were standardized, but one member of the technical staff left and was replaced between doing the studies at factory I and II.

The reaction time test (RT) warrants some explanation, as two other tests were derived from it. The instructions given to the subject prior to doing the reaction time tests were to squeeze the hand grip as quickly and tightly as possible. This enabled not only the RT to be studied, but also the speed with which the grip strength was mobilized. The output from the handgrip dynamometer was divided into two, one going directly to the oscilloscope, the other via a differentiator.

The output from the differentiator, therefore, represented the rate of change in hand grip and the peak of the output, the maximum rate of change (v), i.e., the maximum slope of the undifferentiated output. The time to build up to the maximum rate of change, i.e., highest value of v expressed as a frequency (Hz) was measured using the output of the differentiator (Fig. 1). The maximum rate of change (v) achievable by a subject is determined by his grip strength, but the time taken to reach the maximum rate of change may be influenced by nerve function and selective loss of fibers.

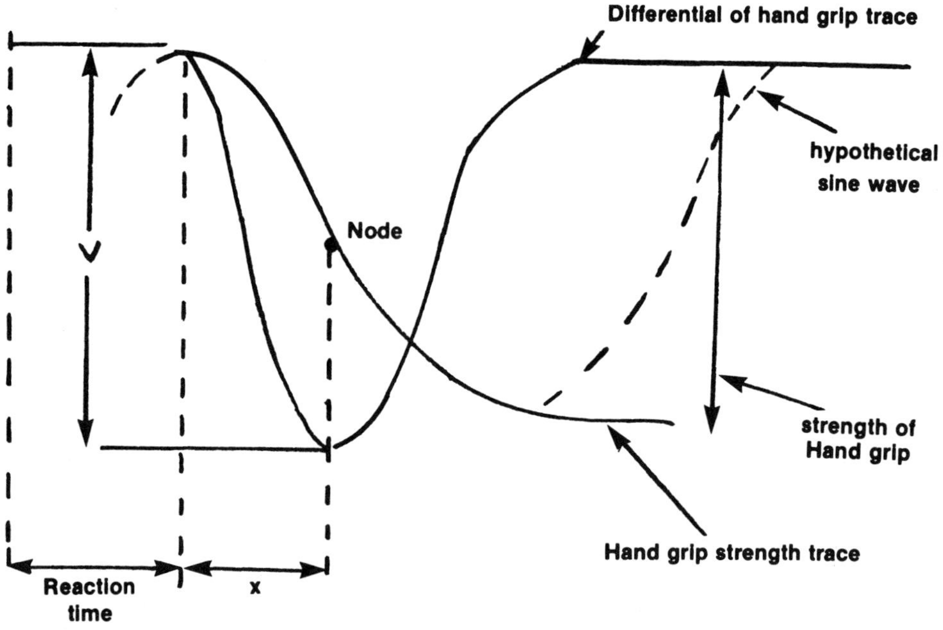

Fig. 1. In calculating the frequency of the hand grip response to the reaction time stimulus, it was assumed that the maximum rate of change in hand grip occurred at the node of a hypothetical sine wave. The time from the crest to this node, i.e. from the start of the reaction to the point where the maximum rate of change in grip was observed is equal to 0.25 of a wave length. The frequency in Hz was therefore calculated as four times the time between the initiation of the reaction and the maximum slope, i.e. X.

RESULTS

Table 3 summarizes the results of the six performance tests. The adding test identifies the office controls, but the results on the other workers are remarkably similar. The tapping test results are very similar in all groups, although there is some indication that both controls achieve higher tap rates than the exposed and that there may be a factory difference. Grip strength differs little between groups, apart from the office controls, who have a low grip length. Endurance is shorter in the office controls and shows a marked factory difference, which could be due to motivation. The results of the one-hole test show little difference between groups. The reaction times indicate that the exposed group at factory I may be faster than the other groups.

When the correlation coefficients between the biochemical and performance parameters were derived, out of 114 possible correlations, 12 were significant, 6 in the expected direction and 6 in the opposite direction (Table 4). What is considered to be the right and wrong direction is possibly debatable, but, nevertheless, the results beg the question as to what expected correlations mean against this background.

TABLE 3. Results of Performance Tests (Mean Values) on Workers at Factory I and Factory II

	Factory I			*Factory II*	
Variable	*EXP*	*CO*	*CM*	*EXP*	*CM*
Adding Test (min)	*19.1*	*14.7*	*19.6*	*21.8*	*19.0*
No. of Errors	*6.6*	*4.4*	*8.9*	*8.0*	*7.6*
Tapping (DP) Total No. of Taps	*129.5*	*133.9*	*134.6*	*147.2*	*144.6*
Tapping (SP) Total No. of Taps	*348.5*	*357.6*	*345.3*	*361.4*	*370.6*
Grip (kg)	*63.0*	*58.1*	*60.1*	*61.3*	*63.2*
Endurance (sec)	*39.7*	*30.7*	*37.9*	*27.2*	*24.1*
One-hole Test Total No. of Pins	*290.1*	*294.1*	*289.7*	*295.9*	*291.7*
RT (msec)	*119.1*	*127.2*	*136.6*	*134.3*	*133.7*
Max Slope (V)	*7.3*	*6.7*	*6.8*	*7.2*	*7.4*
Frequency (Hz)	*3.3*	*3.4*	*3.3*	*3.6*	*3.3*

DP = double plate tapping, 60 seconds
SP = single plate tapping, 60 seconds
V = volts
RT = reaction time

TABLE 4. Correlation Coefficients Obtained between Biochemical Parameters and Results of the Performance Tests (correlations below 0.2 have not been included)

Variable	PbB		PbU		ALA	
	FI	FII	FI	FII	FI	FII
Adding Test						
Time				$-.39^{*}$		
No. of Errors						$-.33$
Tapping Double Plates (60 sec)						
No. of Latches				$-.22$		
No. of Bridge Hits					$.32^{+**}$	$.26$
Tapping Single Plate (60 sec)			$.34^{**}$			
Grip Strength				$-.39^{+**}$		
Endurance				$.28$		
One-Hole Test						
Total Pins	$.25^{*}$	$.25$		$-.31$		
Total Grasp				$-.36^{*}$		
Total Positioning	$.35^{+**}$			$-.51^{**}$		
Reaction Time (arm)						
Maximum Slope (arm)		$-.23$		$-.32$		
Frequency (arm)		$.28$				$.31$
Reaction Time (leg)			$-.32^{**}$			
Maximum Slope (leg)		$-.21$		$-.35^{+*}$		
Frequency (leg)		$.37^{+*}$				
Reaction Time (visual)				$.20$		
Maximum Slope (visual)		$.26$		$-.35^{+*}$		
Frequency (visual)				$-.32$		

* = Significant at the .05 level.
** = Significant at the .01 level.
\+ = Correlation is in the expected direction.

Table 5 presents the product moment correlation coefficients for nerve function and blood lead at factory II. The tests identify age as a modifier of nerve function, although they do not do so every time, indicating that the sample size is small or that their reliability leaves something to be desired. Blood lead, however, has no significant influence on nerve function in this group.

TABLE 5. Correlation Coefficients between Median and Ulnar Nerve Functions and Age, Length of Employment (L of E), and Blood Lead (PbB) at Factory II for the Exposed (EXP) and Control (CM) Groups

		MCV		AMP-W		AMP-E		E/W	
		EXP	CM	EXP	CM	EXP	CM	EXP	CM
MEDIAN	Age	$-.60^{+***}$			$-.31$			$-.23$	$.67^{***}$
	L of E	$-.44^{+**}$			$-.39$			$-.40^{+*}$	
	PbB			$-.24$		$-.24$			
ULNAR	Age	$-.36^{+*}$		$-.50^{+**}$	$-.21$	$-.49^{+**}$	$-.29$		$.26$
	L of E		$.58$						
	PbB	$-.22$				$-.22$	$.22$		

MCV = Motor conduction velocity.
AMP-W = EMG amplitude at the wrist.
AMP-E = EMG amplitude at the elbow.
E/W = Ratio of elbow to wrist amplitude.
** = significant at .05*
*** - significant at .01*
**** = significant at .001*
(Mitran, 1978, p. 214.)

The influence of nerve function on the performance tests in factory II are given in Table 6. The reaction time picture is possibly the most interesting to look at. Of a possible 144 correlations, 10 are significant, and of these, 9 are in the expected direction. Four significant correlations appear in the reaction time to a visual stimulus, 3 in the exposed group, and 1 in the control group. Of the other significant correlations, 4 are in the exposed group and 2 in the control.

There is, therefore, some evidence that visual reaction time, in particular, may be influenced by motor nerve conduction velocity. Although NCV is lowered by lead exposure there appears to be no direct relationship between PbB values and reaction time. Only 3 significant correlations out of 160 correlations appeared between the other performance tests and nerve function, 2 in the control and 1 in the exposed group.

TABLE 6. Correlation Coefficients between Motor Nerve Conduction Velocity (MCV), Amplitude at the Wrist (AMP-W) and elbow (AMP-E) for the Median and Ulnar Nerves and the Reaction Time (RT) of Workers at Factory II

		MCV		AMP-W		AMP-E		E/W	
		EXP	CM	EXP	CM	EXP	CM	EXP	CM
MEDIAN	RT (arm)		.41						
	Max Slope (V)			$.41^{+*}$					
	Hz				$.43^{+*}$		.39		
	RT (leg)								
	Max Slope (V)					$.41^{+*}$			
	RT (visual)	$-.44^{+**}$	-.41	-.30		-.30			
ULNAR	RT (arm)							.33	
	Hz	$-.43^{*}$						-.31	
	RT (leg)		$-.42^{+*}$				-.38	$-.46^{+**}$	-.3
	RT (visual)	$-.41^{+**}$	$-.45^{+*}$					$-.34^{+*}$	
	Hz					-.31		.35	

* = significant at the .05 level
** = significant at the .01 level
\+ = correlation in the expected direction

DISCUSSION

Those performance tests used in this study which were sensitive to learned skill or physical training were able to identify the control group obtained from office staff. The results of the one-hole test, a test requiring eye/hand coordination, were remarkably consistent between groups, as was the adding test, and neither appeared to be sensitive to the influence of nerve function. The reaction time results did not show that PbB level is an influencing factor, but nerve function may possibly be so, particularly with a visual stimulus.

The assessment of the meaning of significant correlations against the overall statistical picture is important (Table 4). Significant correlations are likely to be meaningless unless they fit into a trend and preferably one that fits the physiological assumptions that are being made. Bearing the above in mind, the number of correct correlations between function tests and reaction time is encouraging, as it indicates that reaction time may be a useful indicator or test of changes in nerve function. The lack of a relationship between PbB and reaction time may indicate that lead is not acting directly on the nerves, but via a number of intermediate steps, and during the process, the relationship between PbB and nerve function changes or becomes blurred.

There is, therefore, some indication that performance tests involving the higher centers, e.g., adding, one-hole test, tapping, are not influenced by NCV or the state of the peripheral nervous system, as indicated by the function tests applied to these subjects. On the other hand, reaction time tests may be sensitive to the functional state of the peripheral nervous system. If this is so, any influence of lead over the range met in this study on nervous activity may be purely peripheral. This is an interesting point, because if it is the case, test of intelligence and higher nervous function could be used in matching groups that are to be investigated. We had high hopes of the tremor tests, but the straight comparison that has been done on the basis of exposed and control groups shows no difference between them.

The results we have, therefore, do not indicate that a mean PbB value of 50 g/100 ml has any significant influence on performance and the question arises as to why some studies should show such an effect (Repko, 1976; Haeninen *et al.*, 1978) and others not (Johnson, personal communication; McNeil *et al.*, 1975). At the moment, I do not know, but I believe the explanation may lie in the design of the experiments.

When presented with results such as those described in this paper, there are considerable difficulties of interpretation. Interpretation hinges on two cardinal assumptions. First, there is no stratification in terms of ability, intelligence, or other factors which are likely to result in some workers being more exposed to lead than others and which at the same time are going to influence the results of performance tests (Fig. 2). Second, the control group represents a duplicate of the exposed groups as they would be if they were not exposed to lead.

When using personnel from different factories, we are also making the assumption that the factories operate in the same social milieu and environment, and that they recruit from the same segment of the population as regards ability and life style. These assumptions are so fundamental and the economic and social implications of findings that indicate that low level lead exposure is in any way injurious to health or wellbeing so great, that I believe they must be proved before making assertions about the effects of lead on people's ability to perform certain selected tests.

How does one select the fully matched control populations, how does one make sure that those men doing the dirty jobs with the high blood leads would be the equal of their lower blood lead colleagues but for the presence of lead? I do not know, but I believe it may be possible to test the hypothesis that ability stratification may be responsible for the differences observed between lead workers and non-exposed workers with different blood lead values.

For example, if ability plays a part in determining blood lead values and if high blood lead reduces ability, then with time,

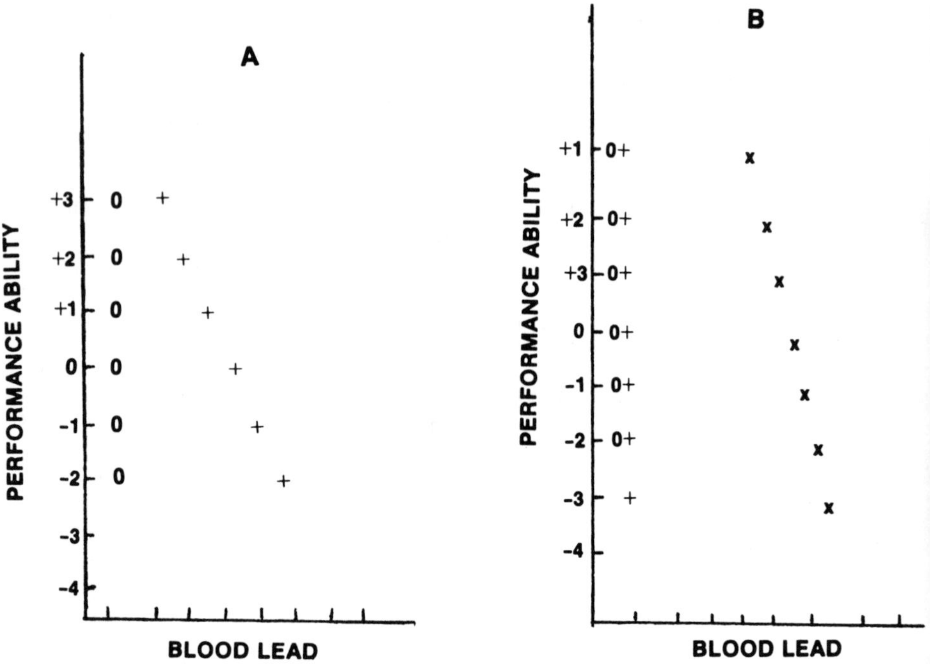

Fig. 2. If the ability of people as measured by performance and intelligence tests is a factor in determining lead exposure levels and hence blood lead levels, tests of performance on a non-exposed group (0) and exposed group (X) may well show the pattern indicated in A. It would be incorrect, however, to ascribe the association between performance and PbB to the adverse effects of lead. If the exposed group were not identical to the control group (O) before exposure as indicated by (+) in B, then the response shown (X) could be interpreted as indicating that lead exposure has a detrimental effect on performance. The relationship observed, however, is due to the two groups not being matched in terms of performance ability.

high blood lead workers should tend to move up to even higher levels or show an increase in the number of interventions taken by management. If, as is likely, environmental lead levels have been coming down in the factories, the high blood lead workers may show a different response to the changing environment by remaining relatively higher.

In a recent paper by Haenninen and coworkers (1978), she observed that over a period of years the maximum PbB observed was highly correlated with the time weighted average for the exposure period, so indicating that workers with high PbB may be consistently high over a number of years. It may be possible, therefore, to determine how high and low PbB groups respond to changes in environmental lead levels.

Haenninen's observation indicates another test of the hypothesis which might be applied, namely to do performance tests on new employees and determine if the blood lead values eventually achieved can be predicted from the results of the tests. It would be important to include those tests which are believed to show the deleterious effects of lead.

Another method of testing the hypothesis which may be feasible is to compare workers in different factories which have different mean PbB levels. The factories would have to be matched in terms of the social milieu within which they operate and the ability spectrum they attract. In addition to lead, factories with other environmental factors which are inert towards the nervous system, but unpleasant, should be selected. If the labor force is stratified in any way and the contaminant has no effect on the nervous system, the dose response relationships should move horizontally, as indicated in Figure 3. A movement down would indicate a possible influence of the contaminants on the nervous system. No association between performance and inert but unpleasant contaminants may indicate that stratification may not be a problem in studies of lead workers.

The problems of matching groups of workers must not be underestimated. In 1962, the British Medical Research Council reported on a study of acclimatization to heat. Three groups of soldiers had been selected by matching for height, weight, and rectal temperature and sweat rate when exposed to heat in a climatic chamber. The three groups of soldiers were given quite different training for a number of weeks and then subjected to "performance tests."

The results were difficult to explain, but when the intelligence tests of the three groups were compared, it was found that one group had a significantly and markedly lower score. Randomizing a physical and physiological variable did not result in the randomizing of a psychological variable. When studying performance, the burden of proof that the groups being studied are matched and that groups are randomly selected for factors influencing performance lies with the experimenter.

Many years ago, the International Biological Program was established by the International Council of Scientific Unions. The Human Adaptability Section set itself a number of tasks, some of which involved measuring certain aspects of human performance. To do this, scientists gathered together to decide what to measure, how to measure it, and how to design the experimental programs. Having done this, groups of scientists gathered in different laboratories to evaluate the methods and be trained in the techniques.

Roy Shephard's book (1978), "Human Physiological Work Capacity," which is an account of the studies of work capacity across social group, race, sex, climatic zone, and altitude, makes salutary reading for anybody who is trying to investigate small, possibly nonexistent, differences between small groups of workers within a factory. Such studies, I believe, are likely to produce data which cannot be interpreted or validated unless the two important assumptions already mentioned are valid.

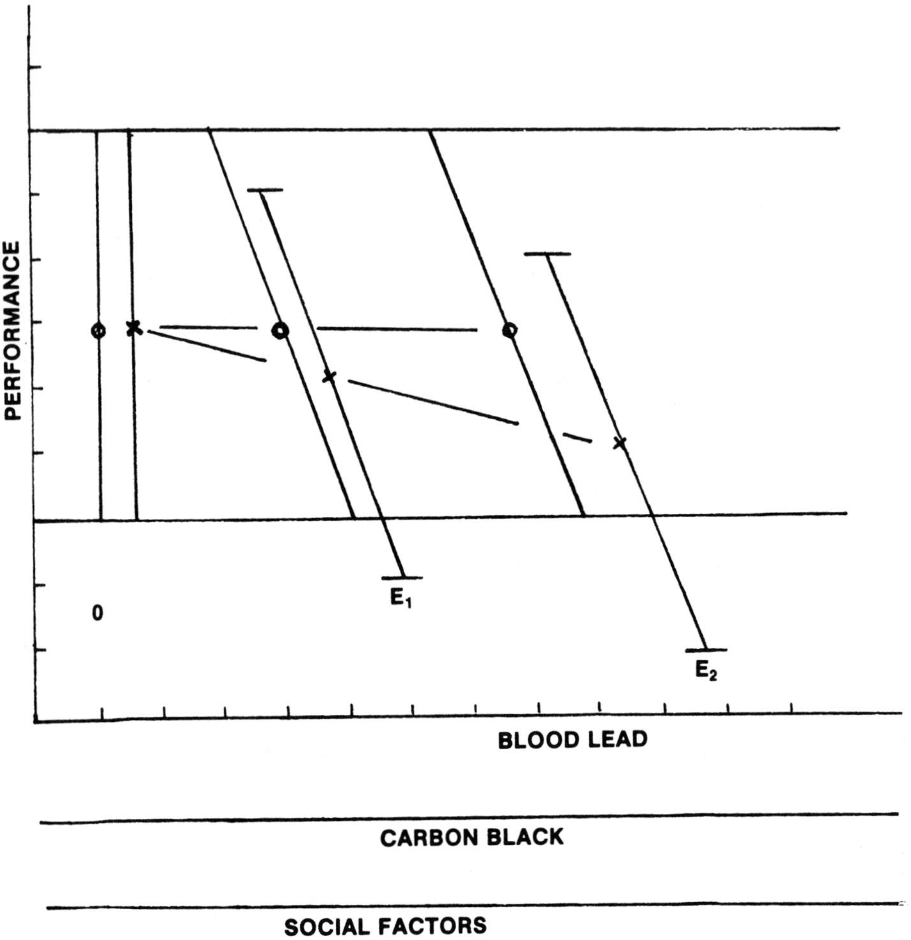

Fig. 3. Test of the ability stratification hypothesis. If ability stratification is responsible for the suggested relationship between blood lead and performance, tests of matched population's with differing mean PbB should show a horizontal movement of the performance range as indicated (O) at E_1 and E_2. This horizontal movement might also be observed if other environmental factors are used for plotting performance against instead of lead. A decrease in performance as shown (X) with increase in exposure to lead (E_1 & E_2) would support the view that lead, or something associated with it, is adversely affecting performance.

The problem of interpreting contradictory evidence and the need for standardization of methods used in monitoring population groups when studying the lead tolerance of man was emphasized by Professor Zielhuis three years ago at Dubrovnik (Zielhuis, 1975); information is required on what should be regarded as a "normal" value, e.g., MCV, RT, and what "normal" environmental factors affect them, e.g., alcohol consumption, smoking, and disease. An agreed methodology which can be validated is required for such studies. Factors which at the moment are thought to be able to influence results, such as ability stratification, must be investigated and shown to be relevant or not relevant to the design of investigations directed towards the measurement of lead tolerance in humans or, indeed, his tolerance of any neurotoxic environmental contaminant.

REFERENCES

Archibald, J. (1976). A clinical and biochemical study of early lead poisoning. *Progress Report.* Department of Occupational Health, University of Manchester, England.

British Medical Research Council (1962). *Acclimatization to Heat,* APRC Project P-21, pp. 20-22. MRC 61/827, England.

Griffin, T. B., Coulston, F., and Wills, H. (1975). Biological and clinical effects of continuous exposure to airborne particulate lead. *Arh Hig Rada Toksikol 26* (Supp), 191-207.

Haenninen, H., Hernberg, S., Mantere, P., Vesanto, R., and Jalkanen, M. (1978). Psychological performance of subjects with low exposure to lead. *J. Occ. Med. 20,* 683-689.

Johnson, B., personal communication.

McNeil, J. L., Ptasnik, J. A., and Croft, D. B. (1975). Evaluation of longterm effects of elevated blood lead concentrations in asymptomatic children. *Arh Hig Rada Toksikol 26* (Supp), 97-117.

Mitran, E. (1978). "Effects of Acceptable Levels of Lead Adsorption on Human Performance in Relation to Biochemical and Neurophysiological Changes," Ph.D. Thesis. University of London, England.

Repko, J. (1976). "Behavioral Methods and Results in the Evaluation of Workers Occupationally Exposed to Inorganic Lead in U.S. Battery Manufacturing Industries." Report prepared for the 2nd International Workshop on Occupational Lead Exposure: Re-evaluation of permissible limits. Amsterdam, Netherlands.

Schwarz, K. (1975). Panel discussion: International symposium on environmental lead research. *Arh Hig Rada Toksikol 26* (Supp), 234.

Shephard, R. J. (1978). Human Physiological Work Capacity. *In* International Biological Programme 15. University of Cambridge Press, England.

Zielhuis, R. L. (1975). Research needs for the future: International symposium on environmental lead research. *Arh Hig Rada Toksikol 26* (Supp), 263-268.

DISCUSSION OF PAPER BY DR. GEOFFREY CROCKFORD

Shiro Tanaka, Niosh: I understand that the new employees are put into one exposure area and kept there for about three years. Have you tried to use the employee as his own control and follow his performance? I suppose in five or six months his blood lead level may reach a plateau or some stable area, during which time, if you can document that his performance goes down, probably you can eliminate the effect of aging in that short period. Have you tried this?

Geoffrey Crockford, London School of Hygiene & Tropical Medicine: No, we haven't. What I was trying to indicate is that this is one of the controls of this hypotheses. This is the sort of thing which I think one needs to do to check if there's anything to it. If you can rule out stratification as being at all important, this, in fact, simplifies the use of performance tests on exposed people.

Grandjean: I agree with the speaker that prospective studies would absolutely be needed in this field. However, there's some indication that your very provocative hypothesis is not correct. When Wechsler, in the 1940's and early 1950's, made his intelligence tests, he did not recognize the fact that they might change the test results. Later on he realized that some of the test results might be sensitive to organic brain damage. He then divided the tests into so-called "hold" tests and "don't hold" tests. If you look at the results of the lead workers, you'll find that the "don't hold" tests, the tests that are sensitive to organic brain damage, will correlate very closely with the blood lead level, the ZPP or the ALAD, while the "hold" tests will not show such close relationship. If your workers had shown the stratification that you talk about, you wouldn't find such difference.

Crockford: Well, I think it's possible that you're right. I'm not suggesting that what I put forward is correct. Correct me if I'm wrong, but I think this is just one study on one group of workers. The point I was trying to make is that because of the world that we're living in, you're quite likely to get that type of result. Now, if you had gone to a factory, where, for example, some relatively inert material, such as carbon black, which made certain jobs unpleasant, that, in a sense, would have been the control for the experiments that have been done. I think that unti you can show that the group of lead exposed workers were normal before their exposure to lead in some way, you can't really draw any conclusions from just one set of results. A pattern has to come out. So, if in succeeding years, other groups of workers come up with the same results, then the pattern's there. But at the moment, I guess you're well aware, people who have looked at performance in lead workers are coming out with quite different results. Some say that they can find no relationship, as we have done, and others say they have found a relationship to some particular performance work. So there's no pattern emerging at the moment.

MORTALITY IN EMPLOYEES OF LEAD PRODUCTION FACILITIES AND LEAD BATTERY PLANTS, 1971-1975

W. Clark Cooper, M.D.

Berkeley, California

INTRODUCTION

In 1975, Cooper and Gaffey reported on the deaths that occurred in a cohort of 7,032 workers in 16 U.S. lead production facilities and battery plants during the years 1947-1970. Lead absorption was known to have been greatly in excess of accepted standards in many members of the cohort. There were 1,356 reported deaths; death certificates were obtained in 1,267. The standardized mortality ratio (SMR) for all causes was 107 for lead production workers and 99 for battery plant workers. Deaths from neoplasms were in slight excess in the former, but not increased in battery plants. There were no excess deaths from kidney tumors. The SMR for cardiovascular-renal disease was 96 for the production plants and 101 for battery plant workers. There was no excess in deaths due either to stroke or hypertensive heart disease. However, deaths classified as "other hypertensive disease" and "chronic nephritis or renal sclerosis" were higher than expected in both populations.

In the present report, the approximately 5,400 members of the cohort who were surviving on December 31, 1970 were observed for an additional five year period, January 1, 1971 through December 31, 1975 (Cooper, 1978).

METHODS

The selection of the original study population was summarized in the reports by Cooper and Gaffey (1975, 1976). Microfilmed personnel records from 10 battery plants and six lead production facilities (referred to at times as "smelters") provided information on 24,494 individuals who had worked in these plants at some

ISBN 0-12-460520-6

time between January 1, 1946 and December 31, 1970. After the exclusion of 950 female employees, approximately 16,000 individuals with periods of employment less than one year, and approximately 350 with missing birth or hire dates, there remained a study population of 7,032. After follow-up, including submission of names and social security numbers to the Social Security Administration, the vital status of the participants as of December 31, 1970 was reported as 5,837 alive, 1,356 dead, and 289 unknown.

The current study is based upon follow-up for an additional five-year period of the 5,837 men reported alive and the 289 men not located as of December 31, 1970. Information received after completion of the first study led to adjustments of the population, since some not located or thought dead were found to have been alive and others were found deceased. The actual population included in the current study was 5,490 men.

Lists of those thought alive on December 31, 1970 were sent to each of the participating plants, with requests that an indication be made for each man as to whether (1) he was still working, (2) he had terminated employment, with date of termination, and (3) he was known to have died, with date and place of death, if known.

After receipt of the foregoing, the names and social security numbers of men whose vital status had not been determined were sent to the Social Security Administration (SSA). From SSA, information was obtained as to whether they (1) were presumably alive, based on continuing payments from an employer or continuing benefits being paid out, or (2) had been reported dead. Others were reported as vital status unknown. For all men who were reported dead, either by the plant or by SSA, requests were sent to states where death occurred or the state of last residence to obtain death certificates.

Causes of death were coded from death certificates, and these were all reviewed by a nosologist in the Office of Vital Statistics of the State of California. Causes were coded to the Seventh Revision of the International Classification of Disease, Injuries, and Causes of Death (ICD, 1955), because this had been used in the previous report. This involved translation of categories used in the Eighth Revision (1965) into appropriate 1955 counterparts.

The percentages of all observed and certified deaths due to selected groups of causes were determined and compared with the percentages expected if there had been the same distribution by cause that occurred in U.S. male deaths in 1973. Adjustments were made for age, using the method of Guralnick (1963). Adjustments were not made by race. Standardized mortality ratios (SMR's) also were calculated, as described in 1975, using the age and cause-specific death rates for the male populations of the United States in 1973, the mid-year of the study period.

The statistical significance of the deviation of each SMR from an expected value of 100 was tested by a method derived from Chin Leong Chiang (1961). The formula for determining the standard error of the SMR was:

$$SE = \sqrt{\frac{100 \times SMR}{\text{expected deaths}}}$$

which follows from the formulas presented in the cited paper. If an observed SMR differed from 100 by more than 1.96 standard errors, it was regarded as significant at the 5% level; if it differed by more than 2.57 standard errors, it was regarded as significant at the 1% level. SMR's based on fewer than 5 observed cases were not tested for significance.

RESULTS

Revision of study population. Although the follow-up began as a study of 5,387 workers believed to be alive as of December 31, 1970, plus 289 of unknown status as of that date, additional information obtained in the follow-up modified this. Records on two men were lost, 58 additional deaths prior to January 1, 1971 were located, and three men previously reported as having been dead were found to have been alive. A comparison between the numbers reported in 1975 and the revised population is shown in Table 1.

The additional deaths prior to 1971 located in this search will not be included in the present analysis. The causes are summarized as follows: 13 malignant neoplasms (1 buccal, 5 gastrointestinal, 2 respiratory, 1 genital, 0 urinary, 0 leukemia, 3 lymphoma, 1 skin); 29 cardiovascular-renal disease (2 cerebrovascular, 23 arteriosclerotic heart disease, 1 other hypertensive disease, 1 chronic nephritis, 2 other), 4 influenza and pneumonia, 2 cirrhosis of liver, 4 senility and ill-defined, 2 suicide, 4 remaining.

The modification of the population resulted in there being 5,490 available for follow-up as of January 1, 1971. Of these, 1,967 were in lead production facilities and 3,523 in battery plants.

TABLE 1. Status of Study Population as of January 1, 1971

	As Shown in First Report	*As Revised for Second Report*
Total Number	*7,032*	*7,030*
Alive	*5,387*	*5,490*
Dead	*1,356*	*1,411*
Total Known	*6,743*	*6,901*
Status Unknown	*289*	*129*

TABLE 2. Distribution of Study Populations by Year of Hire

	Cumulative Percent	
Hired Before	*Production Facilities*	*Battery Plants*
1920	*0.3*	*2.5*
1925	*1.1*	*7.3*
1930	*2.6*	*12.8*
1935	*4.0*	*14.3*
1940	*7.2*	*18.7*
1945	*12.5*	*44.8*
1950	*33.8*	*54.9*
1955	*56.9*	*62.1*
1960	*67.3*	*68.7*
1965	*81.7*	*79.5*
1970	*100.0*	*100.0*

Description of the study population. Table 2 summarizes the distribution of the study population by year of hire. It is of interest to notice the difference in the distribution of peak hire periods for the two study populations. At the beginning of the 1971-1975 observation period, 45% of the battery plant workers and 12.5% of the lead production facility workers had been hired before 1945.

Table 3 summarizes the distribution of the revised study population by age at time of entry into the new follow-up period, January 1, 1971. The mean age for smelter workers was 46.9 years, that for battery plants 44.2 years. Nevertheless, the battery plant had 18.4% of its members aged 65 years or more, compared with 9.3% of the smelter population. These divergent patterns of age distribution can affect mortality analysis, even when adjustments are made for age, as pointed out by Sterling (1964) and by Gaffey (1976).

Table 4 summarizes the distribution of the population by years employed in a lead-producing or lead-using industry. Known gaps in employment are omitted, as well as years following termination of employment. It can be seen that 21.1% of the smelter workers and 26% of the battery plant workers had had 20 or more years of employment as of the beginning of the observation period; 36.8 and 35.6%, respectively, had had 10 or more years of employment.

It was not possible to do an analysis of racial distribution of the populations because this information does not appear in plant records. In the previous report (Cooper and Gaffey, 1975), evidence obtained from plant managers indicated that the smelters probably had about 30% black employees, compared with less than 10% in the battery plants. The former were therefore in excess of the national average in the population (approximately 11%), while the latter were below it. As will be shown later in an analysis

TABLE 3. Distribution of Study Populations by Age as of January 1, 1971

Age -- Equal to or More Than	Cumulative Percentage	
	Production Facilities	Battery Plants
30 Yrs.	89.4	89.7
35	79.7	79.2
40	67.0	70.9
45	52.7	63.6
50	37.7	50.5
55	25.3	37.9
60	15.8	26.7
65	9.3	18.4
70	4.4	11.2
75	2.7	5.7
80	1.2	2.2
85	0.3	0.3

TABLE 4. Distribution of Revised 1971 Study Population by Years Employed, as of January 1, 1971

Years Employed (all before Jan. 1, 1971)	Smelter Workers		Battery Workers	
	No.	%	No.	%
0-4	869	44.6	1,180	33.4
5-9	362	18.6	740	20.9
10-14	153	7.9	460	13.0
15-19	151	7.8	232	6.6
20-24	172	8.8	149	4.2
25-29	98	5.0	217	6.1
30-34	65	3.3	255	7.2
35-39	47	2.4	186	5.3
40-44	27	1.3	87	2.5
45-49	5	0.3	26	0.7
50-54	0	-	1	<0.01
Total Known	1,949	100.00	3,533	100.0
Total Unknown	55	-	82	-
TOTAL	2,004		3,615	

of deaths, this was supported by the fact that 35.4% of the deaths reported in lead production facilities were identified on death certificates as being in Blacks, while only 7.05% of the deaths reported in battery plants were in black employees.

Follow-up. Due to changes of ownership of some of the plants, with transfer and consolidation of records, information from original sources was less complete than needed. There were 4,394 names submitted to SAA for follow-up. Of the 5,490 men established as being alive on January 1, 1971, 609 were not located as of December 31, 1975. The percentage of follow-up was thus relatively low, being only 88.9%. That for lead production facilities was 87.2% and that for battery plants 89.8%. For the 491 deaths, certificates were obtained for 436 (Table 5).

Analysis of deaths. The age and racial distribution of the deaths observed in the study is shown in Table 6. It can be seen that 39 of 110 certified deaths, or 35.4% of the deaths in lead production facilities, were in black workers; the corresponding figures in battery plants were 23 of 326 deaths or 7.1%. These figures can be contrasted with the national average, where in 1973, 11.2% of all deaths in males aged 20 years or more were in blacks.

TABLE 5. Follow-up of 5,490 Workers Alive January 1, 1971

Vital Status	Production Facilities	Battery Plants	Total
Alive 1/1/71	1,967	3,523	5,490
Alive 12/31/75	1,590	2,800	4,390
Dead 12/31/75	127	364	491
Known 12/31/75	1,717 (87%)	3,164 (90%)	4,881 (89%)
Unknown 12/31/75	250 (13%)	359 (10%)	609 (11%)

TABLE 6. Race and Age of Certified Deaths

	Production Facilities	Battery Plants
Certified Deaths	110	326
Percent Black[a]	35%	7%
Mean Age, Black	58.8 years	62.2 years
Mean Age, White	63.8 years	69.9 years
Mean Age, Total	61.7 years	68.9 years

[a]U.S. Average = 11%.

TABLE 7. Percentages of 436 Certified Deaths in Smelter and Battery Plant Workers, 1971-75, Due to Selected Causes, Compared with Percentages Among Deaths in U.S. Males 20 or More Years of Age, for 1973 (proportional mortality)

Cause of Death (1955 ICD No.)	U.S. Males aged 20 yrs. or more	Smelter Workers		Battery Plant Workers	
		Expected[a]	Observed	Expected[a]	Observed
	percent	percent	percent	percent	percent
Tuberculosis (001-019)	0.26	0.38	0	0.25	0.31
Malignant Neoplasms (140-205)	18.29	21.33	16.36	19.43	21.80
Buccal & pharyngeal (140-148)	0.54	0.33	0.91	0.47	0.31
Digestive (150-159)	4.94	5.24	2.73	5.21	4.60
Respiratory (160-164)	6.07	7.29	8.20	6.51	7.36
Genital (170-179)	1.92	1.51	1.82	2.43	1.23
Urinary (180-181)	1.02	1.03	0	1.12	0.92
Leukemia (204)	0.73	0.70	0	0.78	0.92
Lymphoma (200-203-205)	1.00	1.11	0	1.08	0.92
Other sites	2.07	4.12	2.7	1.73	5.52
Diabetes Mellitus (260)	1.51	1.52	1.82	1.50	1.53
Cardiovascular-Renal (330-334, 400-468, 592-594)	45.26	49.52	49.10	52.94	53.99
Cerebrovascular (330-334)	9.25	7.48	9.10	9.45	9.20
Rheumatic (400-402; 410-416)	0.68	0.73	1.82	0.64	0.61
Ischemic heart disease (420)	31.68	37.62	30.91	39.24	37.12
Hypertensive heart disease (440-443)	0.53	0.50	0.91	0.55	0.31
Other hypertensive disease (444-447)	0.36	0.32	0.91	0.34	0.31
Chr. & unspec. nephritis (592-594)	0.34	0.35	0	0.32	0.31
Other	2.42	2.52	3.64	2.40	3.37

Table 7 (Continued)

Cause of Death (1955 ICD No.)	U.S. Males aged 20 yrs. or more	Smelter Workers		Battery Plant Workers	
		Expected[a]	Observed	Expected[a]	Observed
	percent	percent	percent	percent	percent
Influenza and Pneumonia (480-493)	3.00	2.61	3.64	3.02	2.45
Cirrhosis of Liver (581)	2.07	3.17	1.82	1.85	2.15
Senility and Ill-Defined (780-795)	1.46	1.56	3.64	1.30	4.91
Motor Vehicle Accidents (810-835)	3.45	2.43	3.64	1.67	1.53
Other Accidents & Ext. Causes (800-802, 840-962)	3.66	3.34	1.82	2.42	1.53
Suicides (963, 970-979)	1.72	1.79	3.64	1.12	0.31
Homicide (964, 980-985)	1.46	1.28	1.82	0.72	1.23
Residual (all others)	17.86	11.07	12.72	13.82	8.31
Total	100.00	100.00	100.00	100.00	100.00
Based on This No. of Deaths	1,031,966	110		326	

[a]Adjusted for distribution of ages at time of death in certified deaths.

TABLE 8. Standardized Mortality Ratios. Observed and Expected Deaths by Cause, 1971-1975 in Lead Production Facilities and Lead Battery Plants

Cause of Death (ICD No., 1955)	Lead Production Facilities Deaths			Lead Battery Plants Deaths		
	Obs.	Exp.	SMR[b]	Obs.	Exp.	SMR[b]
All Causes	127	118.1	108	364	310.76	117**
All Causes with Death Certificates	110	118.1	108	326	310.76	117
Malignant Neoplasms (140-205	18	23.4	89	71	60.9	136*
Buccal cavity and pharynx (140-148)	1	0.76	--	1	1.81	--
Digestive organs & peritoneum (150-159)	3	6.12	--	15	16.47	102
Respiratory system (160-164)	9	8.51	121	24	20.95	128
Genital organs (170-179)	2	1.81	--	4	6.02	--
Urinary organs (180-181)	0	1.22	--	3	3.39	--
Leukemia, aleukemia (204)	0	0.85	--	3	2.18	--
Lymphoma (200-203, 205)	0	1.81	--	3	3.23	--
Other sites	3	2.33	--	18	6.85	293**

TABLE 8 (Continued)

	Lead Production Facilities			Lead Battery Plants		
	Deaths			Deaths		
	Obs.	Exp.	SMR[b]	Obs.	Exp.	SMR[b]
Major Cardiovascular & Renal Disease (330-334, 400-468, 592-594)	54	60.33	104	176	169.04	116
Vascular lesions--central nervous system (330-334)	10	8.85	131	30	27.52	122
Arteriosclerotic heart disease (420)	34	43.69	90	121	119.97	113
Hypertensive heart disease (440-443)	1	0.56	--	1	1.63	--
Other hypertensive disease (444-447)	1	0.38	--	1	1.09	--
Chronic nephritis, renal sclerosis (592-594)	0	0.41	--	1	1.05	--
Other	8	6.64	139	22	16.98	145
Influenza and Pneumonia (480-493)	4	3.04	--	8	8.90	101
Cirrhosis of Liver (581)	2	3.52	--	7	6.83	124
Symptoms, Senility & Ill-Defined (780-795)	4	1.84	--	16	4.28	418**
Motor Vehicle Accidents (810-835)	2	3.22	--	5	6.04	93
Other Accidents & External Causes (800-802, 840-962)	2	4.13	--	5	8.42	66
Suicide (963, 970-979)	4	2.19	--	1	4.18	--
Homicide (964, 980-985)	2	1.70	--	4	2.83	--
All Other Causes (residual)	18	14.73	141	33	39.34	94
Number of workers		1,717			3,164	
Number of person-years		8,286			14,967	

[a]Only those with over 5 observed listed; adjusted, except "all causes" by +15.45% for 17 uncertified deaths. [b]Only those with over 5 observed listed; adjusted, except "all causes" by +11.66% for 38 uncertified deaths. *Significantly different from 100 at 0.05 level. **Significantly different from 100 at 0.01 level.

Proportionate mortality ratios. In Table 7 the percentages of all deaths attributed to a number of specific causes have been calculated and compared with the percentages reported for U.S. males aged 20 or more at death for the year 1973. Appropriate adjustments were made for ages of death of the study population (using the method of Guralnick, 1962). There were few instances where the percentage in the study population deviated appreciably from the national average.

Standardized mortality ratios. Table 8 shows the observed and expected deaths for lead production facilities and battery plant workers. These will be summarized separately.

Lead smelter workers had an SMR for "all causes" combined of 108. There were no statistically significant deviations above 100 for any specific causes, although malignancies of the respiratory tract showed an adjusted SMR of 121, major cardiovascular-renal disease an SMR of 104, and vascular lesions of the central nervous system 131.

Battery plant workers showed an SMR for all causes combined of 117 (significantly higher than 100). Malignant neoplasms were also in slight excess (SMR = 136). This was accounted for largely by malignant neoplasms of the respiratory system (SMR = 128) and malignant neoplasms of "other or unspecified sites" (SMR = 293). The latter will be listed in more detail later. Major cardiovascular renal disease showed an SMR of 116, vascular lesions of the central nervous system 122, arteriosclerotic heart disease 113, and the category of senility and ill-defined conditions a statistically high SMR of 418. This too will be broken down individually as to causes involved.

Standardized mortality ratios as related to hire dates. In Table 9 the mortality for lead production facility workers hired in 1945 or earlier is compared with that for those hired on or after January 1, 1946. The only apparent differences were in major cardiovascular renal disease, where the pre- and post-1946 hires showed SMR's of 121 and 91, respectively, explained largely by differences in deaths attributed to arteriosclerotic heart disease. There was no difference in malignant neoplasms.

Table 10 provides a similar analysis for battery plant workers. Men hired before 1946 showed a higher overall SMR (125 as compared with 80) and for "major cardiovascular renal disease" and "senility and ill-defined causes." There was no difference in malignant neoplasms, those hired after January 1, 1946 showing somewhat higher SMR's for all malignant neoplasms and for respiratory tract malignancies.

Standardized mortality ratios as related to duration of employment. Tables 11 through 14 show SMR's calculated for lead production facility employees and for battery plant employees, comparing first those with less than 10 years or more than 20 years employment (as of January 1, 1971). There were no consistent patterns of

TABLE 9. Standardized Mortality Ratios. Observed and Expected Deaths by Cause, 1971-1975 in Lead Production Facilities in Men Hired Before January 1, 1946 Compared with Men Hired on and after January 1, 1946

Cause of Death (ICD No., 1955)	Hire Date Before 1/1/46 Deaths			Hire Date 1/1/46 or Later Deaths		
	Obs.	Exp.	SMR[a]	Obs.	Exp.	SMR[b]
All Causes	49	45.33	108	78	72.79	107
All Causes with Death Certificates	40	45.33	108	70	72.79	107
Malignant Neoplasms (140-148)	6	8.88	83	12	14.52	93
Buccal cavity and pharynx (140-148)	0	0.25	--	1	0.50	--
Digestive organs & peritoneum (150-159)	1	2.46	--	2	3.67	--
Respiratory system (160-164)	3	3.00	--	6	5.51	122
Genital organs (170-179)	1	0.97	--	1	0.84	--
Urinary organs (180-181)	0	0.51	--	0	0.70	--
Leukemia, aleukemia (204)	0	0.31	--	0	0.52	--
Lymphoma (200-203, 205)	0	0.44	--	0	0.87	--
Other sites	1	0.94	--	2	1.91	--

Major Cardiovascular & Renal Disease (330-334, 400-468, 592-594)	26	26.15	121	28	34.18	91
Vascular lesions--central nervous system (330-334)	3	4.45	--	7	4.39	177
Arteriosclerotic heart disease (420)	19	18.35	127	15	25.34	66*
Hypertensive heart disease (440-443)	0	0.26	--	1	0.31	--
Other hypertensive disease (444-447)	1	0.17	--	0	0.21	--
Chronic nephritis, renal sclerosis (592-594)	0	0.15	--	0	0.26	--
Other	3	2.77	--	5	3.67	152
Influenza and Pneumonia (480-493)	1	1.41	--	3	1.63	--
Cirrhosis of Liver (581)	0	0.76	--	2	2.76	--
Symptoms, Senility & Ill-Defined (780-795)	2	0.57	--	2	1.28	--
Motor Vehicle Accidents (810-835)	0	0.47	--	4	2.75	--
Other Accidents & External Causes (800-802, 840-962)	1	0.87	--	1	3.26	--
Suicide (963, 970-979)	1	0.38	--	3	1.81	--
Homicide (064, 980-985)	0	0.14	--	2	1.56	--
All Other Causes (residual)	3	4.98	--	12	10.68	125
Number of workers		253			1,464	
Number of person-years		1,147			7,139	

*[a]Only those with over 5 observed listed; adjusted, except "all causes " by +22.5% for 9 uncertified deaths. [b]Only those with over 5 observed listed; adjusted, except "all causes" by +11.43% for 8 uncertified deaths. *Significantly different from 100 at 0.05 level. **Significantly different from 100 at 0.01 level.*

TABLE 10. Standardized Mortality Ratios. Observed and Expected Deaths by Cause, 1971-1975 in Lead Battery Plant Employees Hired Before January 1, 1946 Compared with Those Hired on and after January 1, 1946

Cause of Death (ICD No., 1955)	Hire Date Before 1/1/46 Deaths			Hire Date 1/1/46 or Later Deaths		
	Obs.	Exp.	SMR[a]	Obs.	Exp.	SMR[b]
All Causes	319	254.67	125**	45	56.08	80
All Causes with Death Certificates	285	254.67	125**	40	56.08	80
Malignant Neoplasms (140-205)	58	50.47	128	13	10.39	141
Buccal cavity and pharynx (140-148)	0	1.53	--	1	0.38	--
Digestive organs & peritoneum (150-159)	14	13.97	112	1	2.50	--
Respiratory system (160-164)	18	17.02	118	6	3.93	172
Genital organs (170-179)	4	5.57	--	0	0.45	--
Urinary organs (180-181)	3	2.92	--	0	0.46	--
Leukemia, aleukemia (204)	2	1.76	--	1	0.42	--
Lymphoma (200-203, 205)	2	2.53	--	1	0.70	--
Other sites	15	5.27	317**	3	1.55	--

Major Cardiovascular & Renal Disease (330-334, 400-468, 592-594)	168	145.95	128**	8	25.13	39**
Vascular lesions--central nervous system (330-334)	29	24.81	131	1	2.70	--
Arteriosclerotic heart disease (420)	117	102.72	127*	4	17.25	--
Hypertensive heart disease (440-443)	1	1.42	--	0	0.21	--
Other hypertensive disease (444-447)	1	0.93	--	0	0.15	--
Chronic nephritis, renal sclerosis (592-594)	1	0.84	--	0	0.21	--
Other	19	15.23	139	3	2.63	--
Influenza and Pneumonia (480-493)	8	7.74	115	0	1.11	--
Cirrhosis of Liver (581)	4	4.29	--	3	2.54	--
Symptoms, Senility & Ill-Defined (780-795)	16	3.14	569**	0	1.14	--
Motor Vehicle Accidents (810-835)	5	2.73	204	0	3.32	--
Other Accidents & External Causes (800-802, 840-962)	3	4.88	--	2	3.54	--
Suicide (963, 970-979)	0	2.19	--	1	2.00	--
Homicide (964, 980-985)	0	0.83	--	4	2.00	--
All Other Causes (residual)	23	32.45	79	9	6.89	147
Number of workers		1,488			1,676	
Number of person-years		6,673			8,294	

[a]*Only those with over 5 observed listed; adjusted, except "all causes" by +11.54% for 33 uncertified deaths.* [b]*Only those with over 5 observed listed; adjusted, except "all causes" by +12.5% for 5 uncertified deaths.* **Significantly different from 100 at 0.05 level.* ***Significantly different from 100 at 0.01 level.*

TABLE 11. *Standardized Mortality Ratios. Observed and Expected Deaths by Cause, 1971-1975 in Lead Production Facilities in Men with Less than 10 Years Employment, Compared with Men with 10 or More Years of Employment*

Cause of Death (ICD No., 1955)	Less Than 10 Years Deaths			10 or More Years Deaths		
	Obs.	Exp.	SMR[a]	Obs.	Exp.	SMR[b]
All Causes	45	46.34	97	82	71.90	114
All Causes with Death Certificates	40	46.34	97	70	71.90	114
Malignant Neoplasms (140-205)	6	8.97	75	12	14.45	97
Buccal cavity and pharynx (140-148)	1	0.31	--	0	0.45	--
Digestive organs & peritoneum (150-159)	2	2.24	--	1	3.88	--
Respiratory system (160-164)	2	3.37	--	7	5.14	159
Genital organs (170-179)	0	0.51	--	2	1.31	--
Urinary organs (180-181)	0	0.43	--	0	0.79	--
Leukemia, aleukemia (204)	0	0.34	--	0	0.49	--
Lymphoma (200-203, 205)	0	0.56	--	0	0.75	--
Other sites	1	1.21	--	2	1.64	--

Major Cardiovascular & Renal Disease (330-334, 400-468, 592-594)	13	21.02	70	41	39.37	122
Vascular lesions--central nervous system (330-334)	2	2.70	--	8	6.16	152
Arteriosclerotic heart disease (420)	6	15.54	44**	28	28.19	116
Hypertensive heart disease (440-443)	1	0.19	--	0	0.38	--
Other hypertensive disease (444-447)	0	0.13	--	1	0.25	--
Chronic nephritis, renal sclerosis (592-594)	0	0.17	--	0	0.24	--
Other	4	2.29	--	4	4.15	--
Influenza and Pneumonia (480-493)	1	1.03	--	3	2.01	--
Cirrhosis of Liver (581)	2	1.81	--	0	1.71	--
Symptoms, Senility & Ill-Defined (780-795)	1	0.84	--	3	1.00	--
Motor Vehicle Accidents (810-835)	4	2.07	--	0	1.15	--
Other Accidents & External Causes (800-802, 840-962)	1	2.33	--	1	1.80	--
Suicide (963, 970-979)	2	1.29	--	2	0.84	--
Homicide (964, 980-985)	1	1.21	--	1	0.49	--
All Other Causes (residual)	9	5.77	175	7	9.08	90
Number of workers		1,069			648	
Number of person-years		5,244			3,041	

*[a]Only those with over 5 observed listed; adjusted, except "all causes" by +12.5% for 5 uncertified deaths. [b]Only those with over 5 observed; adjusted, except "all causes" by +17.14% for 12 uncertified deaths. *Significantly different from 100 at 0.05 level. **Significantly different from 100 at 0.01 level.*

TABLE 12. Standardized Mortality Ratios. Observed and Expected Deaths by Cause, 1971-1975 in Lead Production Facilities in Men with Less than 20 Years Employment, Compared with Men with 20 or More Years Employment

Cause of Death (ICD No., 1955)	Less than 20 Years Deaths			20 Years or More Deaths		
	Obs.	Exp.	SMR[a]	Obs.	Exp.	SMR[b]
All Causes	70	65.59	107	57	52.45	109
All Causes with Death Certificates	62	65.59	107	48	52.45	109
Malignant Neoplasms (140-205)	10	12.87	88	8	10.51	90
Buccal cavity and pharynx (140-148)	1	0.44	--	0	0.32	--
Digestive organs & peritoneum (150-159)	2	3.27	69	1	2.86	--
Respiratory system (160-164)	4	4.80	--	5	3.71	160
Genital organs (170-179)	1	0.80	--	1	1.00	--
Urinary organs (180-181)	0	0.63	--	0	0.59	--
Leukemia, aleukemia (204)	0	0.48	--	0	0.35	--
Lymphoma (200-203, 205)	0	0.77	--	0	0.53	--
Other sites	2	1.68	--	1	1.15	--

Major Cardiovascular & Renal Disease (330-334, 400-468, 592-594)	25	30.89	92	29	29.38	118
Vascular lesions--central nervous system (330-334)	6	4.12	165	4	4.71	101
Arteriosclerotic heart disease (420)	13	22.72	64*	21	20.93	119
Hypertensive heart disease (440-443)	1	0.28	--	0	0.28	--
Other hypertensive disease (444-447)	0	0.19	--	1	0.18	--
Chronic nephritis, renal sclerosis (592-594)	0	0.24	--	0	0.17	--
Other	5	3.34	169	3	2.49	--
Influenza and Pneumonia (480-493)	2	1.51	--	2	1.52	--
Cyrrhosis of Liver (581)	2	2.40	--	0	1.12	--
Symptoms, Senility & Ill-Defined (780-795)	2	1.14	--	2	0.70	--
Motor Vehicle Accidents (810-835)	4	2.54	--	0	0.68	--
Other Accidents & External Causes (800-802, 840-962)	1	2.97	--	1	1.16	--
Suicide (963, 970-979)	3	1.64	--	1	0.55	--
Homicide (964, 980-985)	2	1.45	--	0	0.25	--
All Other Causes (residual	11	8.14	143	5	6.58	90
Number of workers		1,337			380	
Number of person-years		6,514			1,771	

[a]Only those with over 5 observed listed; adjusted, except "all causes" by +12.9% for 8 uncertified deaths. [b]Only those with over 5 observed listed; adjusted, except "all causes" by +18.75% for 9 uncertified deaths. *Significantly different from 100 at 0.05 level. **Significantly different from 100 at 0.01 level.

TABLE 13. Standardized Mortality Ratios. Observed and Expected Deaths by Cause, 1971-1975 in Lead Battery Plants, in Men with Less than 10 Years Employment, Compared with Men with 10 or More Years Employment

Cause of Death (ICD No., 1955)	Less than 10 Years Deaths			10 Years or More Deaths		
	Obs.	Exp.	SMR[a]	Obs.	Exp.	SMR[b]
All Causes	77	75.30	102	287	234.98	122
All Causes with Death Certificates	68	75.30	102	258	234.98	122
Malignant Neoplasms (140-205)	19	14.39	149	52	46.41	125
Buccal cavity and pharynx (140-148)	0	0.49	--	1	1.32	--
Digestive organs & peritoneum (150-159)	3	3.65	--	12	12.80	105
Respiratory system (160-164)	8	5.27	172	16	15.67	114
Genital organs (170-179)	2	0.94	--	2	5.06	--
Urinary organs (180-181)	1	0.71	--	2	2.68	--
Leukemia, aleukemia (204)	1	0.55	--	2	1.68	--
Lymphoma (200-203, 205)	2	0.88	--	1	2.35	--
Other sites	2	1.90	--	16	4.85	229

Major Cardiovascular & Renal Disease (330-334, 400-468, 592-594)	*27*	*35.11*	*87*	*149*	*133.67*	*125***
Vascular lesions--central nervous system (330-334)	*6*	*4.82*	*140*	*24*	*22.63*	*118*
Arteriosclerotic heart disease (420)	*15*	*25.64*	*67***	*106*	*94.13*	*126**
Hypertensive heart disease (440-443)	*0*	*0.32*	--	*1*	*1.30*	--
Other hypertensive disease (444-447)	*0*	*0.22*	--	*1*	*0.86*	--
Chronic nephritis, renal sclerosis (592-594)	*0*	*0.27*	--	*1*	*0.77*	--
Other	*6*	*3.84*	*177*	*16*	*14.00*	*127*
Influenza and Pneumonia (480-493)	*0*	*1.87*	--	*8*	*7.04*	*126*
Cirrhosis of Liver (581)	*2*	*2.66*	--	*4*	*4.16*	--
Symptoms, Senility & Ill-Defined (780-795)	*4*	*1.32*	--	*12*	*2.95*	*453***
Motor Vehicle Accidents (810-835)	*0*	*3.25*	--	*5*	*2.79*	*199*
Other Accidents & External Causes (800-802, 840-962)	*3*	*3.63*	--	*2*	*4.78*	--
Suicide (963, 970-979)	*1*	*1.99*	--	*0*	*2.19*	--
Homicide (964, 980-985)	*4*	*1.86*	--	*0*	*0.97*	--
All Other Causes (residual)	*10*			*26*		
Number of workers		*1,638*			*1,526*	
Number of person-years		*8,022*			*6,946*	

*[a]Only those with over 5 observed listed; adjusted, except "all causes" by +13.2% for 9 uncertified deaths. [b]Only those with over 5 observed listed; adjusted, except "all causes" by +11.2% for 29 uncertified deaths. *Significantly different from 100 at 0.05 level. **Significantly different from 100 at 0.01 level.*

TABLE 14. Standardized Mortality Ratios. Observed and Expected Deaths by Cause, 1971-1975 in Lead Battery Plants, in Men with Less than 20 Years Employment, Compared with Men with 20 or More Years Employment

Cause of Death (ICD No., 1955)	Less than 20 Years Deaths			20 Years or More Deaths		
	Obs.	Exp.	SMR[a]	Obs.	Exp.	SMR[b]
All Causes	149	136.19	109	215	174.37	123**
All Causes with Death Certificates	137	136.19	109	189	174.37	123**
Malignant Neoplasms (140-205)	34	26.92	137	37	33.91	124
Buccal cavity and pharynx (140-148)	1	0.88	--	0	0.92	--
Digestive organs & peritoneum (150-159)	8	6.99	124	7	9.48	84
Respiratory system (160-164)	13	9.83	144	11	11.11	113
Genital organs (170-179)	2	1.98	--	2	4.03	--
Urinary organs (180-181)	1	1.38	--	2	2.01	--
Leukemia, aleukemia (204)	1	0.97	--	2	1.21	--
Lymphoma (200-203, 205)	2	1.54	--	1	1.68	--
Other sites	6	3.35	288	12	3.47	510

Major Cardiovascular & Renal Disease (330-334, 400-468, 592-594)	66	67.64	107	110	101.34	124*
Vascular lesions--central nervous system (330-334)	15	9.68	169	15	17.81	96
Arteriosclerotic heart disease (420)	40	49.23	88	81	70.67	131*
Hypertensive heart disease (440-443)	1	0.63	--	0	1.00	--
Other hypertensive disease (444-447)	0	0.42	--	1	0.66	--
Chronic nephritis, renal sclerosis (592-594)	0	0.48	--	1	0.57	--
Other	10	7.20	151	12	10.63	128
Influenza and Pneumonia (480-493)	1	3.37	--	7	4.53	145
Cirrhosis of Liver (581)	5	4.27	127	2	2.55	--
Symptoms, Senility & Ill-Defined (780-795)	9	2.19	447	7	2.09	381
Motor Vehicle Accidents (810-835)	1	4.34	--	4	1.70	--
Other Accidents & External Causes (800-802, 840-962)	3	5.25	--	2	3.16	--
Suicide (963, 970-979)	1	2.84	--	0	1.35	--
Homicide (964, 980-985)	4	2.36	--	0	0.47	--
All Other Causes (residual)	13	17.01	83	20	23.27	98
Number of workers		2,262			902	
Number of person-years		10,960			4,008	

[a]Only those with over 5 observed listed; adjusted, except "all causes" by +8.8% for 12 uncertified deaths. [b]Only those with over 5 observed listed; adjusted, except "all causes" by +13.78% for 26 uncertified deaths. *Significantly different from 100 at 0.05 level. **Significantly different from 100 at 0.01 level.

association other than SMR's for cardiovascular-renal disease and some of its subdivisions, which were consistently higher for those with longer employment. Malignant neoplasms, on the other hand, showed no such relationships.

Standardized mortality ratios as related to estimated exposure to lead. Job titles were categorized into groups estimated as having high, medium, or low exposures to lead. These were used to divide the lead battery production employees and the battery plant workers into groups estimated as having had high, medium, or low exposures. Unfortunately, these estimates involved recent jobs and exposures and almost certainly did not give a valid estimate of lead exposures in the past. There were no consistent associations with any causes of death.

Special consideration of malignant neoplasms of other and unspecified sites. To determine whether this group of diagnoses involved some major primary site of malignancies, the causes on death certificates were reviewed as shown in Table 15. There were 3 tumors of the skin, 2 of the brain, 3 of the thyroid, 1 of bone, 1 of connective tissue, and 9 where primary site was unknown.

Special consideration of "senility or ill-defined conditions." This group of causes, which were particularly high in battery plants, represented deaths listed on death certificates as "not determined," "cause unknown," and "natural causes" (Table 16).

Exposure to lead. The report by Cooper and Gaffey (1975) contained detailed information on urinary and blood lead concentrations. There was relatively little such data obtained before 1960, but by 1970 all but one of the plants included in the study had a program of biological monitoring. It was impossible to relate causes of death to lead levels, because systematic surveillance had begun too recently for most of the deceased workers to have information in their records. In the 1947-1970 study, for example, only 9.7% of the deceased workers had recorded urinary lead values, and only 1.7% had recorded blood lead values.

Review of available data, however, showed that there had been a considerable number with absorption above currently acceptable standards. As shown in Tables 17 and 18, there were many individuals whose average urinary lead levels exceeded 200 μg/L and whose average blood lead levels exceeded 70 μg/100 g or even 100 μg/100 g. One can assume that when surveillance had begun, environmental controls also were improved. There is no doubt that absorption of lead had been even greater in the early days of exposure of these men. For example, 34 smelter workers and 294 battery plant workers had begun employment before 1920.

TABLE 15. Malignant Neoplasms of "Other and Unspecified Sites" in Battery Plants

1955 ICD No.	
190.9	Malignant melanoma
191.3	Squamous carcinoma of right face
191.5	Squamous carcinoma of anus
193.0	Pneumonia, due to frontal glioma
193.9	Metastatic carcinoma of the brain
194	Metastatic carcinoma of thyroid
194	Carcinoma of laryngeal structure due to Ca thyroid and Ca thyroid cartilage
196.4	Carcinomatosis, due to carcinoma right humerus
197.9	Carcinomatosis, due to liposarcoma
199	Carcinomatosis, primary site unknown
199	Cardiorespiratory arrest, due to metastatic squamous cell carcinoma
199	Respiratory failure due to metastatic carcinoma
199	Metastatic carcinoma
199	Asphyxiation due to pneumonia, due to generalized carcinomatosis
199	Respiratory failure due to pneumonia due to carcinoma, metastatis
199	Metastatic squamous cell carcinoma, due to primary site undetermined
199	Squamous cell carcinoma, metastatic, primary unknown
199	Carcinoma of liver (origin unknown)
	Summary
	Skin 3; brain 2; thyroid 2; bone 1; connective tissue 1; unknown 9

DISCUSSION

This additional five years of observation of cohorts of lead production facility employees and lead battery plant workers showed that they continue to have an overall mortality somewhat higher than expected for populations developed from employed groups. They not only do not demonstrate the "healthy worker effect," which would lead one to expect standardized mortality ratios in the range of 80 to 90, but they are actually slightly above 100.

Deaths from malignant neoplasms. Studies in experimental animals are consistent in demonstrating that when administered in dosages close to those that produce fatal lead intoxication,

TABLE 16. Deaths Attributed to "Senility or Ill-Defined Conditions" (ICD 1955: 780-795)

Age	Cause of Death as Given on Death Certificate
	Battery Plants
55	Not determined
56	Bronchial aspiration--gastric contents
59	Not determined
61	Not determined
63	Presumably natural disease
63	Found dead highway
64	Not determined
62	Presumably natural disease
65	Presumably natural disease
65	Presumably natural disease
74	Cardiogenic shock, fluid overload, respiratory failure
76	Presumably natural disease
78	Not determined
80	Presumably natural disease
84	Not determined
86	Not determined
	Lead Production Facility Workers
44	Presumably natural causes
53	Natural cause
67	Presumably natural disease
71	Cause unknown

TABLE 17. Number of Workers with Ten or More PbU Determinations Whose Mean Values Equalled or Exceeded Indicated Concentrations (μg/L) (Uncorrected)

Type of Plant	Numbers of Workers with over 10 determinations	Number with mean values			
		≥150	≥200	≥250	≥300 μg/liter
Smelters	497	289	164	70	27
Battery plants	1,053	249	59	17	7
Total	1,550	538	223	87	34

TABLE 18. Number of Workers Whose Mean PbB Values Over a Working Lifetime Equaled or Exceeded the Indicated Concentrations

Type of Plant	*No. of Workers*	*Number with mean values* ≥*40*	≥*70*	≥*80*	≥*100 μg/100 g*
Smelters	*534*	*457*	*138*	*83*	*25*
Battery plants	*1316*	*976*	*350*	*105*	*52*
Total	*1850*	*1433*	*488*	*188*	*77*

parenterally administered lead salts produce benign and malignant tumors of the renal cortex in rats (Zollinger, 1953; Tonz, 1957; Matthews and Walpole, 1958; Roe *et al.*, 1965; Coogan *et al.*, 1972). Studies in which lead salts were added to the diet also produced kidney tumors (Boyland, 1962; van Esch *et al.*, 1962; Mao and Molnar, 1967; Zawirska and Medras, 1968; van Esch and Kroes, 1969; Coogan *et al.*, 1972). The evidence that lead is capable of producing primary tumors of organs other than the kidney is very weak (Zawirska and Medras, 1968, 1972; Oyasu *et al.*, 1970; Stoner *et al.*, 1976). There is one study by Kobayachi and Okamoto (1974) which supports the hypothesis that lead oxide is co-carcinogenic with benzo(a)pyrene, possibly attributable to the inhibition of detoxifying enzymes.

Epidemiologic studies in man have never demonstrated a convincing excess of malignancies attributable to lead. Dingwall-Fordyce and Lane (1963) reported the mortality in 425 pensioners assembled from lead industries in the United Kingdom. Members of the cohort had to have had at least 25 years of service to be eligible, and nearly all had reached 65 years of age. Based on the numbers of observed and expected malignancies in groups with low, negligible, and heavy exposures, the authors concluded that they found "no evidence to suggest that malignant disease was associated with lead absorption." Malcolm in 1971 reported further observations of the same population. They stated that deaths from cancer in pensioners totalled 19 where 27 were expected, and that in employed workers, there were 33 such deaths with 34 expected. Primary sites were not given. Although the report contained few details, it is probable that any excess of cancer deaths would have been apparent.

In the previous report of the current population by Cooper and Gaffey (1975), and Cooper (1976), smelter workers were found to have some excess of deaths from malignant neoplasms (69 where 54.95 were expected), while battery plant workers had 186 deaths where 180.34 were expected. In both groups this was largely explained by modest excesses in cancers of the digestive organs and respiratory tract. There was no excess of kidney tumors in either population.

The conclusion from the earlier study was that there was inadequate evidence to support lead as an etiologic agent for the small excess of cancer deaths because of the absence of correlation with onset, duration, or level of exposure, and the presence of other possible carcinogens in the work environment of the smelters. The five-years of additional observation produced inexplicably different results. Cancer deaths from malignancy in the smelters were below expected (SMR 89), with the only site in excess being the respiratory tract, where 9 deaths were observed with 8.51 expected (adjusted SMR of 121).

In the battery plants, on the other hand, there were 71 deaths from malignant neoplasms, with 60.9 expected, giving an adjusted SMR of 136 (significant at the 95% confidence level). The excess was largely explained by a slight excess of cancers of the respiratory system (SMR 128) and of "other and unspecified sites" (SMR 293). The latter consisted of 9 with unknown primary site, 3 skin cancers, 2 brain tumors, 2 thyroid tumors, 1 bone tumor, and 1 connective tissue tumor. There was only 1 kidney tumor in the battery plant workers, included in urinary tract tumors.

Again, as in 1947-1970, no consistent relationships could be found between those hired before 1946, those with 10 or more years of employment, 20 or more years of employment, or those known to have worked in jobs with high exposure, and the incidence of cancer.

The updated study thus did not support a carcinogenic role for lead, although it left unanswered questions. The relatively small excess of tumors was found in only one of the cohorts. No kidney tumors were found. A large part of the excess resulted from respiratory tract tumors, where the absence of smoking histories prevented adjustment for this critical factor. An excess of heavy smokers in this work force could easily account for the relatively small excess of lung cancer, as pointed out by Lundin *et al.* (1969) in their study of uranium miners. Also, most of the workers in the study lived in urban areas, which could have been a factor. Against an association with lead was the failure to demonstrate any internal associations within either cohort that related to lead exposures.

Cardiovascular and renal disease. Deaths from major cardiovascular and renal diseases were slightly higher than expected in both cohorts. This could reflect the aging of the populations. It was most notable in the battery plant workers, where it was observed that the average age at time of death was 69.4 years. Vascular lesions of the central nervous system, i.e., cases that would commonly be called "stroke," were in slight excess in both smelter workers and battery plant workers, although the excesses were not statistically significant. There was no excess comparable with that observed by Dingwall-Fordyce and Lane (1963), who reported 24 deaths from cerebral hemorrhage, cerebral thrombosis, and cerebral arteriosclerosis in a heavily exposed group, where 9.3 were expected.

Cooper and Gaffey (1975) showed in 1947-1970 an excess of deaths from "other hypertensive disease" and "chronic nephritis and other renal sclerosis," categories consistent with the known toxic effects of lead upon the kidney. In the 1947-1970 period, there were 14 deaths in smelters and 27 deaths in battery plants from these categories, where 4.7 and 15.8 had been expected. In the current follow-up period, however, these excesses were not duplicated. There were 2 such deaths in the lead production facilities where 0.79 were expected, and 2 in lead battery plants where 2.72 were expected. We have no explanation for the difference.

General considerations. It is extremely important that the relatively small differences in the standardized mortality ratios between the study cohorts and the general population not be overinterpreted. Also, differences between lead production facilities and battery plants, and between various subcohorts may reflect differences in the age distributions of the groups as well as different environmental influences. As Sterling (1964) pointed out, the factors that produce the "healthy worker effect" lead to working populations not being comparable to the general population. The differences are not constant or predictable. As Sterling puts it:

> "In comparison with the mortality or morbidity in the general population, a group of industrial employees is likely to exhibit low mortality and morbidity from all causes, during their earlier and middle years (i.e., from approximately 20 to 60). However, when this population ages, it contains a number of excessively prolonged survivors, when contrasted with the general population or with the population in the same residence area. This will make for a comparatively higher mortality during this period. Especially will this be true for such categories as cancer and cardiovascular diseases, since failing to die from many of the ordinary ills and obstacles, they will tend to die with degenerative diseases...."

While it is ordinarily assumed that the age adjustments involved in calculating SMR's correct for this factor, the relationships are actually too complex for one to be sure of this. It is essential, therefore, when comparing the SMR of one population with the SMR of another, that one be sure of age comparability. That the age of lead production facility workers was strikingly different from that of lead battery plant workers is shown in Table 3 and by the higher average age of those who died. Further analyses are planned to elucidate the contribution of these factors.

SUMMARY

This is a continuation of a study of 7,032 men who had been employed for one or more years in lead production facilities or lead battery plants in the United States. Mortality during the period 1947-1970 was reported by Cooper and Gaffey in 1975. The present report describes the mortality of 5,490 members of the cohort who were surviving as of January 1, 1971 during a subsequent follow-up period, 1971-1975. There were 491 deaths during 22,253 person-years of observation. Total mortality was slightly higher than expected, standardized mortality ratios (SMR's) being 108 and 117 for smelters and battery plants, respectively. There was a slight deficit in malignant neoplasms in smelters (SMR = 89) and a slight excess in battery plants (SMR = 136). A major portion of the latter was based on cancers of unknown primary site. SMR's for lung cancer were 121 and 128 in smelters and battery plants, respectively.

There was no demonstrable correlation between elevated SMR's and either duration or intensity of exposure to lead. Deaths from hypertensive disease or from chronic nephritis and renal sclerosis, which has been found to be in significant excess in the period 1947-1970, were fewer than expected during the period 1971-1975. Review of blood and urine lead concentrations indicated that many members of the cohort had had lead exposures greatly exceeding current standards. Because of differences in the age distributions of the two main cohorts and between various sub-cohorts, it was pointed out that differences in SMR's should not be overinterpreted.

Acknowledgment

This study was sponsored by the International Lead Zinc Research Organization, Inc., and performed by Equitable Environmental Health, Inc. The present address of the author is: W. Clark Cooper, M.D., 2150 Shattuck Avenue (Suite 401), Berkeley, California 94704.

REFERENCES

Boyland, E., Dukes, C. E., Grove, P. L., and Mitchley, B. D. (1962). The induction of renal tumours by feeding lead acetate to rats. *Brit. J. Cancer 16,* 283-288.

Chiang, C. L. (1961). Standard error of the age-adjusted death rate. *Vit. Stat. Spec. Rept. 47,* 275-285.

Coogan, P., Stein, L., Hsu, G., and Hass, G. (1972). The tumorigenic action of lead in rats. *Lab. Invest. 26,* 473 (Abstract).

Cooper, W. C. (1976). Cancer mortality patterns in the lead industry. *Ann. N.Y. Acad. Sci. 271,* 250-259.

Cooper, W. C. (1978). Mortality in workers in lead production facilities and lead battery plants during the period 1971-1975. Report to the International Lead Zinc Research Organization, January 31, 1978, New York.

Cooper, W. C., and Gaffey, W. R. (1975). Mortality of lead workers. *J. Occup. Med. 17,* 100-107.

Cooper, W. C., and Gaffey, W. R. (1975). Mortality study of lead workers. *Arh. Hig. Rad. Toksikol.* (Zagreb) 26, 209-229.

Dingwall-Fordyce, I., and Lane, R. E. (1963). A follow-up study of lead workers. *Brit. J. Ind. Med. 20,* 313-315.

Guralnick, L. (1963). Mortality by occupation and cause of death among men 20-64 years of age: U.S. 1950. *Vit. Stat. Spec. Rept. 53(3).*

Gaffey, W. R. (1976). A critique of the standardized mortality ratio. *J. Occup. Med. 18,* 157-160.

Kobayashi, N., and Okamoto, T. (1974). Effects of lead oxide on the induction of lung tumors in Syrian hamsters. *J. Nat. Cancer Inst. 52,* 1605-1607.

Lundin, F. E., Jr., Lloyd, J. W., Smith, E. M., Archer, V. E., and Holaday, D. E. (1969). Mortality of uranium miners in relation to radiation exposure, hard-rock mining, and cigarette smoking--1950 through September 1967. *Health Phys. 16,* 571-578.

Malcolm, D. (1971). Prevention of long-term sequelae following absorption of lead. *Arch. Environ. Health 23,* 23, 292-298.

Mao, P., and Molnar, J. J. (1967). The fine structure and histochemistry of lead-induced renal tumors in rats. *Mer. J. Path. 50,* 571-603.

Matthews, J. J., and Walpole, A. (1958). Tumours of the liver and kidney induced in Wistar rats with 4'-fluoro-4-aminodiphenyl. *Brit. J. Cancer 12,* 234-241.

Oyasu, R., Battifora, H. A., Clasen, R. A., McDonald, J. H., and Hass, G. M. (1970). Induction of cerebral gliomas in rats with dietary lead subacetate and 2-acetylaminofluorene. *Cancer Res. 30,* 1248-1261.

Roe, R. J. C., Boyland, E., Dukes, C. C., and Mitchley, B. C. (1965). Failure of testosterone or xanthopterin to influence the induction of renal neoplasms by lead in rats. *Brit. J. Cancer 19,* 860-866.

Sterling, T. D. (1964). Epidemiology of disease associated with lead. *Arch. Environ. Health 8,* 333-348.

Stoner, G. D., Shimkin, M. B., Troxell, M. C., Thompson, T. L., and Terry, L. S. (1976). Test for carcinogenicity of metallic compounds by the pulmonary tumor response in strain A mice. *Cancer Res. 36,* 1744-1747.

Tonz, O. (1957). Renal tumors in rats with lead phosphate. *Z. Ges. Exper. Med. 123,* 361-377.

Van Esch, G. J., Van Genderen, H., and Vink, H. H. (1962). The induction of renal tumors by feeding of lead acetate to rats. *Brit. J. Cancer 16,* 289-297.

Zawirska, B., and Medras, K. (1972). Role of the kidneys in disorders of porphyrin metabolism during carcinogenesis induced by lead acetate. *Arch. Immunol. Therap. Exptl. 20,* 257-272.
Zawirska, B., and Medras, K. (1968). Tumoren und Storungen des Prophyrinstoffwechsels bei Ratten mit chronischer experimenteller Bleiintoxikation. *Zbl. allg. Path. 111,* 1-2.
Zollinger, H. U. (1953). Durch chronische Bleivergiftung erzeugte Nierendadenome und carcinome bei Ratten und ihre Beziehungen zu den entsprechenden Neubildungen des Menschen. *Virchows Arch. 323,* 694-710.

DISCUSSION OF PAPER BY DR. W. CLARK COOPER

Betsy T. Kagey, Downstate Medical Center: I have a question on the idea of healthy worker effect. I'm not sure I followed your population. You stated that these workers had at least one year of exposure and were hired during this particular period of time, or up to 1970. Was your study population hired continually up to 1970?

Dr. W. Clark Cooper: Not necessarily. The population included individuals who were working as of the first of January, 1945 and others who were hired later. But Bill Gaffey did not start his period of observation until the first of January, 1946, so that individuals who were working as of January 1, 1946, or who were hired subsequently, had to have worked for at least one year. This is why I separated out those hired before 1945 and those hired after.

Kagey: Would those hired after 1945 influence the results, the SMR's? Wouldn't they lower, or what would you expect to see, as an SMR?

Cooper: Certainly they should be the ones who should exhibit the healthy worker effect, if it occurred. To some extent, this was true before this period of observation began. That effect was demonstrable, I think, in the previous report. By the time this study began, the great majority of the people had been employed for 5, 10 or 15 years or more. Are you inferring that you should be able to demonstrate the healthy worker effect in the more recent hires?

Kagey: Yes. Was your population exposed to other materials, for example, cadmium in the operations? You just sort of briefly talked about lead.

Cooper: There were other exposures. In the primary smelters there were definitely opportunities for other exposures, arsenic and cadmium. In the battery plants, there were some other low level arsenic exposures in some areas. I think they were well below even the action level of the present standard. There were also some sulfuric acid exposures in some of the battery plants. There were other exposures in the smelter.

Kagey: Did you take any previous work history of those employed at these particular plants and where they worked prior to their entry?

Cooper: No, the cohort was generated from personnel records. Practically none of the personnel records, particularly in the early days, contained this kind of detailed information.

Michael K. Williams, Chloride Automotive: This study was of workers from 1971-1975. Your previous study was up to 1971. I wonder why you don't combine the results to get larger numbers?

Cooper: The first study was 1947 through 1970. Your question is the same question that ILZRO keeps asking me. Why not pool the entire results. It's been entirely a technical and computer problem, in that we cannot just add the numbers in the person years in making the calculations. We do plan to do this, that is, combine them over the entire period. I still believe it's desirable to break down the deaths in various time periods. I think we can learn things from that, too. But we do intend to combine them at the next go around.

Paul B. Hammond, University of Cincinnati: I noticed that you made a special point of the fact that the excess deaths in the malignancy category did not involve renal tumors.

Cooper: Yes.

Hammond: Is there a good basis for believing that because you can demonstrate renal tumors in rats and mice that, therefore, the tumor you would see in man would necessarily express itself as a renal tumor? For example, there is another study demonstrating gliomas in mice, and I just don't quite see the inference, if any.

Cooper: Clearly, no one can get up and say that animal experiments necessarily determine the site of tumors in other species. I still feel that since in animals the kidney is the locus for pathologic effects other than tumor and in man the kidney has been demonstrated as being a site of pathologic effects of lead, that it is of interest that you don't find any. Clearly, one would be on very shaky ground, in fact, it would be indefensible to say that you have to find tumors at the same site. But we did think it was an interest that we found neither an excess of brain tumors nor an excess of kidney tumors, the only two that had been described in animals.

HEALTH STUDY OF A LEAD-EXPOSED POPULATION

M. Fugaš
M. Šarić

Institute for Medical Research and Occupational Health
Zagreb, Yugoslavia

The objective of the study was to learn more about possible general health effects caused by a prolonged exposure to lead. The study was carried out on a population living around a lead smelter, exposed to lead for several generations, showing all biological changes indicative of lead exposure, but no clinical symptoms of lead poisoning (U.S. EPA, 1977; Fugaš *et al.*, 1976). A population living in an area of similar geographic and climatic characteristics and of similar urbanization level, but without a large lead source, served as a control group.

Both areas are river valleys about 500 m above sea level, surrounded by mountains up to 1600 m high. According to the 1973 census, the number of inhabitants in the country including the exposed area was about 24,000 and in the control area 15,200. As already reported (Fugaš *et al.*, 1976), the population in the control area was relatively older than in the exposed area. The most frequent occupation of men in the control group and of those in the exposed group who are not employed in the lead smeltery is wood exploitation and building.

Several different approaches were used in the study: retrospective analysis of vital statistics data, hospital treatment data, and causes of death, and health study of family samples by means of a questionnaire, clinical, and laboratory examinations. To define exposure to lead of the investigated population, a survey of environmental contamination by lead was carried out comprising determination of lead in air, depositions, household dust, drinking water, and food.

ISBN 0-12-460520-6

ENVIRONMENTAL CONTAMINATION BY LEAD

Sampling and Analysis

Weekly samples of airborne lead were taken at five sites in the exposed area continuously over the period of investigation. Samples were collected on membrane filters from about 14 m^3 of air. In the control area, 24-hour samples were collected continuously over the year on millipore filters from about 200 m^3 of air. Monthly samples of deposited dust were collected at four sites in the exposed and at one site in the control area over a one year period. The samples were collected in plastic containers of 1.5 liter volume with a 10 cm diameter opening.

Composite samples of household dust were collected in randomly selected homes of the examined families. In each home, dust was collected from the floor of the living room and the kitchen by means of a pump with a special adapter, membrane filters serving as a sampling surface. Thirty-five samples were collected in the in the exposed area and 12 in the control.

Samples of drinking water were collected in 5 liter plastic containers from five sites in each area (public and domestic water supplies, wells, and springs). Samples of various foodstuffs were repeatedly collected in both the exposed and the control areas in foodstores or in the households of examined families, either in the kitchen from food prepared for a meal, or from the garden. The raw foodstuffs were washed under tap water in order to simulate the conditions of preparing food for cooking.

All dust samples were analyzed by atomic absorption spectrophotometry after being treated with nitric acid, evaporated to dryness, and redissolved in 1% EDTA solution (pH 8). Water samples were analyzed in the same way, after interfering ions were removed by ion exchange (Korbisch and Sorio, 1975). Samples of foodstuffs were digested in a autoclave and determined by flameless atomic absorption spectrophotometry (standard addition method).

RESULTS AND DISCUSSION

Summarized results of environmental measurements are shown in Table 1. Yearly cycles of monthly averages of airborne and deposited lead by sites are shown in Figures 1 and 2.

The concentration of lead in air of the exposed area (Fig. 1) shows a marked winter maximum due to frequent calm periods with temperature inversions in winter and rainy periods in summer. Yearly averages did not change very much in the last three years. It was shown earlier (U.S. EPA, 1977; Fugaš *et al.*, 1973) that due to the indoor/outdoor concentration ratio of lead in air, the weighted average exposure to lead in air of the general population

TABLE 1. *Mean Concentrations of Lead in Environmental Media*

		Exposed Area			Control Area			
Environmental Medium	Units	Mean ($\overline{C}_E$)	Range	Comments	Mean ($\overline{C}_C$)	Range	Comments	Ratio $\overline{C}_E/\overline{C}_C$
Air	$\mu g/m^3$	17	0.7-115	Five year measurements at five sites	0.10	0.09-0.76	One year measurements at one site	170:1
Deposit	mg/m^2/month	173	5-612	One year measurements at four sites	2.3	0.3-7.5	One year measurements at one site	75:1
Household dust	$\mu g/g$	4.077	1.460-9.650	Thirty-five samples	158	50-310	Twelve samples	25:1
Drinking water	$\mu g/g$	10.7	7.4-13.2	Five different water sources	1.9	0.3-2.6	Five different water sources	6:1
Daily food	$\mu g/g$	2.52		Weighted average for daily consumed food	0.44		Weighted average for daily consumed food	6:1

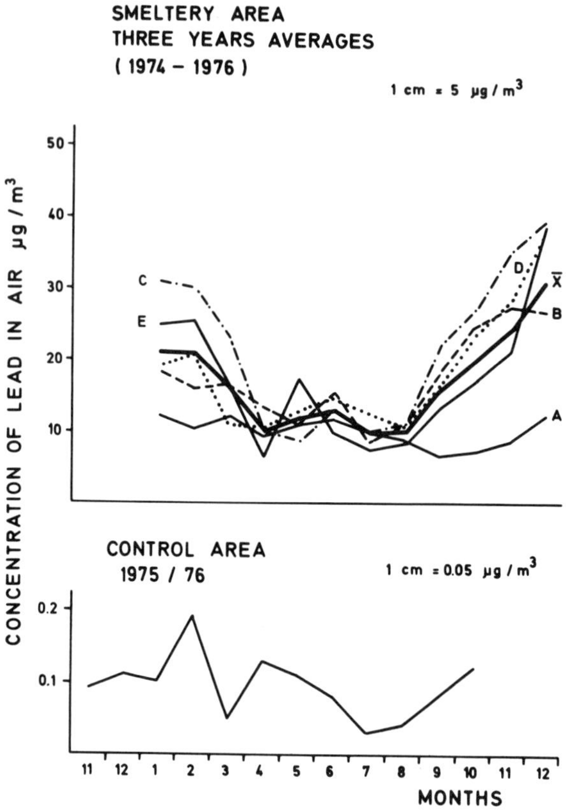

Fig. 1. Yearly cycles of mean monthly concentrations of lead in air by sites.

in the smeltery area is about 50% to 80% of the concentration measured outdoors (depending on subjects' occupation and season) and that the MMD of lead particles in the exposed area is 2.0 to 2.5 μm (less than 5% below 0.5 μm and about 95% below 5 μm). The yearly cycles of deposited lead is less pronounced (Fig. 2).

There was a marked difference in lead content of all environmental media between the exposed and the control area, but the concentration of lead in water of the exposed area was, nevertheless, below the WHO standard (50 μg/l) and cannot be considered as a serious source of lead intake. The average daily lead intake by food calculated from the composition of the diet obtained by a questionnaire (Fugaš *et al.*, 1976) and from the average lead content of various foodstuffs amounted to 2.9 mg for the exposed and to 0.51 mg for the control area.

Using the obtained data on lead in air, water, and diet, and supposing that 10% of the ingested lead is absorbed while 50% of

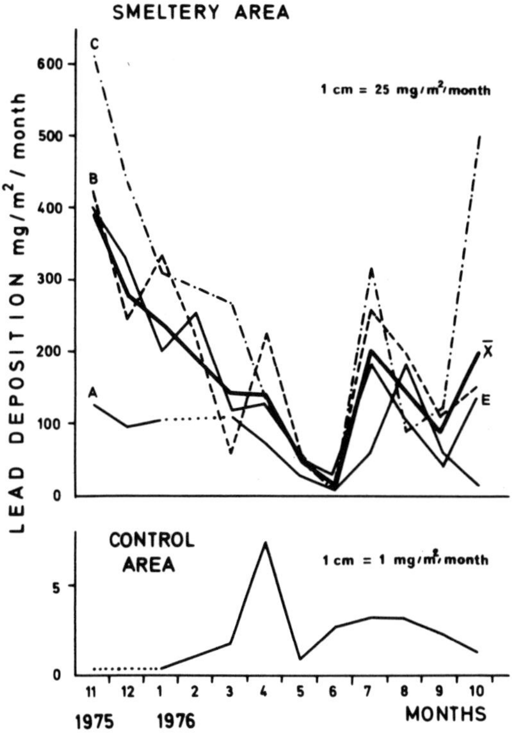

Fig. 2. Yearly cycles of monthly lead deposition by sites.

the inhaled lead is retained and completely absorbed, as suggested by Zielhuis (1973), the lead absorption of the exposed population is estimated to be about 8 to 10 times higher than that of the control group. The contribution of air lead to the total lead absorbed may amount to 30% in the exposed and to 1.5% in the control area. Thus, oral lead intake is the main source of lead body burden of the control population, while in the exposed area, airborne lead plays an important role.

RETROSPECTIVE STUDY

Analysis of Vital Statistics Data

The data on birth rate, infant mortality, general mortality, natural population increase, rate of total and spontaneous abortion, and fetal mortality collected in the lead smelter area and in the control area were analyzed comparatively for the period from

1966 to 1973. The results of the analysis were presented and discussed earlier (Fugaš *et al.*, 1976).

Analysis of Hospital Treatment Data

Data about hospital treatment of the inhabitants from both the smeltery and the control area were collected for the same period. The rate of selected groups of diseases treated in hospitals was presented earlier (Fugaš *et al.*, 1976). A more detailed analysis on hospital treatment is shown in Tables 2 and 3. The rate of hospital treatment was higher in the exposed than in the control area. This can be due partly to slightly better facilities for hospital treatment in the smeltery area, although the health insurance system is practically the same in both areas. On the other hand, the population from the control area was older than the exposed population.

The structure of diseases treated in hospitals indicates that there were some differences in the specific hospital morbidity rate. As already stressed (Fugaš *et al.*, 1976), in the interpretation of

TABLE 2. Rate of Hospital Treatment for Both Smelter and Control Areas by Sex (1966-1973)

Year		Rate of treated per 1000	
		Smeltery Area	Control Area
1966	m.	140.5	85.1
	f.	164.7	90.2
1967	m.	101.6	79.7
	f.	131.3	86.5
1968	m.	115.3	63.9
	f.	137.9	84.6
1969	m.	100.1	67.8
	f.	126.8	79.5
1970	m.	110.0	67.5
	f.	121.7	78.1
1971	m.	98.9	67.4
	f.	118.5	81.4
1972	m.	107.0	64.6
	f.	133.7	83.9
1973	m.	114.1	66.2
	f.	130.9	89.1
$\overline{X}_m$		111	70.2
$\overline{X}_f$		133	84.2
$\overline{X}_{tot}$		122	77

TABLE 3. Hospital Treatment Rate of Selected Disease for the Inhabitants of Both Smeltery(S) and Control (C) Areas 1966-1973

Diseases (Groups of Diseases)	All Inhabitants S	All Inhabitants C	Males S	Males C	Females S	Females C
1. Malignant neoplasm (140-209)	4.7	4.7	5.0	4.3	4.3	5.1
2. Diseases of the thyroid gland (240-246)	2.9	2.2	0.6	0.5	5.2	3.8
3. Diabetes mellitus (250)	1.0	0.6	0.9	0.6	1.2	0.5
4. Diseases of other endocrine glands (251-258)	0.2	0.1	0.2	0.2	0.2	0.0
5. Other metabolic diseases (270-279)	0.3	0.1	0.3	0.2	0.4	0.1
6. Diseases of the blood and blood-forming organs (280-289)	1.1	0.5	1.0	0.6	1.2	0.3
7. Psychoses (290-299)	1.0	1.5	0.9	1.7	1.1	1.4
8. Alcoholism (303)	1.0	0.5	1.8	1.0	0.3	0.0
9. Mental retardation (310-315)	0.1	0.1	0.1	0.2	0.1	0.0
10. Hypertensive disease (400-404)	0.9	0.5	0.7	0.3	1.2	0.6
11. Ischemic heart disease (410-414)	3.3	1.4	3.0	1.6	3.6	1.2
12. Cerebrovascular disease (430-438)	1.4	1.0	1.4	1.2	1.4	0.8
13. Diseases of arteries, arteriolas and capillaries (440-448), diseases of veins and lymphatics, and other diseases of circulatory system (450-458)	4.3	1.9	4.4	1.7	4.3	2.0
14. Ulcer of stomach, duodenum, peptic, gastrojejunal (531-534)	2.1	1.4	3.5	2.3	0.8	0.7
15. Nephritis and nephrosis (580-584)	1.1	0.5	1.3	0.6	0.9	0.4
16. Other diseases of urinary system (590-599)	5.1	2.0	3.3	1.6	6.8	2.4
17. Abortion (640-645)					13.3	10.8
18. Complications of pregnancy, childbirth and the puerperium excluding abortion (630-639, 651-661, 670-678)					6.8	3.3
19. Diseases of the musculoskeletal system and connective tissue (710-738)	4.2	2.4	4.9	2.8	3.4	2.0
20. Congenital anomalies (740-759)	1.5	1.5	1.8	1.5	1.2	1.4

[a]International classification codes.

these data one should be aware that diseases treated in hospitals do not reflect a general morbidity pattern. This is particularly true for the diseases which are mainly treated by general practitioners and in outpatient departments. A different attitude of general practitioners toward hospital treatment may also have some influence on the specific hospital treatment rate.

Analysis of Causes of Death

On the basis of health statistics data, diseases as cause of death were analyzed in both the smeltery and the control area for the 1966-1973 period. The collected data were classified by age groups and calculated as mean rate per 1000 inhabitants for the analyzed period. In the classification of the causes of death, the International Classification of Diseases was used.

Tables 4 and 5 show the mean age and specific death rate by diseases and groups of diseases for males and females, respectively. Symptoms and inadequately defined conditions were mostly equally represented as cause of death in the compared populations, which might indicate that there was no significant difference in medical-diagnostic criteria for establishing causes of death.

HEALTH STUDY IN FAMILY SAMPLES

In the smeltery area, two population groups were selected for assessment of health status. Group I consisted of 100 families with a father occupationally exposed to lead. Group II consisted of 100 families from the same area, but father was not employed in the smeltery. Group III consisted of 95 families from the control area. In all the three groups, there was at least one child of school age (7-15 years of age) per family. The program of the investigation was described earlier (Fugas *et al.*, 1976).

The number of families summoned for examination and those examined (according to family groups), as well as the number of families that underwent special laboratory examinations in which the level of lead exposure (absorption) and its biological effect were determined, are shown in Table 6. The response of the selected subjects to examinations was satisfactory.

The age of the compared parents is shown in Table 7, and the age of children in Table 8. Although efforts were made to form groups with approximately the same age distribution, boys in group III were a little younger (Table 8). The difference in arithmetic means between male subjects in group II and those in two other groups was also statistically significant. The frequency distribution analysis showed, however, that most male subjects between the 10th and 90th percentiles, in all the three groups were in the same age range and that there was little difference in medians

TABLE 4. Mean Age Specific Death Rate by Diseases for the Male Inhabitants in Both Smelter (S) and Control (C) Areas 1966-1973

Age Groups	0-15		15-25		25-45		45-65		65-75		>75	
Groups of Diseases	S	C	S	C	S	C	S	C	S	C	S	C
Tuberculosis 0.10-0.19	0	0	0	0	0.09	0	0.28	0.12	1.36	0.27	0	2.49
Intestinal infections, bacterial, viral and parasitic diseases, 001-136	0.10	0.09	0	0	0	0.16	0.38	0	0	0.81	0.94	0
Malignant neoplasm 140-239	0.10	0.09	0.14	0.28	0.53	0.08	5.45	3.12	11.56	11.61	17.86	17.43
Diabetes mellitus, 250	0	0	0.07	0	0.04	0	0	0	0.68	0	0	0
Diseases of the nervous system, 320-358	0.05	0	0	0	0	0	0	0	0	0	0	0
Hypertensive disease 400-404	0	0	0	0	0.04	0.08	0.19	0.12	0.34	0	0	0.83
Ischemic heart disease 410-414	0	0	0	0	0.04	0.40	1.79	1.32	5.10	5.13	15.04	4.98
Other forms of heart disease, 420-429	0	0	0	0	0	0	0.56	0.36	0	0.81	2.82	5.81
Cerebrovascular disease, 430-438	0.05	0	0	0	0.04	0.32	0.94	1.44	11.56	8.91	17.86	22.41
Other diseases of circulatory system 390-458	0	0	0	0.14	0.04	0.08	1.50	1.08	8.16	8.10	30.08	41.50
Influenza and pneumonia 470-486	0.05	0	0	0	0.04	0.16	0.94	0.84	2.38	2.43	7.52	10.79
Diseases of the digestive system and others up to 560	0.10	0	0	0.14	0	0.08	1.79	2.16	5.78	7.02	23.50	14.94
Nephritis and nephrosis 580-584	0	0	0	0	0.04	0.16	0.66	0.84	1.70	0.54	0.94	0
Certain causes of perinatal mortality, 760-779	0.52	0.9	0	0	0	0	0	0	0	0	0	0
Symptoms and ill-defined conditions, 780-796	0.21	0.09	0.14	0.14	0.18	0.08	0.38	0.24	1.36	0.27	5.64	4.98
Other diseases up to 796	0.21	0.81	0	0.28	0.70	0.88	2.07	2.64	13.26	7.29	18.80	18.26
Accidents, poisoning and violence, N 800-N 999	0.52	0.9	0.78	1.40	0.88	1.76	3.48	1.32	6.46	2.70	9.40	4.15

TABLE 5. Mean Age Specific Death Rate by Diseases for the Female Inhabitants in Both Smeltery (S) and Control (C) Areas 1966-1973

Age Groups / Groups of Disease	0-15		15-25		25-45		45-65		65-75		>75	
	S	C	S	C	S	C	S	C	S	C	S	C
Tuberculosis 0.10-0.19	0	0	0	0	0.09	0	0.16	0.09	0.25	0	0	1.59
Intestinal infections, bacterial viral and parasitic diseases, 001-136	0	0	0.08	0	0	0.08	0.40	0	0	0.23	0.54	0
Malignant neoplasm 140-239	0.05	0	0	0.15	0.38	0	2.16	2.88	4.08	4.83	8.64	5.83
Diabetes mellitus, 250	0	0	0	0	0	0	0.08	0	0.50	0	0	0
Diseases of the nervous system, 320-358	0	0	0	0	0	0	0	0	0	0	0	0
Hypertensive disease 400-404	0	0	0	0	0	0.08	0.24	0	1.00	0.92	1.08	0
Ischemic heart disease 410-414	0	0	0	0	0.14	0.08	0.80	0.45	3.75	2.99	16.74	12.19
Other forms of heart disease, 420-429	0	0	0	0	0	0	0.16	0	0	0.92	0.54	6.36
Cerebrovascular disease, 430-438	0	0.09	0.08	0	0.05	0.24	0.96	0.72	7.50	7.13	19.44	24.38
Other diseases of circulatory system 390-458	0	0	0	0	0.05	0	0.80	0.45	6.00	8.05	24.84	38.69
Influenza and pneumonia 470-486	0.05	0.18	0	0	0	0	0.56	0.63	0.25	3.91	5.94	9.54
Diseases of the digestive system and others up to 560	0.10	0.27	0.08	0	0	0	0.48	0.45	1.50	4.14	9.18	12.72
Nephritis and nephrosis 580-584	0	0	0	0	0	0	0.40	0.09	0.50	0.23	2.16	1.59
Certain causes of perinatal mortality, 760-779	0.97	1.26	0	0	0	0	0.08	0	0	0	0	0
Symptoms and ill-defined conditions, 780-796	0.27	0	0	0	0	0	0.08	0.27	0.25	0	5.94	5.30
Other diseases up to 796	0.22	0.45	0.08	0	0.14	0.24	1.04	1.35	4.75	3.68	14.58	6.89
Accidents, poisoning and violence, N 800-N 999	0.70	0.27	0.56	0.45	0.24	0.08	0.64	0.45	1.00	0.69	7.56	1.59

TABLE 6. Number of Summoned and Examined Families

Group	No. of Families Summoned for Examination	Responded	No. of Families With All Members Examined	No. of Families Tested for Lead Exposure
I	100	90	85	36
II	100	73	69	32
III	95	88	80	30

TABLE 7. Age Distribution in Parents

Fathers

	I		II		III	
Age Range	N	%	N	%	N	%
26-30	3	3.5	1	1.5	1	1.2
31-35	12	14.1	10	14.5	11	12.9
36-40	27	31.8	16	23.2	27	31.8
41-45	27	31.8	22	31.9	29	34.1
46-50	11	12.9	12	17.4	9	10.6
51-55	5	5.9	2	2.9	4	4.7
56-60	-	-	3	4.4	3	3.5
61-65	-	-	3	4.4	1	1.2
50% younger than	40		42		41	
Between 10th and 90th percentile	33-48		33-52		33-49	

Mothers

	I		II		III	
Age Range	N	%	N	%	N	%
26-30	9	10.6	7	10.1	6	6.9
31-35	19	22.4	16	23.2	23	26.4
36-40	26	30.6	17	24.6	28	32.2
41-45	22	25.9	15	21.7	11	12.6
46-50	8	9.4	13	18.8	15	17.2
51-55	1	1.2	1	1.5	3	3.5
56-60	-	-	-	-	1	1.2
50% younger than	37		38		38	
Between 10th and 90th percentile	31-45		31-47		31-48	

TABLE 8. *Age Distribution in Children*

	Boys			Girls		
	Group			Group		
Age Range	I	II	III	I	II	III
7-9	7	4	17	12	5	9
10-12	13	16	26	18	15	19
13-15	23	14	8	12	14	7
	43	34	51	42	34	35
50% younger than	12.5	12.0	10.5	11.0	12.2	10.7
Between 10th and 90th percentile	9.0-15.3	9.1-15.0	7.6-13.4	8.3-14.2	9.0-15.0	8.2-14.0

(Table 7). Therefore, the groups may be considered comparable.

More than 50% of males in group I represent an autochtonous population that has lived in the area where the smeltery is located for generations. Group II, from the same area, showed an even higher percentage of autochthonous population (over 60%). However, the highest percentage of autochthonous population is found in group III, about 90%. The highest percentage of autochthonous population is also found in female group III (83%), and the lowest in group I (57%).

The compared groups showed no essential differences in socioeconomic characteristics. All males in group I were smeltery workers, and those in group II and III predominantly worked in wood exploitation and building industry. Nineteen percent (19%) of males in group III were miners. About 50% of females in groups I and III were housewives. The percentage of housewives in group II was even higher (61%).

The percentage of smokers in the compared groups of males ranged from 63.5% (group I) to 40% (group III). The highest percentage of female smokers was found in group II (23%), while in two other female groups, the percentage of smokers was 12%.

The largest alcohol consumption was recorded among males in group III, and the lowest in group I. On the other hand, the largest female alcohol consumers came from group I, and the lowest from group III. Females, however, even if they consumed alcohol, consumed little (1-2 glasses of wine a day).

Objective assessment of lead exposure, i.e., lead absorption level, based on blood lead level and biological indices, is shown for the three groups in Tables 9, 10, and 11. The analysis of

TABLE 9. Analysis of Biological Indices of Lead Exposure

Male Inhabitants

Biological Index	*Group I* *N*	*x̄*	*SD*	*Group II* *N*	*x̄*	*SD*	*Group III* *N*	*x̄*	*SD*
E /N×10⁶/	83	4.15	0.33	69	4.25	0.34	84	4.50	0.32
Hb /g/100 ml/	84	13.9	1.40	68	14.6	1.45	85	15.2	1.58
Hct /cm³/100 ml/	36	42.4	2.07	33	44.0	1.83	30	44.6	1.86
BpE /N/10⁶E/	35	1700		33	459		30	160	
Rtc /N/10³E/	34	17	6.41	31	13	4.75	30	9	4.20
PbB /µg/100 ml/	36	93.1	12.1	33	65.8	13.8	31	34.2	10.6
EPP /µg/100 ml/	19	492	168	15	143	90.5	11	27.2	29.6
ALAD /u/1 ml E/	36	13.0	13.9	33	33.9	20.5	30	142	35.7
CPU /µg/100 ml/	36	144	139	32	26.7	26.3	30	13.2	5.92
ALAU /mg/100 ml/	36	3.72	2.19	32	0.80	0.59	30	0.38	0.12

Note: Methods of determination are described in the Final Report ILZRO Project LH-171.

TABLE 10. Analysis of Biological Indices of Lead Exposure

Female Inhabitants

Biological Index	*Group I* *N*	*x̄*	*SD*	*Group II* *N*	*x̄*	*SD*	*Group III* *N*	*x̄*	*SD*
E /N×10⁶/	85	3.93	0.29	67	3.9	0.32	86	4.02	0.32
Hb /g/100 ml/	85	13.1	1.26	67	13.2	1.31	87	13.5	1.45
Hct /cm³/100 ml/	36	41.3	2.11	33	41.6	1.32	30	41.2	1.94
BpE /N/10⁶E/	36	342		32	681		30	233	
Rtc /N/10³E/	35	16	14	30	15	6.8	29	10	4.4
PbB /µg/100 ml/	36	48.0	12.4	33	52.7	10.7	30	22.8	6.05
EPP /µg/100 ml/	19	118	11	15	112	90.6	11	13.9	3.17
ALAD /u/1 ml E/	36	54.2	29.6	33	48.0	25.4	30	172	36.4
CPU /µg/100 ml/	36	18.8	8.92	33	18.9	13.6	29	11.2	4.60
ALAU mg/100 ml/	36	0.64	0.27	33	0.68	0.26	29	0.39	0.15

laboratory findings confirms the supposed differences in lead exposure level among male groups. In females, there was a significant difference in laboratory findings between those living in the exposed and in the control areas, but the difference in findings between group I and group II was not statistically significant. A comparison of the same variables in children showed similar results as in females. Lead exposure of females and children in each of the three groups was also similar (Table 12).

Data on family and personal medical history did not show any particular differences between the compared groups. However, some interesting findings concern subjective complaints of the examined subjects. They are shown in Table 13 for males and Table 14 for females.

TABLE 12. Statistically Significant Differences in Biological Indices of Lead Exposure ($p < 0.05$)

Biological Index	Males Groups I/II	Males Groups I/III	Males Groups II/III	Females Groups I/II	Females Groups I/III	Females Groups II/III	Children Groups I/II	Children Groups I/III	Children Groups II/III
E		I<III	II<III					I<III	II<III
Hb	I<II	I<III	II<III		I<III				
Hct	I<II	I<III						I<III	
BpE	I>II	I>III							
Rtc	I>II	I>III	II>III		I>III	II>III		I>III	II>III
PbB	I>II	I>III	II>III		I>III	II>III		I>III	II>III
EPP	I>II	I>III	II>III		I>III	II>III	I<II	I>III	II>III
ALAD	I<II	I<III	II<III		I<III	II<III		I<III	II<III
CPU	I>II	I>III	II>III		I>III	II>III		I>III	II>III
ALAU	I>II	I>III	II>III		I>III	II>III		I>III	II>III

TABLE 11. Analysis of Biological Indices of Lead Exposure

Children	Group I			Group II			Group III		
Biological Index	N	$\bar{x}$	SD	N	$\bar{x}$	SD	N	$\bar{x}$	SD
E /N×10^6/	85	3.93	2.77	69	3.87	0.33	85	4.03	0.32
Hb /g/100 ml/	85	13.2	1.15	69	13.0	1.21	87	13.3	1.4
Hct /cm^3/100 ml/	36	41.1	1.7	34	41.0	1.2	30	41.3	1.7
BpE /N/10^{6}E/	36	341		33	412		31	93.5	
Rtc /N/10^{3}E/	33	11	4.29	33	11	4.28	29	8	1.55
PbB /μg/100 ml/	36	52.8	12.2	34	57.2	11.7	30	26.2	5.0
EPP /μg/100 ml/	19	97.5	75.2	14	189	112	13	17.0	8.40
ALAD /u/1 ml E/	36	46.1	25.6	34	40.0	21.6	30	178	19.3
CPU /μg/100 ml/	36	17.5	12.3	34	19.6	11.1	30	11.3	3.63
ALAU /mg/100 ml/	36	0.68	0.44	34	0.72	0.35	30	0.41	0.17

A clinical examination of the compared groups did not show any marked differences. The mean value of the systolic and diastolic blood pressure in all compared groups depended on the age as well as on the body mass. Taking into consideration these two parameters (age and body mass), the mean values of blood pressure were somewhat higher in males and females in group III than in the groups from the smeltery area (groups I and II). In the latter two groups, the mean blood pressure values were practically the same.

Among the compared groups, there were practically no differences in the ECG findings in males. The highest number of borderline and pathological findings in females was found in group III, while the other two groups did not differ in that respect, but a more detailed analysis of the ECG findings showed a rather large diversity.

The rate of sedimentation of erythrocytes, leukocyte count, level of cholesterol, sugar in blood, SGPT, alkaline phosphatase activity, creatinine and iron in serum did not show a significant difference among the groups of males. SGOT was on the average higher in groups I and II than in group III, while the average number of leukocytes was the highest in males from group III.

Females in group III also had the highest average number of leukocytes. SGPT was on the average lower in females of groups I and II than in group III. Leukocytes, SGOT and SGPT findings in groups of children differed in the same way as in males and females. In children in groups I and II, the mean value of sugar in blood was almost identical but higher than in children in group III. However, in spite of the noted differences, the values of laboratory findings were within normal limits.

The analysis of the measured anthropometric characteristics was carried out with regard to sex and groups of subjects (group I, group II, group III). The examined groups showed no signifi-

TABLE 13. Subjective Complaints

Question	Answer	Males, Group I N	I %	II N	II %	III N	III %	χ^2 I-II	I-III	II-III
Weakness	Never	32	37.7	32	46.4	58	67.4	3.86	15.23^{xx}	9.53^{xx}
	Occasionally	44	51.8	35	50.7	23	26.7			
	Frequently	9	10.6	2	2.9	5	5.8			
Fatigue	Never	12	14.1	25	37.7	26	30.2	12.14^{xx}	6.42^{x}	1.71
	Occasionally	56	65.9	36	52.2	46	53.5			
	Frequently	17	20.0	7	10.1	14	16.3			
Irritability	Never	9	10.6	12	17.4	12	14.0	1.52	0.65	0.41
	Occasionally	66	77.7	50	72.5	66	76.7			
	Frequently	10	11.8	7	10.1	8	9.3			
Pains in muscles and joints	Never	21	24.7	30	43.5	26	30.6	6.08^{x}	2.18	3.54
	Occasionally	50	58.8	31	44.9	51	60.0			
	Frequently	14	16.5	8	11.6	8	9.4			
Insomnia	Never	52	61.2	41	59.4	55	64.0	0.05	0.48	0.62
	Occasionally	27	31.8	23	33.3	27	31.4			
	Frequently	6	7.06	5	7.3	4	4.7			
Headache	Never	33	38.8	25	36.2	49	57.0	3.26	6.77^{x}	7.01^{x}
	Occasionally	42	49.4	41	59.4	33	38.4			
	Frequently	10	11.8	3	4.4	4	4.7			
Dizziness	Never	50	58.8	46	66.7	58	67.4	1.85	2.66	0.02
	Occasionally	31	36.5	22	31.9	27	31.4			
	Frequently	4	4.7	1	1.5	1	1.2			
Do you have pains in your stomach?	Yes	26	30.6	13	18.8	16	18.6	2.78	3.31	0.001

TABLE 14. *Subjective Complaints*

Question	Answer	Females Group I N	I %	II N	II %	III N	III %	χ^2 I-II	χ^2 I-III	χ^2 II-III
Weakness	Never	28	33.3	22	31.9	51	58.6	1.49	11.09^{xx}	11.62^{xx}
	Occasionally	55	65.5	44	63.8	35	40.2			
	Frequently	1	1.2	3	4.4	1	1.2			
Fatigue	Never	11	13.1	4	5.8	38	43.7	5.63	21.25^{xx}	28.19^{xx}
	Occasionally	63	75.0	62	89.9	46	52.9			
	Frequently	10	11.9	3	4.4	3	3.5			
Irritability	Never	9	10.7	5	7.3	17	19.5	0.00	3.36	4.82
	Occasionally	68	81.0	60	87.0	66	75.9			
	Frequently	7	8.3	4	5.8	4	4.6			
Pains in muscles and joints	Never	26	31.0	24	34.8	48	55.2	0.31	10.40^{xx}	6.48^{x}
	Occasionally	52	61.9	41	59.4	36	41.4			
	Frequently	6	7.1	4	5.8	3	3.5			
Insomnia	Never	45	53.6	31	44.9	52	59.8	1.13	3.42	6.07
	Occasionally	37	44.1	36	52.2	29	33.3			
	Frequently	2	2.4	2	2.9	6	6.9			

Table 14 (Cont'd)

Headache	Never	8	9.6	9	13.0	27	31.0	2.21	11.92^{xx}	8.33^{x}
	Occasionally	63	75.9	55	79.7	51	58.6			
	Frequently	12	14.5	5	7.3	9	10.3			
Dizziness	Never	33	39.3	19	27.5	46	52.9		3.17	
	Occasionally	51	60.7	49	71.0	41	47.1			
	Frequently	0	0.0	1	1.5	0	0.0			
Do you have pains in your stomach?	Yes	28	33.3	36	52.2	16	18.4	5.53^{x}	4.99^{x}	19.76^{xx}
Menstruation	Irregular	13	15.5	12	17.4	14	16.9	0.10	0.06	0.01
	Painful	37	44.1	26	37.7	35	42.2	0.63	0.06	0.32
	Strong	45	54.9	33	51.6	53	64.6	0.16	1.62	2.54
Spontaneous abortions	Yes	18	21.3	18	26.1	21	24.1	1.32	0.43	1.85
Artificial abortions	Yes	35	41.2	22	31.9	30	34.5	3.25	0.84	1.85
Do you take contraceptive pills?	Yes	15	17.9	18	26.5	11	13.3	1.64	0.67	4.21^{x}

^{x}Significant at the 95% level. ^{xx}Significant at the 99% level.

cant statistical heterogeneities. The finding suggests that the examined samples belong to an identical population. No difference with regard to anthropometric dimensions in children was established either. They also showed, with regard to the values at a particular age, an identical trend of changes in variables.

DISCUSSION

A summary evaluation of the results of the retrospective analysis of vital statistics data, hospital treatments, and death causes, as well as of the data on health status of the examined and compared families, indicates a possibility that some of the analyzed parameters are related to lead exposure. It should be added, however, that the results obtained by different approaches are not sufficiently consistent.

The retrospective analysis of vital statistics data indicates that the abortion rate, including spontaneous abortions, is higher in the smeltery area than in the control area. Data on hospital treatment also indicate that abortions, as well as complications in pregnancy, delivery, and puerperium, are more frequent in the smeltery area than in the control area. On the other hand, anamnestic data collected during examinations of the selected female samples do not indicate any differences in the number of spontaneous abortions between the compared groups from the exposed and the control area.

Nevertheless, in evaluation of these parameters more significance should probably be attributed to the data from the retrospective analysis. Since the anamnestic data collected during examinations refer to a relatively small number of women and since the family selection criterion required at least one child (7-15 years of age) per family, the samples might have been biased.

General mortality was higher in the control than in the exposed area. This is due, on one hand, to a somewhat lower infant mortality in the smeltery area, and, on the other hand, to the fact that the population in the control area was structurally older.

A comparison of the data on treatment in hospitals and on mortality causes (standardized with regard to age) shows that hypertensive disease, ischemic heart disease, and nephritis were more frequent in the exposed than in the control area, both as a subject of treatment in hospitals and as a mortality cause.

Hospital treatment of ischemic heart disease in females from the exposed area was proportionally more frequent than in males. As a cause of death, it was also somewhat more frequent in females in the exposed area than in the control area, but in contrast to hospital treatment, the disease was a more frequent cause of death in males than in females in the exposed area.

In the oldest age groups, cerebrovascular diseases and other diseases of the circulation were a more frequent cause of death in the control area than in the exposed area, while they were treated in hospital proportionally more often in the exposed area than in the control area.

Other diseases, which according to the statistics on hospital treatment were more frequent in the exposed area than in the control area, were not simultaneously a more frequent cause of death in the exposed area than in the control area. Taking into consideration the character of these diseases, this is quite logical, because, as a rule, they are not death-causing.

Clinical examination of the families did not indicate any significant differences among the compared groups either with regard to the prevalence of some chronic diseases or with regard to the family anamnesis. Distribution of arterial blood pressure did not show higher values in groups exposed to lead. The values were even proportionally higher in the control group.

From the presented data, it is certainly hard to draw any firm conclusions on the relationship between lead exposure and occurrence and frequency of individual chronic diseases. Nevertheless, the fact that in the analysis of hospital treatment and of causes of death some diseases stand out as more frequent in the exposed area than in the control area indicates a possibility that the development and frequency of these diseases are influenced by lead exposure. This relates particularly to ischemic heart disease, hypertensive heart disease, and the chronic kidney disease.

These findings again underline the question which has been a subject of discussions and controversial opinions in the literature: does lead contribute to the development of degenerative diseases of blood vessels and does it cause chronic kidney disease? In the case of blood vessel damages and their possible connection with lead exposure, there was, again, a certain discordance between the findings from the retrospective analysis and the results obtained in the examination of the compared groups of families.

In addition to the fact that the groups of examined subjects are too small to allow evaluation of chronic diseases in individuals on the basis of the examination, a fact should be borne in mind that a possible lead effect, e.g., regarding hypertension and chronic kidney disease, is manifested as a delayed effect. Taking into consideration the age of our subjects, it is, consequently, too early, perhaps, to expect such effect.

On the other hand, it ought to be stressed that by means of a retrospective analysis it was not possible to single out those who had actually been exposed to higher lead concentrations in the exposed area (with regard to their place of residence in relation to the smeltery and the place of work). The analysis included all the inhabitants of the commune, including those who were actually not exposed to lead in concentrations any higher than those in the control area. In this way, the exposed group was, to a certain degree, "diluted," which might have had a certain influence on the results, under the assumption that frequency of individual diseases is partly related to lead exposure. Otherwise, on the basis of the collected data on socio-economic and other characteristics of the mentioned differences in age structure, there were no other significant differences between the compared areas.

In relation to the presented results of the examination, we also ought to mention the expected higher frequency (with regard to the level of lead exposure) of subjective complaints in the group of males occupationally exposed to lead (as compared to the other two examined male groups and especially to the control area group). These complaints can certainly be related partly to elevated lead absorption. In this respect, wives of subjects occupationally ex-
population, it appears that, apart from lead exposure and already mentioned differences in age structure, there were no other significant differences in age structure, there were no other significant differences between the compared areas.
regard to health, at least as far as it can be concluded from the collected data.

Since in addition to the males occupationally exposed to lead, other groups of subjects from the smeltery area (smeltery workers' wives, their children, males without occupational lead exposure, their wives and children) also showed elevated lead absorption, a question can be asked whether the population has not, to a certain extent, become adapted to lead exposure, so that the frequency of symptoms and signs accompanying elevated lead absorption is lower than would be expected. The possible existence of such a adaptation mechanism has already been suggested (Ricklin, 1952; Carow and Leist, 1961; Stockinger, 1975). However, such an "adaptation" does not necessarily mean that those with an increased lead exposure become more resistant to some of the possible chronic adverse health effects of lead.

REFERENCES

Carow, G., and Leist, J. (1961). Beobachtungen über Veränderungen der Laboratoriumsbefunde bei Bleiarbeiten in einem Metallbetrieb. *Zentralbl. Arbeitsmed. Arbeitschutz 11*, 289.

Fugaš, M., Wilder, B., Pauković, R., Hršak, J., and Steiner-Škreb, D. (1973). Concentration levels and particle size distribution of lead in the air of an urban and an industrial area as a basis for the calculation of population exposure. Proceedings of the International Symposium on Environmental Health Aspects of Lead held in Amsterdam, 1972, p. 961. Commission of the European Communities, Luxembourg.

Fugaš, M., Markičević, A., Prpić-Majić, D., Rudan, P., Seničar, Lj., Sušnik, J., and Šarić, M. (1976). Health study of a lead-exposed population. *Arh. Hig. Rada. Toksikol. 26* (Suppl.), 119.

Korbisch, J., and Sorio, A. (1975). Determination of cadmium, copper, and lead in natural waters after anion-exchange separation. *Anal. Chim. Acta 76,* 393.
Ricklin, W. (1952). Beitrag zur Bewertung des Bleigehaltes im Blut. *Z. Unfallmed. und Berufskr. 45,* 141.

Stockinger, H. E. (1975). Usefulness of biologic and air standards for lead. *J. Occup. Med. 17,* 108.
U.S. Environmental Protection Agency (1977). Biological significance of some metals as air pollutants, Part I: Lead. Environmental Health Effects Research Series 600/1.77 041. Research Triangle Park.
Zielhuis, R. L. (1973). Lead absorptions and public health, an appraisal of hazards. Proceedings of the International Symposium Environmental Health Aspects of Lead held in Amsterdam, 1972, p. 631. Commission of European Communities, Luxembourg.

DISCUSSION OF PAPER BY PROFESSOR MARKO ŠARIĆ AND DR. MIRKA FUGAŠ

Maurice A. Shapiro, Graduate School of Public Health, University of Pittsburgh: I wonder whether you have studied learning disabilities in these two populations. I understand that the exposed population has been exposed for generations. Is there any information about the difference in learning capabilities among the children in the exposed area and the nonexposed area?

Professor Marko Šarić, Institute for Medical Research and Occupational Health, Zagreb: Well, we haven't done any particular study concerning this question, but we think that there are no essential differences in these two populations as far as the standard of education and everything else is concerned. We haven't done any particular studies just to see whether there are differences in learning ability.

Malcolm: I was interested in the finding of this excess of ischemic heart disease, because most of the studies on populations of lead workers, not on total populations, don't seem to show this finding. Also, the epidemiological studies of soft water areas have shown an excess of ischemic heart disease, and it has been suggested this might be due to some plumbosolvency, although I'm not really convinced by that. It seems to me that the difference here may well be a question of the very long exposure, particularly the heavy exposure of children. If you remember, the Australian work of Nye, Henderson *et al.*, the children who got lead poisoning in Queensland often died of renal failure in their 30's. I wonder if you looked at this effect. The other thing is, are there any other exposures of this group other than lead which might be related to ischemic heart disease?

Sarić: Well, of course, there is a certain exposure to cadmium in the smelter area. As far as I know, in the case of children from Queensland, a rather high percentage of those children developed acute lead intoxication in the early stages of their lives, and, in the interpretation of the results, this was considered a very important fact. In our study, in spite of rather high lead in blood levels and other indicators of increased lead absorption, we had no information that clinical lead poisoning had occurred in these children. So I don't know what is going to happen. Maybe, in the meantime, the conditions also will improve. I agree that it will be very interesting to follow this especially young population to see whether they will develop more chronic renal kidney disease. Concerning the water supply and the softness and hardness, maybe Dr. Fugaš can answer the question better.

Dr. Mirka Fugaš, Institute for Medical Research, Yugoslav Academy of Science and Arts: The water was mostly hard. It was rather soft in only one small area.

Dr. Harvey C. Gonick, University of California: I would like to address myself to another component to the question concerning the increased hypertensive and renal incidence in the hospitalized population. I am sure that given the interest in Balkan Nephropathy in your country, you examined the populations to make sure that no one had moved into the area from an endemic area where Balkan Nephropathy existed. I am also sure that you took careful selection to be sure that such individuals were deleted from the study population. However, I should like to ask you whether or not some of the same factors that maybe lead to Balkan Nephropathy may also be superimposing on the effect of chronic exposure to lead? For instance, if well water indeed is the source of elements that may lead to Balkan Nephropathy, was there a part of the population which was exposed to well water, or, conversely, if it's a fungus that is grown in some of the storage bins in the rural areas that are nearby, could some of the population have been exposed to this fungus?

Šaric: Both areas have a rather high standard of living, at least concerning our conditions in Yugoslavia and the hygienic standard is also rather high in both areas. Of course, it is always possible that some other factors might interfere with the results. For that reason, we used several approaches to see whether we get the same or similar tendencies in the results. It is very difficult to make very firm conclusions about our studies, particularly, from hospital treatment data because, as I already mentioned, other factors, for example, the attitude towards hospital treatment, general attitude of the people, and so on, may also have some influence on the results. But we don't think that other metals, well, maybe, with the exception of cadmium of course, to some extent, had an important role, as, maybe, marked differences in hygienic conditions which were more or less the same, rather high. High, I mean, compared with other parts of Yugoslavia. This study was done in Slovenia, which is the more developed part of Yugoslavia.

Bobby G. Wixson, University of Missouri - Rolla: We have worked with three of six lead smelters in the United States for the past 10 years. I have a question and a comment. My question is how far away from the smelter do you have to go before reaching what would be considered normal background concentrations? And, a second question, are you looking at other metals such as zinc, cadmium, copper, that type of thing as well as lead?

Fugaš: Well, this is a river valley. The contamination is dispersed along the valley within a distance of approximately 10 kilometers. We have not measured further on, but as you can see from the data, the concentrations at the more distant sites are already lower. The control area is separated by a high mountain from the contaminated area and the pollution from the smelter cannot reach it. We did measure other metals in air samples but not in blood samples. The concentrations of cadmium and zinc were elevated as compared to urban areas, but much lower than for instance the concentrations of lead.

Wixson: In our studies with one smelter located in a valley, we found that in the zone extending up and down the long axis of the valley there was selective fallout of lead and zinc close to the smelter with the cadmium going out a greater distance. That is what I would assume was the type of situation you would run into.

Jerry F. Stara, U.S. Environmental Protection Agency: Mrs. Fugas, in your summary statement you suggested that perhaps 30 percent of the daily intake of lead is from inhalation. You did not state what is the contribution of lead from water in the smelter area. Is it strictly the drinking water or is it also the water that is used as part of foods, for example, soups, coffee and so on, or do you make any estimates in this respect?

Fugaš: Well, at least 30 percent is just a rough assessment, but I mentioned that lead in water was rather low. In the exposed area, the concentration was about 10 micrograms per liter, which is much less than the WHO standard which is 50 micrograms per liter. Therefore, we didn't consider water as a source of lead body burden. The same water is used for cooking and drinking. There is a public water supply, but many use water from a well. We also took samples from a spring. The concentration was never above 14 or 15 micrograms per liter. It seems that the ground water is rather deep and protected by an impermeable layer so that water was not contaminated by lead.

PARTICLE SIZE, SOLUBILITY, AND BIOCHEMICAL INDICATORS[1]

E. King

National Occupational Hygiene Service, Ltd.
Manchester, England

This study was primarily concerned with the possible effect of the particle size and solubility in industrial airborne lead dust upon its absorption by those exposed to it. Secondarily, the various biochemical parameters of absorption and "effect" were correlated in an attempt to obtain more mathematically sound relationships than hitherto available.

Three groups of workers were studied. Factory 1 gave a group of 19 lead acid battery pasters for whom we had a blood lead history averaging 6.9 years (range 1-13); Factory 2 gave a group of 34 pigment production workers for whom we had a blood lead history averaging 3.7 years (range 1-6); and Factory 3 gave a group of 48 smelters, all with blood lead histories of more than one year (the smelter joined our blood lead analytical service some three years before the survey). All the men were job-stabilized, as evidenced by blood lead history. No cases of poisoning had occurred in the factories for some years, with none ever occurring at Factory 2. There was the possibility of some staff transfer in the 1960's, giving some selection effect in Factory 1, but not in Factory 2 (no transfer due to high blood lead in its history), and not to any significant extent in Factory 3.

The survey was based upon one 8-hour (shift) personal air sample per week for 12 weeks, one urine sample per week for lead (PbU) and δ-aminolevulinic acid (ALAU) analysis, and monthly blood lead, with some FEP analyses. Air samples were by Casella Cascade Impactors at 2.5 lpm, with the four sized slides and the one backing filter being sequentially leached for one hour with water, 0.1 N HCl, 1.0 N HCl, and 10.0 N HCl. The survey data were analyzed for 1st and 2nd degree regression, using all the biochemical

[1]*The material on which this paper is based comes from the ILZRO Project LH-207 and is published in full in the Annals of Occupational Hygiene, Volume 22, pages 213-239, 1979, to which reference should be made for details of the data discussed herein.*

ISBN 0-12-460520-6

parameters and chosen combinations for particle size and solubility for air data in permutation.

Overall, the project failed in its primary purpose, in that it did not identify a specific combination of size and solubility that had such good agreement with the biochemical data that it could be used as a primary standard for industry. This failure could have been partly due to the very wide range of materials and exposures in factories 2 and 3. However, the more likely reason is the fact that while there is a higher percentage absorption of small particles (shown in the survey), the large particles, with a lesser percentage systemic absorption, dominate the overall mass. Indeed, the more soluble of the mid-sized particles may well go into solution in the mid-respiratory tract, with complete absorption. Therefore, for the purpose of this present discussion, only "total" lead in air data are used.

The secondary purpose of the project was more successful and gave a hypothesis which may not only explain some of the anomalies observed in the industrial lead absorption field, but which also has a bearing on the setting of "air" standards. Essentially, we obtained linear relationships between blood FEP (FEP) and ALAU, between FEP and PbU, and between ALAU and PbU. This suggests that all three are linear with another factor, namely, "bioactive" or "bioavailable" lead. This is contrary to an often stated concept that the biochemical responses show increased curvature as their reserves are overcome by linearly increasing blood lead. This latter concept arises from the observed curved relationships, as found in the survey, between ALAU and PbB, and FEP and PbB.

However, in the survey, the relationship of PbU against PbB was also curved and not linear, the curvature following the same pattern of ALAU and FEP. Excepting another hypothesis of non-linear renal excretion, this finding completes the pattern of a single, bioactive fraction of blood lead which increases disproportionately with whole blood lead, and with which biochemical responses and urinary excretion are linear.

It follows from this basic hypothesis that with an increasing "dose" (lead in air) which attempts to increase whole blood lead, we should expect disproportionate (to dose) increase in urinary lead. This, in fact, was found in the survey, with the curved urinary excretions being the "best" patterns observed with lead in air (PbA). It also follows from the hypothesis that blood lead should <u>not</u> increase linearly with "dose," but should attempt to self-limit, as the increasing excretion counters the increasing intake.

Of the three factories in the survey, Factory 1 had too narrow a blood lead range, while Factory 3 was complicated by the mixed fume/dust exposures, with two exposure populations emerging from the particle sizing considerations. Factory 2, however, gave a curved relationship, the data covering a wide exposure range, with the curve giving a y-intercept of the right magnitude, unlike the majority of linear regressions produced of PbB against PbA in this survey and elsewhere. The curve obtained was very similar to that produced by Dr. M. K. Williams, at the Amsterdam Workshop (1976), for grouped data.

This hypothesis explains the observation of many of us with long experience in factories, of very high personal lead in air levels with only marginally excessive blood lead levels. It also explains the relatively small increase in whole blood lead which is often associated with a change from full health to "poisoning" in the lead worker. Further, if such curvature applies at the lower levels of absorption, it could explain the wide range of whole blood lead levels found in closely defined populations.

The implications of this concept and of the survey findings in the field of "standard" setting are considerable. We are all increasingly having to make, or recommend, executive decisions on numerical data rather than on opinions or judgments, and while many of us may regret this, it is a fact of life. With lead, only one of our available parameters is sufficiently stable in an individual and reproducible in a laboratory for that purpose, and this is blood lead.

Whether this is the "right" parameter to use is arguable, and will be debated in the future as it has in the past. Currently, however, it is the primary standard governing the "permissible" absorption of lead by workers. Traditionally, lead in air standards have been derived from it, on the basis of a linear and "reasonably" good relationship. Present evidence is that the relationship is poor and probably non-linear, making the derivation of precise air standards from air/biological data of dubious soundness.

An example of our alternative approach is to be found in the United Kingdom Consultative Document on the Control of Lead at Work (1978). In this, the control of an individual worker is based upon blood lead estimation, as is common practice throughout the world, with 80 μg Pb/100 ml as the maximum. The lead in air is based, however, upon a Hygiene Standard of 0.15 mg Pb/m^3; levels between 0.15 and 0.45 mg/m^3 should be interpreted carefully in the light of the biochemical data and history of those so exposed, provided, of course, that respiratory protection is not in use.

The air data referred to are whole shift personal samples, and not to "static" or "breathing zone" or "vicinity" samples, which tend to give much lower levels. Further, in the U.K., such standards are applied only after "reasonably practicable" control of the dust and fume has been carried out, rather than as an excuse for not so doing.

The importance of the document is that it recognizes that over the whole range of lead compounds, dust sizes, and solubilities, a single lead in air level which cannot be related with any precision to the biological burden it is supposed to control is illogical, impractical, and unnecessary.

REFERENCES

Health and Safety Commission (1978). Consultative Document of Lead at Work: Draft Regulations and Draft Approved Code of Practice. Her Majesty's Stationery Office, London.

Williams, M. K. (1976). "Blood Lead-Air Lead Relationship." Paper prepared for the 2nd International Workshop on Permissible Levels for Occupational Exposure to Inorganic Lead, September 21-23, 1976. University of Amsterdam, The Netherlands.

DISCUSSION OF PAPER BY EDWARD KING

Hammond: I didn't hear the critical number. Did you say that the British standard is to be 0.45 mg/m^3, 0.15 to 0.45, or what?

King: No. What we think we are going to do, at present this is only a consultative document, it is not law, is to have a standard of 0.15 mg/m^3. However, levels between 0.15 and 0.45 mg/m^3 shall be interpreted in the light of biochemical evidence for that population.

Hammond: What is the biochemical evidence that will determine whether the 0.45 is allowable?

King: Essentially blood leads that have been done on all these populations at routine intervals. The populations will be essentially under blood lead control and the air lead measurements are an adjunct to that, rather than a separate issue.

Jerome F. Cole, International Lead Zinc Research Organization, Inc.: In reading through the recent OSHA standard on lead, a great amount of attention is given to your study, as well as other studies relating or not relating, as the case may be, air lead and blood lead. Is one of the reasons that, as OSHA maintains, a very poor relationship between air lead and blood lead is usually seen, is that the air lead and blood lead measurements are only taken at a point in time and no consideration given to the tenure of workers. OSHA contends that because of this there is tendency to underestimate the blood lead from a given air lead because of tenure, the theory being that there is a continuous pouring out of stored lead from tissues and that this builds up over time. I was wondering whether you have any comment on this and whether or not your workers were sufficiently different in tenure to have accounted for so much of the scatter in the data?

King: I'm not sure about this. It looks as though somebody has created a model. No. Of the populations we have studied, and this is one of the points about Britain, our working population is very much more stable. The group of pasters from Factory 1 has accepted that employment is more of a way of life than anything else. They take jobs and remain in them. So this population is made up of very long termers. Factory Number 2 has exactly the same reputation. People join the work force and tend to stay there. There is very small labor turnover. Factory 3, I'm not so sure about. So what we are, in fact, seeing are long term employees. Now on the possible swing, I'm afraid it's the other way around. In 1975, when this study was carried out, there was a great oil price rise. It was the year of our recession. I think you had one here in the U.S. It was the year when production was going down, not up. In Factory Number 1 we, in fact, lost a group of people because their job simply disappeared, and they had to be transferred elsewhere. So if anything, I would suggest the air data we found in that particular year were possibly lower than they were in preceeding years when the production was much higher.

Gonick: I am particularly intrigued by your use of the term "bioavailable" or "bioactive" blood lead. I like it because it has implications which I think are far-reaching. Utilizing that terminology and its implications to examine the relationship between blood lead and urine lead, however, leaves something to be desired. The theoretical reasons for an increase in urinary lead would either be an increased filtered load which reflects then a plasma phenomenon which, unfortunately, we are incapable of measuring, or it may reflect decreased reabsorption at the tubular level because of some other phenomenon which we do not understand. Or it reflects increased tubular secretion because of an increased total body burden, perhaps, including the renal tubular level. The biggest problem, I think, we have in interpreting any interrelationship between total blood lead and urinary lead is a lack of understanding of almost any of the renal physiology that applies to the renal handling of lead. The only recent study is the one by Vander using lead isotopes, looking at clearance data, and, certainly, lead as such seems to be filtered in part. But we know very, very little about that plasma component which is filterable versus that which is nonfilterable. I think this is a whole area which needs to be examined in order to truly understand the implications of your very intriguing statements.

King: In answer to that, I've spent the last 10 years trying to find an accurate method of measuring bioactive lead. I suspect that there is very little of it. There is some evidence, and this is in work published by Dr. Mac Roberts, of Bradford, England, that, in fact, the plasma lead content bears quite a nice relationship with clinical signs, rather better than whole blood lead. However, this is circumstantial evidence, not hard proof.

LEAD TRACERS AND LEAD BALANCES

A. C. Chamberlain
M. J. Heard

Atomic Energy Research Establishment
Harwell, England

BALANCE STUDIES

The overall lead balance of an individual can be described by the equation:

$$I + f_a L = F + U + S \tag{1}$$

and the systemic lead balance by:

$$\begin{aligned} T &= f_g I + f_r L \\ &= F_e + U + S \\ &= U(1 + F_e/U) + S \end{aligned} \tag{2}$$

when the symbols have the following meaning (units in μg/d)

I = dietary intake

L = intake by inhalation

f_g = fractional uptake in gut

f_a = fractional deposition in respiratory tract

f_b = fraction of lead deposited in respiratory tract which is absorbed, either in the lung or in the gut

$f_r = f_a f_b$ = fraction of inhaled lead which is absorbed

$A = f_r L$ = uptake from lung

$G = f_g I$ = uptake from gut

$T = A + G$ = total uptake

F = fecal excretion

F_e = endogenous fecal excretion

U = urinary excretion

S = amount stored

ISBN 0-12-460520-6

Kehoe (1961) gave volunteer subjects experimental doses of lead either orally or by inhalation. In subsidiary inhalation experiments, he estimated f_a from the lead content of inhaled and exhaled air. He was thus able to estimate the term S, the amount of lead transferred to storage.

During the experimental periods, Kehoe was not able to distinguish endogenous fecal lead from unabsorbed dietary lead, but in follow-up periods after the end of exposure, he estimated the excess of urinary and fecal lead compared with pre-exposure periods, which was due to the excretion of some of the stored lead. Kehoe concluded that subject EB excreted about equal amounts of stored lead in urine and feces, but subject MR's endogenous fecal excretion was small.

RADIOACTIVE AND STABLE LEAD TRACERS

Experimental doses of radio-lead can be measured independently of normal dietary lead, and comparison of fecal and urinary excretion of radio-lead given by injection or by inhalation of submicron particles gives an estimate of the ratio F_e/U (Chamberlain *et al.*, 1978). There are also advantages in the use of radioactive aerosols in measuring f_a and f_b, in particular:

1. only short inhalations are needed and the stable lead concentrations can be kept low;
2. deposition in the lung can be measured directly by external gamma ray counting of the activity in the chest or by comparison of the concentrations in inhaled and exhaled air;
3. deposition in the upper respiratory tract and removal by the mucociliary mechanism can be estimated; and
4. the transfer from lung to blood can be followed by analysis of sequential blood samples.

It is a disadvantage of the radioactive method that the most convenient isotope ^{203}Pb has a radioactive half-life of only 52 hours, and it is not possible to follow the uptake and excretion of an experimental dose for longer than about 14 days. The isotope ^{210}Pb, with a 20 year half-life, is useful for animal studies but is not suitable for human experiments. The stable isotope ^{204}Pb used by the UCLA group (Rabinowitz *et al.*, 1976) has complementary advantages. It can only be administered orally and daily doses for months are needed to alter the isotopic ratio in the subject's tissue, but thereafter the biological stay time can be studied for much longer than is possible with ^{203}Pb.

SUMMARY OF RESULTS WORK WITH RADIOACTIVE TRACERS

Deposition in Respiratory Tract

Chamberlain *et al.* (1978) synthesized tetraethyl lead with a ^{203}Pb label, added it to petrol, and burned it in a small (50 cc), four-stroke engine. By varying the dilution of the exhaust with air, particles with mean diameters of 0.02, 0.04, or 0.09 μm were made. The particle sizes were deduced from electron micrographs and also from the Brownian diffusivity. Volunteers inhaled the aerosols for periods of a few minutes in each experiment. The respiratory parameters of length of respiratory cycle and volume per breath were controlled. The former was the most important parameter affecting deposition (Fig. 1). Fractional deposition increased with decrease in particle size (Fig. 2), in agreement with theoretical calculations. Deposition by Brownian diffusion is independent of particle density, and in the size range below 0.5 μm, it is not appropriate to use the mass median equivalent diameter.

Nozaki (1966) generated lead fume and measured the concentration in inhaled and exhaled air by light scattering. For particles with mean (electron microscope) diameter of 0.05 μm, f_a was 0.425. As particle size increased, it fell slightly to 0.393 at 0.1 μm, and then increased to 0.632 at 1 μm diameter. The mass median equivalent diameter of the 1 μm particle was probably about 2.5 μm. Kehoe (1961) generated lead aerosols by burning tetraethyl lead in propane. For an average (electron microscope) diameter of 0.05 μm, μm, f_a was 0.36, and for 0.9 μm, it was 0.46.

In addition to the measurements with radioactive exhaust aerosols, Chamberlain *et al.* (1978) also measured deposition of lead in subjects inhaling the ambient aerosol near motorways and at other places where lead in air (PbA) was in the range 2-10 μg/m^3. This was done by measuring the lead exhaled and comparing it with the concentration inhaled. The fractional deposition in the lung was found to be 0.6 in subjects inhaling fresh exhaust near motorways and 0.5 for other urban atmospheres.

Figure 2 shows that f_a must be expected to vary with the degree of coagulation of the primary exhaust aerosol (diameter about 0.04 μm according to Chamberlain *et al.*) with other exhaust or ambient aerosol particles. In industry, a larger particle size would be expected. If this is in the range 0.1-0.5 μm, f_a will be reduced, possibly to 0.2 or 0.3, but if a considerable fraction of the mass of the particles exceeds about 1 μm diameter (2 to 3 μm mass median equivalent diameter), then f_a will increase again as deposition by impaction becomes more important. Mehani (1966) measured f_a with workers in a battery factory and shipbreaking yard and found values in the range of 0.28-0.7 for different groups of workers, but there were no particle size measurements for comparison.

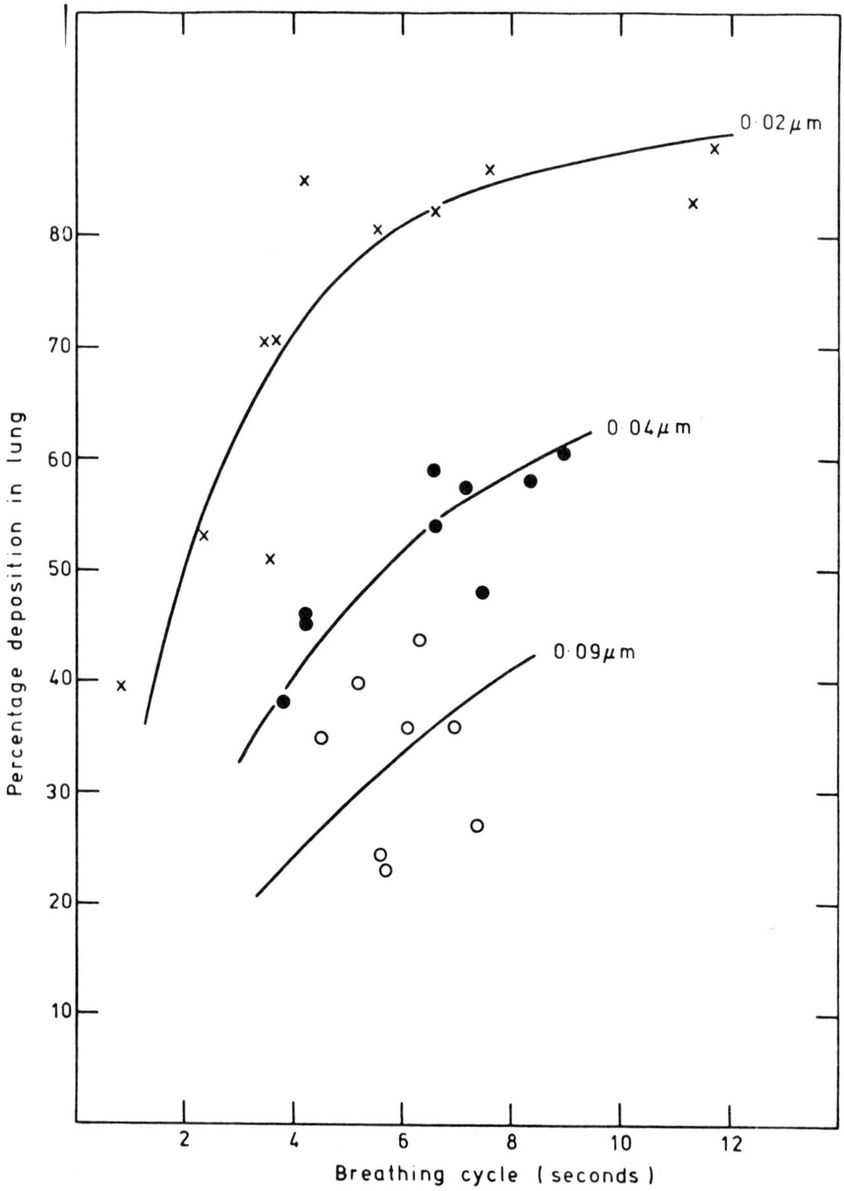

Fig. 1. Deposition in lung of wind tunnel aerosols.

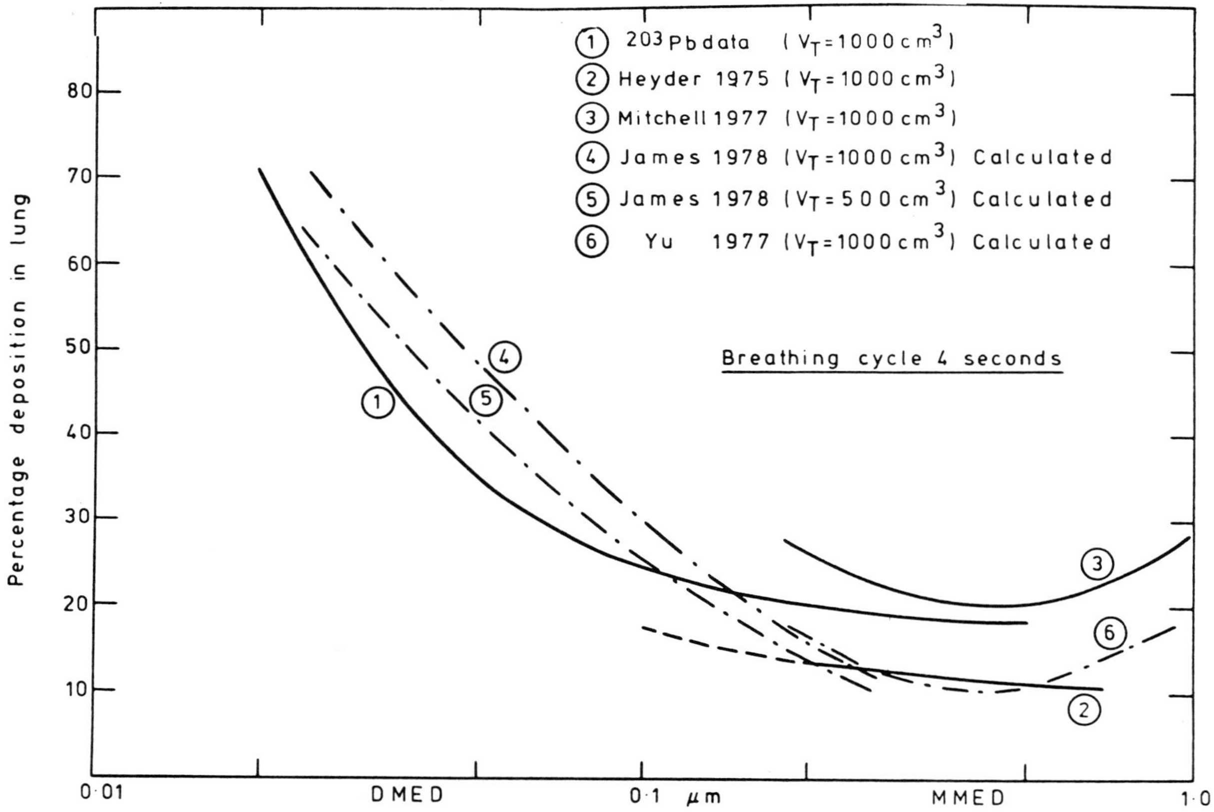

Fig. 2. A comparison of published data on the deposition in the lung of submicron particles.

Uptake from Lung and Appearance in Blood

Chamberlain *et al.* (1978) monitored the removal of ^{203}Pb from the lung by external gamma ray counting and also measured the entry into blood by analyzing sequential venous blood samples. The graph of removal from the lung was not a simple exponential. About 20% of the lung burden was absorbed within 1 hour and 70% within 10 hours. The chemical form of the aerosol (fresh exhaust, exhaust aged in sunlight, lead oxide, lead nitrate aerosols) affected the time scale to some degree, but in all cases, 90% of the lung burden was removed within 4 days. This was confirmed by comparing ^{203}Pb in blood 1-2 days after inhalation and after intravenous injection (Fig. 3). Irrespective of the mode of administration, about 55% of the dose became attached to red cells. If an aerosol of larger particle size had been inhaled, deposition in the upper respiratory tract would have resulted in some mucociliary clearance to the gut, and the fraction absorbed would have been less.

From measurements of ^{203}Pb in red cells and plasma immediately after injection, it was concluded that rapid equilibration occurred (Fig. 4) between red cells, plasma, and extracellular fluid (ECF), as previously described by Stover (1959). From a few hours onward, the fraction in plasma, and by implication also the fraction in ECF, was very small and it was considered likely, though not proved, that the dose had been partitioned 55% in red cells and 45% in other tissues, with an affinity for lead strong enough to maintain a concentration gradient against plasma and ECF. From animal experiments, it is known that 30-50% of intravenous doses of lead is deposited in the skeleton, and it seems probable that much, though not all, of this uptake takes place in the first few hours.

In rats injected with ^{210}Pb and sacrificed serially in groups, Schubert and White (1952) found 2.3% of the dose in the femurs 30 minutes after injection, 3.5% at 1 day, and 3.7% at 6 days. From results of Morgan *et al.* (1977), it is probable that the femurs received about 10% of the total skeletal uptake. Hammond (1971) found 45% of injected ^{210}Pb in the skeleton of rats at 1 day, increasing to 65% at 6 days, and Hackett and Sikov (1977) found 30% in the skeleton at 90 minutes, increasing to 40% at 11 days.

Comparison of fecal excretion after inhalation and after injection showed that not more than a few percent of the submicron ^{203}Pb aerosol was deposited in the upper respiratory tract, but Kehoe (1961) found enhanced fecal excretion in a subject who inhaled lead oxide aerosol with optical diameter 0.9 μm and mass median equivalent diameter of about 2.0 μm. Fugaš and Sarič (1978) have reported a mass median diameter (possibly actually mass median equivalent diameter) of 2.0-2.5 μm for the lead aerosol near a smelter in Yugoslavia. For such particles, f_a would be expected to be about 0.75 (Task Group on Lung Dynamics, 1966), of which pulmonary deposition contributes 0.25 and nasopharyngeal deposition 0.5. Assuming total uptake of the pulmonary fraction and 20% uptake of the nasopharyngeal fraction, f_b would be 0.47 and f_r would

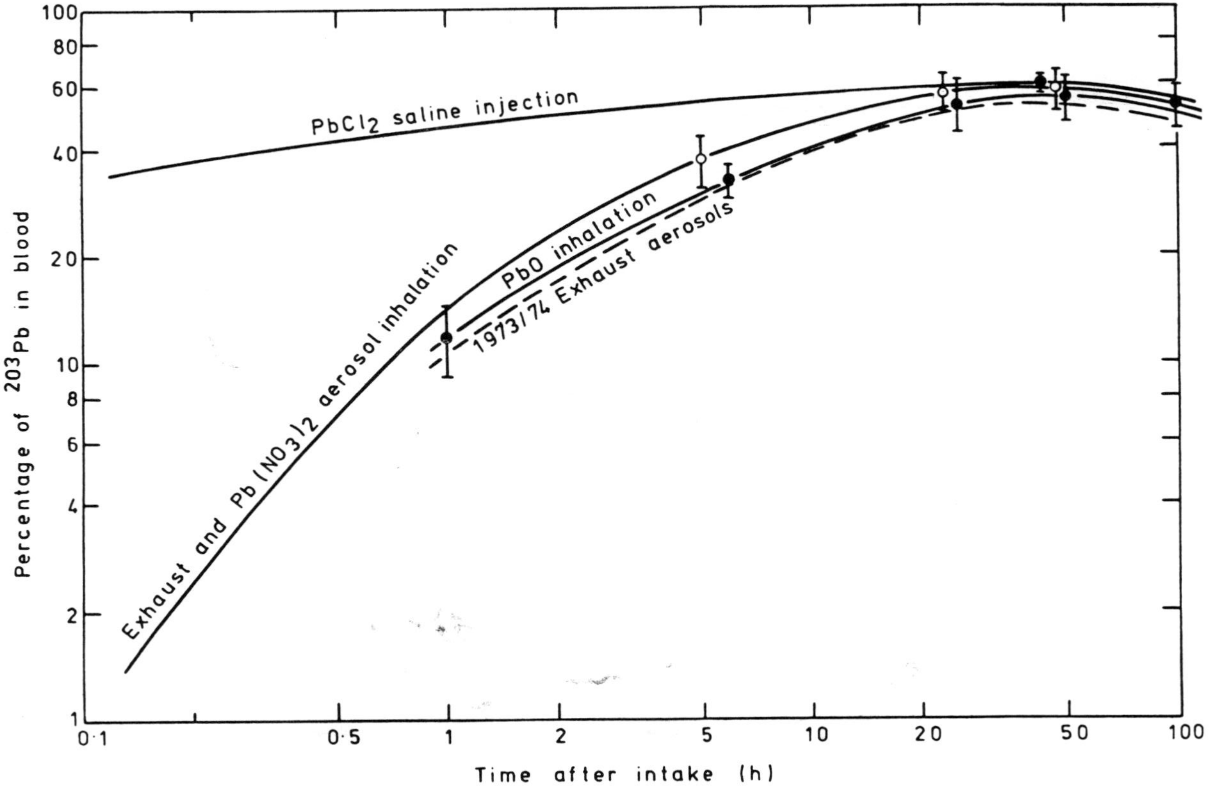

Fig. 3. *Levels of* ^{203}*Pb in blood following inhalation of exhaust, oxide, or nitrate aerosols, or injection of* $PbCl_2$.

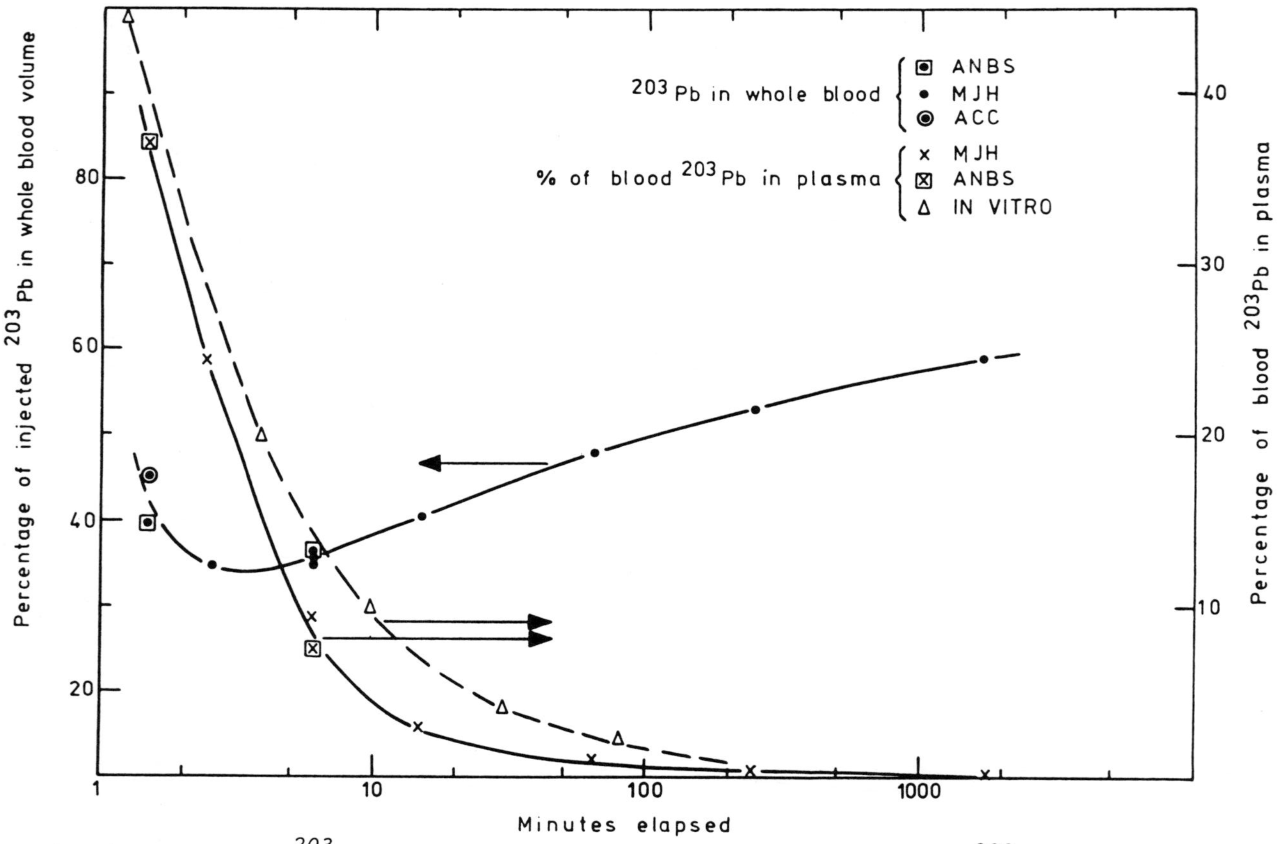

Fig. 4. Levels of ^{203}Pb in blood following intravenous injection of ^{203}Pb Cl_2 in saline.

TABLE 1. *Fractional Deposition in Lung and Uptake of Lead Aerosols*

Aerosol	Particle size (μm)	f_a	f_b	f_r	Reference
Fresh exhaust	0.04	0.5	1	0.5	Chamberlain et al., 1978
Accumulation mode	0.1-1	0.35	1	0.35	WHO, 1977
Industrial	∿2	0.75	0.47	0.35	See text

be 0.35. For purposes of calculation, the values of f_a, f_b, and f_r given in Table 1 will be assumed.

Uptake from Gut

Measurements of the human uptake of orally ingested lead using the radioactive tracer ^{203}Pb or the stable tracer ^{204}Pb have given very variable results, depending on whether the oral dose is taken with food or fasting (Table 2). It can also be expected that the dietary status of the subject, particularly the fat content of the diet, will also affect dietary uptake (Strehlow and Barltrop, 1978). The generally accepted figure of 10% uptake seems to be based mainly on comparisons of urinary and dietary lead in groups

TABLE 2. *Percentage Uptake of Lead from the Gut*

Lead compound	No. of Subjects	Percentage uptake with food	Between meals	Fasting	Reference
$PbCl_2$	3	13		--	Harrison et al., 1969
$PbCl_2$ with alginate	3	13		--	Harrison et al., 1969
PbS	3	--		35	Wetherill et al., 1975
$Pb(NO_3)_2$	3	9.5		34	Wetherill et al., 1975
$PbCl_2$	11	--	21	--	Blake, 1976
PbS	6	6		12	Chamberlain et al., 1978
$PbCl_2$	6	7		45	Chamberlain et al., 1978

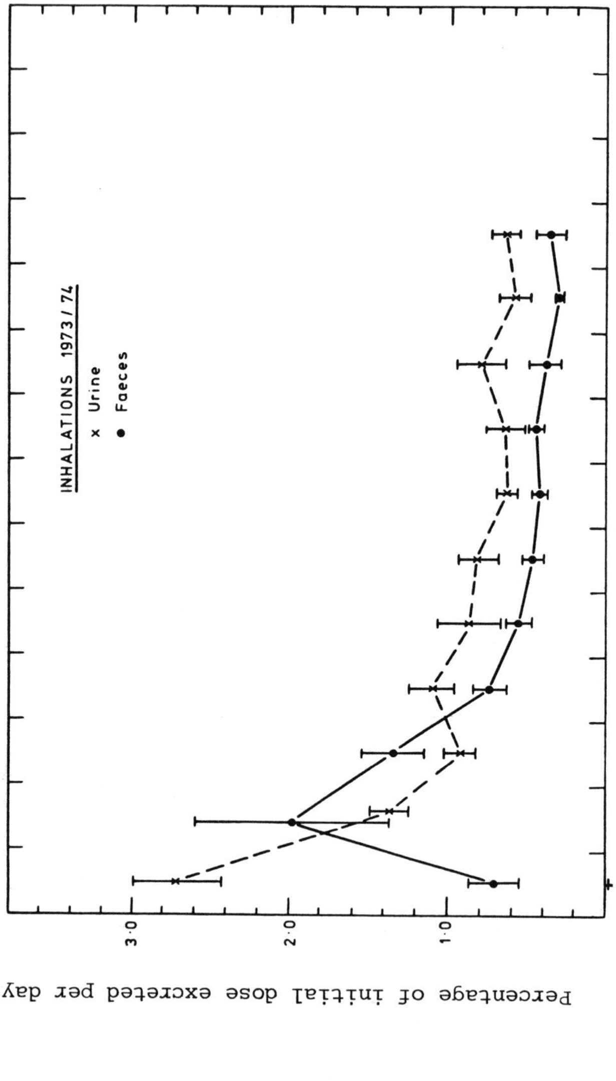
INHALATIONS 1973/74
x Urine
● Faeces
3·0
2·0
1·0
Percentage of initial dose excreted per day

Fig. 5a

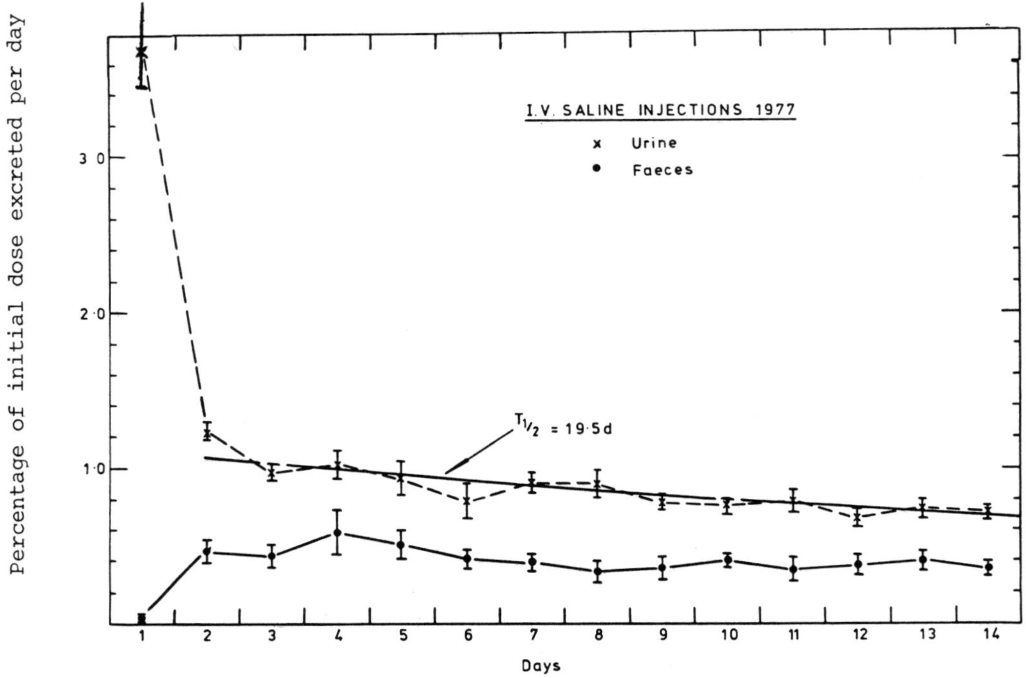

Fig. 5b

Fig. 5a,b. Excretion of ^{203}Pb in urine and faeces following (a) inhalation of motor exhaust aerosols and (b) I.V. injection of $^{203}Pb\ Cl_2$.

of subjects (for example, the data of Kehoe, 1961; Thompson, 1971; and Zurlo *et al.*, 1970). However, as pointed out by WHO (1977), this ignores the possible contribution of inhaled lead to the uptake, and the contributions of endogenous fecal excretion and long-term storage to the disposal of the absorbed lead. If $F_e/U = 0.5$, then the total excretion (urinary plus fecal) is about 15% of the dietary lead.

Excretion

The excretion of ^{203}Pb following inhalation and following injection is shown in Figure 5. Over the 3-14 days post-injection, urinary excretion averaged 0.83% per day of the original dose and fecal excretion 0.41% per day, giving $F_e/U = 0.5$. Following inhalation, F_e/U was 0.6 (Chamberlain *et al.*, 1978). Results from the literature for F_e/U in animals injected with ^{203}Pb or ^{210}Pb are given in Table 3. There is no evidence that F_e/U changes with time after injection. Measurements of F_e/U for lead and other elements in human subjects are summarized in Table 4.

The measurements of Holtzman (1978) refer to ^{210}Pb excreted by subjects who had acquired doses of ^{226}Ra by injection or by oral uptake many years previously. The decay of ^{226}Ra, mainly in the

TABLE 3. Ratio of Endogenous Fecal to Urinary Lead in Animals Injected with Radiolead

Animals				
Species	*No.*	*Period of study (days from injection)*	F_e/U	*Reference*
Baboon	*3*	*10-50*	*0.7*	*Cohen, 1970*
Baboon (immature)	*1*	*366-370*	*1.1*	*Strehlow, 1971*
Dog	*4*	*10-30*	*1.8*	*Hursh, 1973*
Dog	*3*	*1-21*	*3.1*	*Lloyd et al., 1975*
Rat	*3*	*1-16*	*1.2*	*Schubert and White, 1952*
Rat	*10*	*0-1*	*1.0*	*Morgan et al., 1977*
Rat	*1*	*2-5*	*1.7*	
Rat	*5*	*1-14*	*2.2*	*Castellino and Aloj, 1964*
Rat	*3*	*1-78*	*1.3*	*Lucas and Stanford, 1967*
Guinea pig	*1*	*1-20*	*1.1*	*Horiuchi and Horiguchi, 1958*

Table 4. Ratio of Endogenous Fecal to Urinary Excretion in Humans

No. of subjects	Nuclide	Method	Period (days)	F_e/U	Reference
1	^{45}Ca	Injection	1-58	0.55	Bronner et al., 1956
5	^{85}Sr	Ingestion	20-100	0.30	Harrison and Sutton, 1967
3	^{85}Sr	Injection	20-100	0.24	Harrison and Sutton, 1967
11	^{239}Pu	Injection	1-6	1.3	Durbin, 1972
	^{239}Pu	Injection	19-24	1.1	
5	^{203}Pb	Injection	3-14	0.5	Chamberlain et al., 1978
3	^{203}Pb	Inhalation	3-14	0.6	Chamberlain et al., 1978
14[a]	^{210}Pb*	Ingestion	10^4	0.8	Holtzman, 1978
5[b]	^{210}Pb*	Injection	10^4	0.9	

[a]Former radium workers.
[b]Patients receiving ^{226}Ra by injection many years previously.
*^{210}Pb formed as decay product mainly in bone.

bones, gave ^{210}Pb as a daughter product, and this was gradually released by resorption of bone. Thus, the conditions of Holtzman's measurements were very different from those of Chamberlain *et al.*'s work with ^{203}Pb.

The renal clearance rate, defined by:

$$\text{Renal clearance(g/h)} = \frac{^{203}\text{Pb excreted in urine per hour}}{^{203}\text{Pb per g whole blood}}$$

averaged 3.8 g/h compared with 4.5 g/h (4.3 ml/h) found in the earlier series of inhalation experiments (Chamberlain *et al.*, 1975). Kehoe (1961) found an average blood lead (PbB) of 28 μg/100 g and average urinary output (PbU) of 30 μg/dl in 10 normal subjects, giving a renal clearance rate of 4.5 g/h. There is evidence that renal clearance rate increases at higher PbB levels (King, 1979). Measurements by Devoto and Spinazzolo (1973) of PbB and PbU in city and country dwellers, taxi drivers, and lead workers in Sardinia show a slow increase of renal clearance as PbB increases from 19.6 to 78.3 μg/100 g (Table 5).

TABLE 5. Lead in Blood and Urine of Categories of Persons[a]

Category	*A*	*B*	*C*	*D*
No. of subjects	*50*	*50*	*20*	*50*
PbB (μg/100 g)	*19.6*	*21.2*	*27.7*	*78.3*
PbU (μg/d)	*22.0*	*25.0*	*34.3*	*131.7*
Renal clearance (g/h)	*4.7*	*4.9*	*5.2*	*7.0*

[a]*Devoto and Spinazzola, 1973.*

Re-interpretation of Balance Studies with Assumed Value of F_e/U

Kehoe's (1961) balance studies included measurements or estimates of I, L, F, U, and f_a in equation (1), and he was thus able to deduce S. He also appreciated that to solve equation (2), it is necessary to know the ratio F_e/U. Since in the normal way, endogenous fecal lead is masked by unabsorbed dietary lead, Kehoe analyzed the urinary and fecal excretion of two subjects after the end of a lengthy period in which they had taken daily oral doses of lead, in order to estimate how much of the stored lead was excreted via the urinary and fecal routes in the post-experimental period. He found that subject EB excreted approximately equal amounts in urine and feces, whereas subject MR apparently excreted very little endogenous fecal lead. However, this method of deducing endogenous fecal excretion is imprecise, since it is obtained from the difference between the measured fecal excretion and the pre-exposure fecal excretion.

Chamberlain *et al.* (1978) analyzed Kehoe's balance experiments with the assumption that $F_e/U = 0.5$, as found in the Harwell ^{203}Pb experiments. The percentage uptake from the gut was found to be 16% (MR), 12% (EB), and 17% (FC). The percentage of the total uptake (including, in the case of subject FC, the estimated uptake from the lung during his experimental exposure to airborne lead) which was stored during the experimental periods was 40% (MR), 50% (EB), and 38% (FC). Since the experimental periods were many months in duration, it is likely that this storage was mainly in bone.

Table 6 shows an analysis of the uptake, storage, and excretion of three nurses (two females and one male) at St. Mary's Hospital, London, as given by Barltrop and Strehlow (1979). Measurements of I, F, and U for each subject were made for either 3 or 4 periods of three days each. The PbA concentrations to which the subjects were exposed were also measured. In estimating uptake from the lung, it has been assumed that $f_a = 0.5$ and $f_b = 1$. When lead balances are done for a limited period, considerable scatter of results is to be expected. Taking Barltrop and Strehlow's three adult subjects together and averaging the results, the urinary excretion at 48 μg/dl is more than 30% of the dietary lead. There

TABLE 6. Uptake, Excretion, and Storage of Lead (μg/d) in Three Adult Subjects[a]

Subject	*Sex*	*I*	*A*	*F*	*U*	F_e	*S*	*G*	*fg*
AW	*F*	*165*	*10*	*113*	*26*	*13*	*36*		
SMC	*F*	*261*	*11*	*187*	*66*	*33*	*19*		
CD	*M*	*210*	*10*	*248*	*52*	*26*	*-80*		
Mean		*212*	*10*	*183*	*48*	*24*	*-12*	*50*	*0.24*

Notes: $S = I + A - F - U$ $G = U + F_e + S - A$

[a]*Barltrop and Strehlow, 1979.*

is a small negative storage term of -12 μg/dl. The calculated fractional uptake from gut is 0.24.

Relation Between Total Uptake and Blood Lead

Table 7 shows the total uptake of lead by groups for which measurements or estimates of dietary and inhaled lead and of blood lead have been published. Williams *et al.* (1969) equipped workers with personal air samplers to measure their exposure to airborne lead. In calculating the total uptake in Table 7, the ventilation rate at work is assumed to be 10 m^3 per 8 hour shift (Mehani, 1966). Averaged over the 7-day week, this is equivalent to $V_o = 7\ m^3/d$. The ventilation rate away from work, V_a, is taken as 11 m^3/d. The air lead at work, PbA_o, is as given by Williams *et al.*, and the air lead away from work, PbA_a, is assumed to be 1 $\mu g/m^3$. Fractional uptake (f_r) values of 0.35 and 0.5 have been assumed for the industrial and non-industrial exposures.

The dietary lead is taken as 270 μg/d (Thompson, 1971), and f_g as 0.15. For Kehoe's subjects, who received oral doses of lead, and for Zurlo and Griffini's subjects, f_g is again 0.15, and a small contribution from inhaled lead is obtained by putting $PbA_o = 2\ \mu g/m^3$, $f_r = 0.5$, and $V_a = 15\ m^3/d$. Fugas and Saric (1978) measured both dietary and airborne lead in a population living near a smelter and in a control population. As the particle size was about 2 μm, f_r has been taken as 0.35 for the population in the smelter area.

In Figure 6, the mean blood lead of the various groups is plotted against the estimated total uptake (not intake) from diet and air. The full curve is fitted by eye to the points. The dashed curve is that previously derived in the National Academy of Sciences (1972) report. The curvilinear nature of the relation between blood lead and intake is evident and may be related partly, at least, to increase of renal clearance with increase in PbB. A curvilinear relation between PbB and ingestion of lead from

TABLE 7. *Blood Lead Versus Uptake from Gut and Lung*

Subject	PbB μg/dl	Air PbA_o μg/m³	Lead PbA_a μg/m³	Ventilation m³/d V_o	V_a	Uptake (μg/d) A_o	A_a	Food & drink (μg/d) I	Uptake (μg/d) G	Total Uptake, T (=A_o+A_a+G)	References
Machine pasters	74.2	218	1	7[a]	11	530	5	270	40	575	Williams et al., 1969
Hand pasters	63.2	150	1	7	11	370	5	270	40	415	"
Formers	63.0	134	1	7	11	330	5	270	40	375	"
Plastics	28.1	10	1	7	11	25	5	270	40	70	"
Unexposed	28		2	15	15		15	230	35	50	Kehoe, 1961
SW	37		2		15		15	530	80	95	"
MR	55		2		15		15	1230	185	200	"
EB	68		2		15		15	2230	335	350	"
FC	45	150	2	3.2[b]	12	12	160	220	33	205	"
Men (duplicate portions)	12.3							230	35	35	Nordman, 1975
Men (composite analyses)	34.6		2		20		20	505	76	101	Zurlo and Griffini, 1973
Men, in smelter area but not occupationally exposed	65.8		17		20		120	2900	435	562	Fugas and Saric, 1978
Men in control areas	34.2		0.1		20		1	510	76	77	"

[a] Assuming 10 m³ breathed per 8 h shift (Mehani, 1966), 5 shifts per week.
[b] Subject sedentary: 0.6 m³/h for 7.5 h/d, 5 d/week.

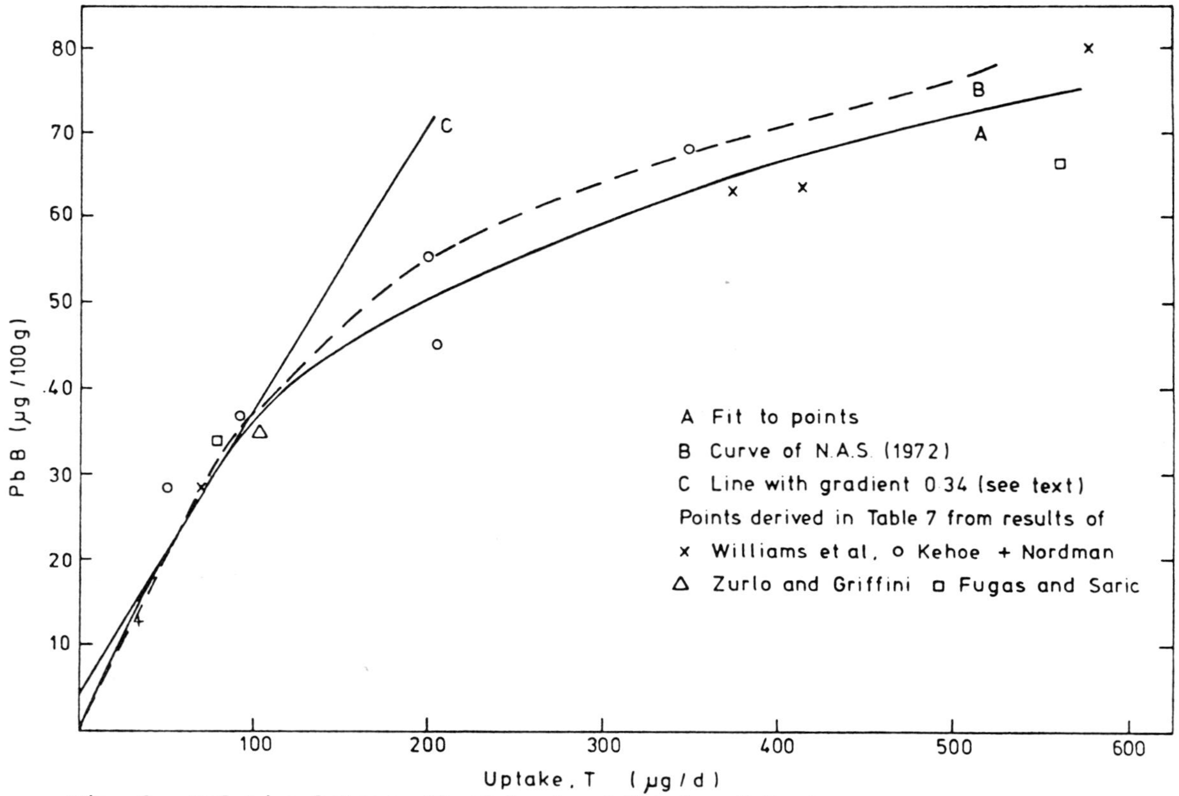

Fig. 6. Relation between Blood Lead and Uptake of Lead

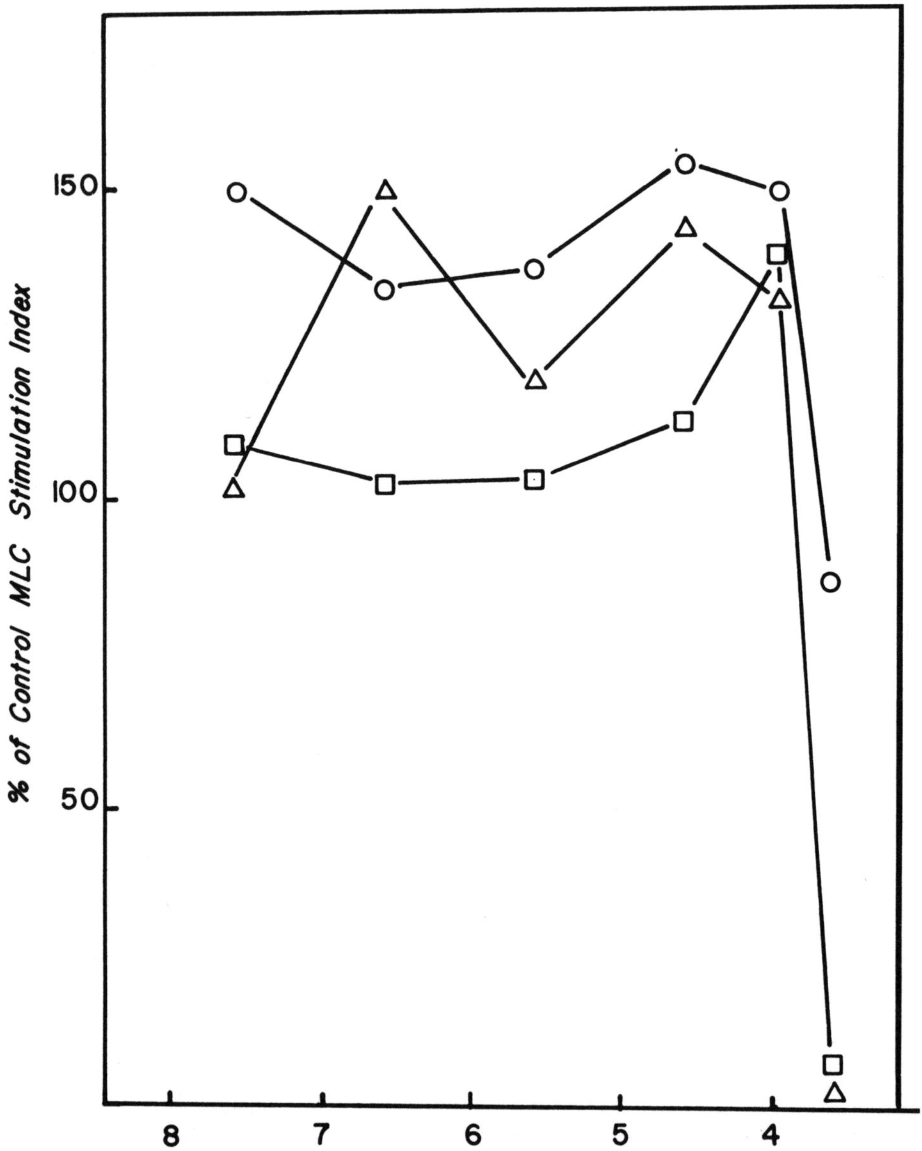
% of Control MLC Stimulation Index
150
100
50
8
7
6
5
4
Molarity PbCl₂ (-Log)

Fig. 7.

plumbo-solvent water supplies has been suggested by Moore *et al.* (1977).

The straight line in Figure 7 is the increase, ΔPbB, in blood lead to be expected from an increase, ΔT, in uptake of lead on the assumptions (Chamberlain *et al.*, 1978):

1. a fraction, 0.55, of the uptake becomes attached to red cells;
2. the biological half-life in blood is 18 days;
3. a factor, 1.3, is to be allowed for long-term resorption and re-entry into blood of some of the lead which is stored in bone; and
4. the mass of blood is 5400 g.

With these assumptions:

$$\frac{\Delta PbB}{\Delta T} = \frac{0.55\times 18\times 1.3}{54\times 0.693} = 0.34\ \mu g/100\ g \text{ per } \mu g/d.$$

The straight line in Figure 6 is drawn with this slope through the point T = 50, PbB = 20, and is close to being a tangent to the curve of PbB against T.

REFERENCES

Barltrop, D., and Strehlow, C. D. (1979). The absorption of ingested lead by children: Proceedings of the 2nd International Symposium on Environmental Lead Research held in Cincinnati (1978). (elsewhere in this publication)

Blake, K. C. H. (1976). Absorption of ^{203}Pb from the gastrointestinal tract of man. *Environ. Res. 11*, 1-4.

Bronner, F., Harris, R. S., Maletskos, C. H., and Benda, C. E. (1958). Studies in calcium metabolism: the fate of intravenously injected radiocalcium in human beings. *J. Clin. Invest. 35*, 1-6.

Castellino, N., and Aloj, S. (1964). Kinetics of the distribution and excretion of lead by rats. *Brit. J. Ind. Med. 21*, 308-314.

Chamberlain, A. C., Clough, W. S., Heard, M. J., Newton, D., Stott, A. N. B., and Wells, A. C. (1975). Uptake of lead by inhalation of motor exhaust. *Proc. Roy. Soc. B., 192*, 77-110.

Chamberlain, A. C., Heard, M. J., Little, P., Newton, D., Wells, A. C., and Wiffen, R. D. (1978). Investigations into Lead from motor vehicles. *AERE Harwell Report* R9198.

Cohen, N. (1970). *The Retention and Distribution of Lead-210 in the Adult Baboon*, Ph.D. Thesis. New York University, New York.

Durbin, P. W. (1972). Plutonium in man. *In* Radiobiology of Plutonium (B. J. Stover and W. S. S. Jee, eds.). University of Utah.

Devoto, G., and Spinazzola, A. (1973). L'effet de la présence du plomb dans l'air, dans l'eau et les denrées alimentaires sur son taux d'absorption chez différentes catégories de personnes. *In* Environmental Health Aspects of Lead, pp. 859-867. Commission of European Communities, Luxembourg.

Fugaš, M., and Sarič, M. (1979). Health study of a lead-exposed population: Proceedings of the 2nd International Symposium on Environmental Lead Research held in Cincinnati (1978). (elsewhere in this publication).

Hackett, P. L., and Sikov, M. R. (1977). Lead toxicity in pregnant rats, pp. 263-264. *Laboratory Report* 2100 Pt. 1, Battelle-Northwest, Richland, Washington.

Hammond, P. B. (1971). The effects of chelating agents on the tissue distribution and excretion of lead. *Toxicol. Appl. Pharmacol. 18,* 296-310.

Harrison, G. E., Carr, T. E. F., Sutton, A., Humphreys, E. R., and Rundo, J. (1969). Effect of alginate on the absorption of lead in man. *Nature 224,* 1115-1116.

Harrison, G. E., and Sutton, A. (1967). Ratio of fecal to urinary clearance of strontium in man. *In* Strontium Metabolism (J. M. A. Lenihan, J. F. Loutit, and J. H. Martin, eds.). Academic Press, New York.

Heyder, J., Armbruster, L., Gebhart, J., Frein, E., and Stahlhofen, W. (1975). Total deposition of aerosol particles in the human respiratory tract for nose and mouth breathing. *J. Aerosol Sci. 6,* 311-328.

Holtzman, R. B. (1978). Application of radiolead to metabolic studies. *In* The Biochemistry of Lead in the Environment, Part B (J. O. Nriagu, ed.). Elsevier, Amsterdam.

Horiuchi, K., and Horiguchi, S. (1958). Studies on the industrial lead poisoning: an experimental study with radioactive lead (radium D). *Oasaka City Med. J. 4,* 159-169.

Hursh, J. B. (1973). Retention of ^{210}Pb in beagle dogs. *Health Phys. 25,* 29-35.

James, A. C. (1978). Lung deposition of sub-micron aerosols calculated as a function of age and breathing rate. *In* Annual Research and Development Report (1977), pp. 71-75. National Radiological Protection Board, Harwell, Oxon.

Kehoe, R. A. (1961). The metabolism of lead in health and disease. *J. Roy. Inst. Pub. Health Hyg. 24,* 81-96, 101-120, 129-143, 177-203.

King, E. (1979). Particle size, solubility, and biochemical indicators. *Ann. Occup. Hyg.* (to be published) (also, a less detailed discussion appears elsewhere in this publication).

Lucas, H. F., and Stanford, J. E. (1967). Excretion and retention of lead-210 in rats. *In* Report ANL 7360, pp. 105-110. Argonne National Laboratory.

Lloyd, R. D., Mays, C. W., Atherton, D. R., and Bruenger, F. W. (1975). ^{210}Pb studies in beagles. *Health Phys. 28,* 575-583.

Mehani, S. (1966). Lead retention by the lungs of lead exposed workers. *Ann. Occup. Hyg. 9,* 165-71.

Mitchell, R. (1977). Lung deposition in freshly excised human lungs. *In* Inhaled Particles, IV, pp. 163-175. Pergamon, London.

Moore, M. R., Meredith, P. A., Campbell, R. C., Goldberg, A., and Pocock, S. J. (1977). Contribution of lead in drinking water to blood lead. *Lancet.*, Sept. 24, p. 661.
Morgan, A., Holmes, A., and Evans, J. C. (1977). Retention, distribution and excretion of lead by the rat after intravenous injection. *Brit. J. Ind. Med. 34,* 37-42.
National Research Council (1972). *Airborne Lead in Perspective.* National Academy of Sciences, Washington, D.C.
Nordman, C. H. (1975). *Environmental Lead Exposure in Finland,* Doctoral Thesis. University of Helsinki, Finland.
Nozaki, K. (1966). Method for studies on inhaled particles in human respiratory system and retention of lead fume. *Ind. Health* (Japan) *4,* 118-128.
Rabinowitz, M., Wetherill, G. W., and Kopple, J. D. (1976). Kinetic analysis of lead metabolism in healthy humans. *J. Clin. Invest. 58,* 260-270.
Schubert, J., and White, M. R. (1952). Effect of sodium and zirconium citrates on distribution and excretion of injected radiolead. *J. Lab. Clin. Med. 39,* 260-266.
Stover, B. J. (1959). Pb^{212} (ThB) tracer studies in adult beagle dogs. *Proc. Soc. Exp. Biol. Med. 100,* 269-272.
Strehlow, C. J. (1971). *The Use of Deciduous Teeth as Indicator of Lead Exposure,* Ph.D. Thesis. New York University.
Strehlow, C. J., and Barltrop, D. (1978). Nutritional status and lead exposure in a multi-racial population. 12th Annual Conference on Trace Substances in Environmental Health, University of Missouri, Columbia.
Task Group on Lung Dynamics (1966). Deposition and retention models for internal dosimetry of the human respiratory tract. *Health Phys. 12,* 173-207.
Thompson, J. A. (1971). Balance between intake and output of lead in normal individuals. *Brit. J. Ind. Med. 28,* 189-194.
Wetherill, G. W., Rabinowitz, M., and Kopple, J. D. (1975). Sources and metabolic pathways of lead in normal humans. *In* Recent Advances in the Assessment of the Health Effects of Environmental Pollution held in Paris (June 1974). Commission of European Communities, Paris.
Williams, M. K., King, E., and Walford, J. (1969). An investigation of lead absorption in an electric accumulator factory with use of personal samplers. *Brit. J. Ind. Med. 26,* 202-216.
World Health Organization (1977). *Environmental Health Criteria 3: Lead.* WHO, Geneva.
Yu, C., and Taulbee, D. (1977). A theory of predicting respiratory tract deposition of inhaled particles in man. *In* Inhaled Particles, IV, pp. 35-49. Pergamon, London.
Zurlo, N., Griffini, A. M., and Vigliani, E. C. (1970). The content of lead in blood and urine of adults living in Milan not occupationally exposed to lead. *Amer. Ind. Hyg. J. 31,* 92-95.
Zurlo, N., and Griffini, A. M. (1973). Le plomb dans les aliments et dans les boissons consommés à Milan. *In* Environmental Health Aspects of Lead, pp. 93-98. Commission of European Communities, Luxembourg.

DISCUSSION OF PAPER BY DR. A. C. CHAMBERLAIN

Vostal: I would like to congratulate you. It's really a very impressive and very excellent type of study, which we wish we had in our hands many years ago. It's really a great possibility of using Pb-203 rather than the elaborate methods using Pb-212 which have been used before. I am also very glad that you have mentioned the paper by Betsy Stover. I think that this is also relevant to our discussion yesterday with Dr. Grandjean, and probably we should really unify what we mean by the biological halftime of Pb-203 in the blood. I assume that you have used a carrier-free Pb-203, and I assume that the increment of the lead exposure by this amount of the lead was practically infinitesimal. So, probably we should not talk about the half-time of lead in blood being 18 or 20 days. I think that we should say the radioactivity halftime of lead in blood which disappears. As you have described, the Betsy Stover data indicate very clearly that there is not only the process of excretion or clearance from the blood, there is a continuing process of exchange with the lead in other compartments, and unless we can measure specific activity changes, probably we should not relate it so directly to the metabolism of the lead in the blood or in the body.

Dr. A. C. Chamberlain, AERE Harwell: Thank you very much for making those points which I should have made during my talk. When we inhaled lead-203, in most cases, we added virtually infinitesimal amounts of carrier. In some of the earlier experiments, in the closed box system, we had carrier because we were not so expert then at making tetraethyl lead. We inhaled up to about a milligram of lead, which was enough to raise our blood lead slightly, from say 30 to 35 or 20 to 25 μg/dl. So, we covered a fairly wide range from infinitesimal stable lead intake to considerable, although not high, stable lead intake. As to the effect of this on the biological half-life, I agree that there is evidence that with heavy exposure the half-life in blood is decreased and renal clearance increased. The 16, 18, or 20 day half-life which we find and which the UCLA people found in long term lead-204 measurements is, to some extent supported by the way blood lead increases when people are moved from one environment to another, in particular a study by Tola *et al.*, in Finland, where they measured the blood lead of people before they entered employment and then subsequently after they entered employment. The way in which blood lead increases with time seems to indicate a biological half-life of the same order as we found don't think it's entirely a radioactive lead factor. It seems to be substantiated in some cases by experiments of Tola's type and also by the data of Griffin *et al.*, from Albany where the time taken for the volunteers blood to respond to the higher concentration seemed to be consistent with the half-life of about 20 days.

Gary Ter Haar, Ethyl Corporation: Dr. Chamberlain, I wonder if based on your measurements and experience in the work you've described, you would speculate on what you believe the alpha is for a mass median equivalent diameter of about 0.2 µm.

Chamberlain: I think that Fig. 1 comparing 0.02, 0.04 and 0.09 µm particles suggests that if your mass median is 0.2 µm and if the particulate size spectrum is fairly narrow, there will be a fairly low lung retention. Certainly, Fig. 2 shows a minimum lung retention at a particle size of from about 0.2 to 0.6 or 0.8 µm. A particle size of 0.2 µm, perhaps, gives a slightly higher deposition than, say, 0.3 or 0.4, but the minimum is somewhere in that area. Now, the value of that minimum is disputed, with retention between 15 percent and 30 or 35 percent and the cause of this variation is not fully understood.

Ter Haar: I think you would agree that that might give you an alpha closer to one.

Chamberlain: It would, yes.

Ter Haar: And this might explain the difference between the observations of the epidemiological studies which generally give lower numbers.

Chamberlain: It certainly might also explain the difference between our observations and, for example, those that have been deduced from Dr. Kehoe's experiments and from industrial experience.

Grandjean: You mentioned that organic lead has a different distribution in the body than inorganic lead. I wonder if any radioactive organic lead was present in the exhaust gases used in your experiment.

Chamberlain: A negligible amount of organic lead was present. This was monitored, and we do not think that when tetraethyl lead is burnt in an engine organic lead is exhausted. We believe that the organic lead which is found in the urban situation is by direct evaporation from petrol. We did check this, and it was so small that it did not affect the results in any way. Incidentally, if there had been organic lead present one would have expected a loss by exhalation after an inhalation session, and this wasn't found.

In our full report (Chamberlain *et al.*, 1978) there is a table showing that at ten days after inhalation, the loss of total body radioactivity equalled the sum of the fecal and urinary excretions, suggesting that loss by sweat, hair, and exhalation was only a few percent of the total.

LEAD IN DRINKING WATER: THE CONTRIBUTION OF HOUSEHOLD TAP WATER TO BLOOD LEAD LEVELS

D. Worth, M.D.
A. Matranga

Tufts-New England Medical Center

M. Lieberman, Ph.D.
E. DeVos
Harvard Graduate School of Education

P. Karelekas
C. Ryan

U.S. Environmental Protection Agency
Boston, Massachusetts

G. Craun

U.S. Environmental Protection Agency
Cincinnati, Ohio

INTRODUCTION

Water as a source of lead poisoning has been known from antiquity. Lead pipe was used extensively in water works by the Romans (Frontinus, 1973). Weston (1920) cites Galen, who described the harmful action of water on large lead service pipes in the Fourth Century A.D., as well as an earlier document by Vitruvius, an architect in the time of Augustus Caesar, who forbade the use of lead pipe for conveying water.

In 1899, the Massachusetts Board of Health reported that during the previous two years it had undertaken an "investigation ... an account of a large number of cases of lead poisoning in several of the towns of the State, which upon investigation (were) found to be due to the fact that the water supplied publicly to these

ISBN 0-12-460520-6

towns was of such a character as to actively attack lead, and cause an appreciable amount of lead to be taken constantly into their systems by persons who habitually used these waters" (Massachusetts State Board of Health, 1899). The State Board of Health reported then that 71 of 136 reporting cities and towns in Massachusetts had lead or lead lined water service pipes in use. The service pipe, or line, connects the interior plumbing to the water main. The mains, major conduits delivering water to many service lines, never were of lead composition.

Among the properties which influence quality of drinking water are hardness and pH. Hardness, is caused by divalent metallic cations, principally calcium, magnesium, strontium, ferrous iron and manganous ions. Hardness in water is derived largely from contact with soil and rock formations. In general, hard waters originate in areas where top soil is thick and limestone is present; soft waters originate where top soil is thin and limestone formations are sparse or absent. Hardness is ordinarily expressed in terms of calcium carbonate ($Ca\ CO_3$). In common usage, water is classified as soft if it contains less than 75 parts per million of hardness as calcium carbonate.

Soft water may be corrosive to metallic surfaces. The pH of water derived from both geologic and atmospheric factors also may contribute a corrosive quality when either high or low (Durfor and Becker, 1964). Natural water supplies in New England tend to be both soft and of a relatively low pH.

In 1920, Weston gave a detailed account of the recognition in Massachusetts and elsewhere of the problems associated with the use of lead pipe to convey drinking water. He stated: "Many water works ... have thousands of dollars invested in lead services. If the water supplying such works be found to attack lead services, the best plan is to replace them gradually ... meanwhile protecting consumers by proper treatment of the water supplied through the existing services." He cited the practice, common in England, of adding calcium carbonate to very soft waters, adding "it may be stated in brief that prevention of corrosion consists in neutralizing the acid components of the water by the addition of alkalies" (Weston, 1920).

Most public water suppliers treat raw water to improve its quality by removing undesirable constituents, such as microbiologic pathogens, inorganic and organical chemical, turbidity, and taste and odor-causing compounds. Treatment consists of variable combinations of screening, coagulation, flocculation, sedimentation, filtration, and disinfection. Adjustment of pH and softening are also practiced to obtain the desired "finished" water quality. Only a few communities in New England have instituted corrosion control as a part of their water treatment.

In a 1924 survey, 51% of 539 cities in the United States used lead or lead-lined service lines to some extent (Donaldson, 1924). In most areas of Massachusetts, policies developed against new use of lead in water supply systems during the late 1920's and 30's, although no concerted effort was made to remove the pipes already in service. Many of these pipes have never been replaced and are found associated primarily with old housing.

STUDY DESCRIPTION

In 1972, we conducted a preliminary study of tap water quality in the Beacon Hill area of Boston and showed that 65% of 54 household samples exceeded the 50 micrograms per liter (μg/l) federal standard for lead, with some samples over 5 times the acceptable limit. On the basis of this information, we proceeded with a large-scale study to further assess the extent and significance of the problem in the metropolitan Boston area.

This study was designed to determine whether a significant relationship exists between lead in household drinking water and blood lead levels of the household residents. Other household sources of lead and demographic factors were also considered. The cities of Boston (Brighton-Allston district), Cambridge, and Somerville were selected for study. The three study areas were chosen primarily because their municipal water department records showed that many houses had lead or lead-lined water service lines in current use.

In a previous report on the quality of the water samples taken in this study, we described the watersheds, the treatment processes, the chemical and physical properties of raw and finished waters, and the range of trace metals in finished water and in tap water after distribution (Karelekas *et al.*, 1976). The water in all three cities was low in lead both before and after treatment and prior to distribution. During the study period, and for many prior years, Boston and Somerville purchased water from the Metropolitan District Commission (MDC), a state agency, after the raw water was treated with chlorine and ammonia. The water distributed in Boston and Somerville was soft, with a pH varying between 6 and 7 during the sampling period, and was expected to be corrosive to lead piping. The study area in Cambridge uses water from a different source. The natural water supply is also soft, with a pH less than 7, but for some years, including the sampling period, the municipal water department has adjusted the pH to levels varying from 6.9 to 8 by addition of sodium hydroxide. There are no known industrial point sources of lead, such as smelters, which might pollute air in the areas. Geographic relationship of the cities to heavily travelled highways is similar, as are traffic patterns within the cities.

Participants were contacted by letters sent from the city governments inviting participation of households in selected districts and by letters sent home with school children. Referrals were accepted from neighborhood health centers of families wishing a household survey. Households in the study included single family homes, two and three family houses, and a few units in apartment buildings. Water samples were collected from 320 households, during the period January, 1974 through January, 1975. Water samples were preserved with 1.5 ml of concentrated nitric acid per quart. Four types of water samples were collected in plastic containers which were rinsed out with tap water prior to being filled:

A Standing Grab Sample (STH_2O) was taken immediately after turning on a cold water tap and rinsing out the sample container. This sample of water was in contact with interior plumbing for an unknown length of time.

A Running Grab Sample ($RUNH_2O$) was taken after the water had run for approximately 4 minutes. This represented water which would have minimal contact with the service line and interior plumbing.

A Composite Sample ($COMH_2O$) was taken to assess the average concentration of lead in water during a one day period. Residents were asked to fill a one gallon sample bottle with one quart of water at each meal.

An Early Morning Sample (EMH_2O) was collected from houses where lead would be expected in drinking water because of lead piping or because previous samples indicated lead in the drinking water. A one quart container was filled early in the morning, at the time when the participant felt the water temperature change from warm to cold. Since water would be expected to warm slightly after standing in interior plumbing, the sample of colder water would have been standing overnight just outside the foundation of the house, in the underground service line. This sample would be expected to have the longest contact time with the service line during 24 hours.

Composition of interior plumbing and of the exterior service line (which can be observed between the foundation and the household meter) were determined by inspection. The local water department record was reviewed for each dwelling to determine whether lead had ever been used in the service line.

Water samples were analyzed by the Water Supply Research Laboratory of the EPA in Cincinnati, Ohio. Analysis for water lead was done by flameless atomic absorption with a heated graphite atomizer (Perkin-Elmer HGA-200) (Caldwell *et al.*, 1974). The detectable limit for lead is higher for Cambridge than for Boston and Somerville (MDC water) because of interferences encountered during analysis; MDC water chloride content is 7 mg and sulfate is 10 mg per liter, causing a suppression of the lead signal by an amount equivalent to 8 μg per liter of lead. Cambridge water, which contains approximately 50 mg of chloride and 30 mg of sulfate per liter, caused a suppression equivalent to 16 μg per liter of lead. These values were added to the instrument reading of 5 μg per liter for acidified distilled water blanks, resulting in minimum detectable levels of 13 μg, and 21 μg per liter for MDC and Cambridge water, respectively. Samples collected at the water treatment plants confirmed records of the Massachusetts Department of public Health showing the lead concentration of water prior to distribution to be at or below detection limits for all three cities.

Dust samples were collected in plastic collection bags from participants' vacuum cleaners. Whenever possible, the entire contents were removed from the machine. If there was no vacuum cleaner or if the machine was empty, no sample was collected.

Dust samples were analyzed by the Water Supply Research Laboratory of the EPA. Each dust sample was processed through a series of stainless steel sieves mounted on a Tyler Rotap shaker and the sifted material was collected in a stainless steel pan. The final screen had a grid size of 0.18 millimeters. Each sample was digested in nitric acid and analyzed with a Perkin-Elmer 403 Atomic Absorption Spectrophotometer.

Tests for lead content of paint were performed on walls and window sills of all rooms used by children aged 6 years or less. Paint analysis was performed by an experienced environmental technician, using 2 portable Model XK-2 Princeton Gamma Tech X-Ray fluorescence lead detectors calibrated every 2-3 days. Presence or absence of teeth marks on window sills and of loose paint on walls and ceilings were recorded.

Blood samples were obtained from all participants by finger-stick, collected in 100 μl capillary tubes, with polyethylene caps to prevent evaporation and contamination. One person with previous experience collected all blood samples. Analysis for blood lead was performed by Environmental Sciences Associates, Burlington, Massachusetts, by anodic stripping voltametry, using the procedure developed by Matson (Matson *et al.*, 1971).

Distance from major highways was measured for each residence on a large scale area map; when distances were short, on-site measurements were made. Traffic density patterns for local, state, and interstate roads were obtained from the agencies compiling such data.

A Residential Traffic Density scale was constructed (TD 50) for the amount of traffic within 50 meters of a highway (Table 1).

TABLE 1. Residential Traffic Density Scale for Within 50 Meters of a Highway

Scale	*Traffic Location and Density*
High	*Within 50 meters of a highway with over 40,000 autos daily*
Intermediate	*Within 50 meters of a highway with over 20,000 autos daily but less than 40,000*
Low	*Within 50 meters of a highway with less than 20,000 autos daily*

A similar 100 meter (TD 100) scale was also used, and each residence was designated high, intermediate, or low density area by each scale.

The head of household or wife provided the following information for each participant: age, sex, educational level attained by head of household, total family income, history of pica for children 6 years of age or younger, and use of tobacco by persons 20 years of age or older.

RESULTS

Table 2 shows the numbers and types of samples obtained.

Water: Descriptive statistics for water lead levels were calculated using an imputed value for below detectable levels. The imputed values are the midpoints of the range below detectable values. Figure 1 shows the frequency of lead pipe, of detectable levels of lead in tap water in any sample, and of levels which exceed the standard for lead in drinking water in any sample. Since many water samples show no detectable lead and because lead was not present prior to distribution, detectable levels of lead in water samples taken at the tap indicate that lead is being picked up somewhere in the distribution system, presumably either in the service line or in the interior plumbing.

Detectable levels of lead as well as levels which exceed the standard occur with greater frequency in Boston and Somerville than in Cambridge. These differences may be explained partially by the less corrosive quality of water in Cambridge and by differ-

TABLE 2. Lead Levels in Water, Dust, and Paint

Sample Source (No.)	*Range*	*Mean*	*S.D.*	*Median*
Running Water (320) Lead mg/l	≤0.013-0.208	0.017	0.021	0.011
Standing Water (320) Lead mg/l	≤0.013-1.508	0.030	0.097	0.013
Composite Water (25) Lead mg/l	≤0.013-0.758	0.078	0.161	0.028
Early Morning Water (131) Lead mg/l	≤0.013-1.108	0.096	0.178	0.048
Dust - mg/g (214)	0.17-99.00	2.58	9.36	1.40
Maximum Paint Lead mg/cm^3 (93)	0.9-58.0	15.0	11.6	14.8

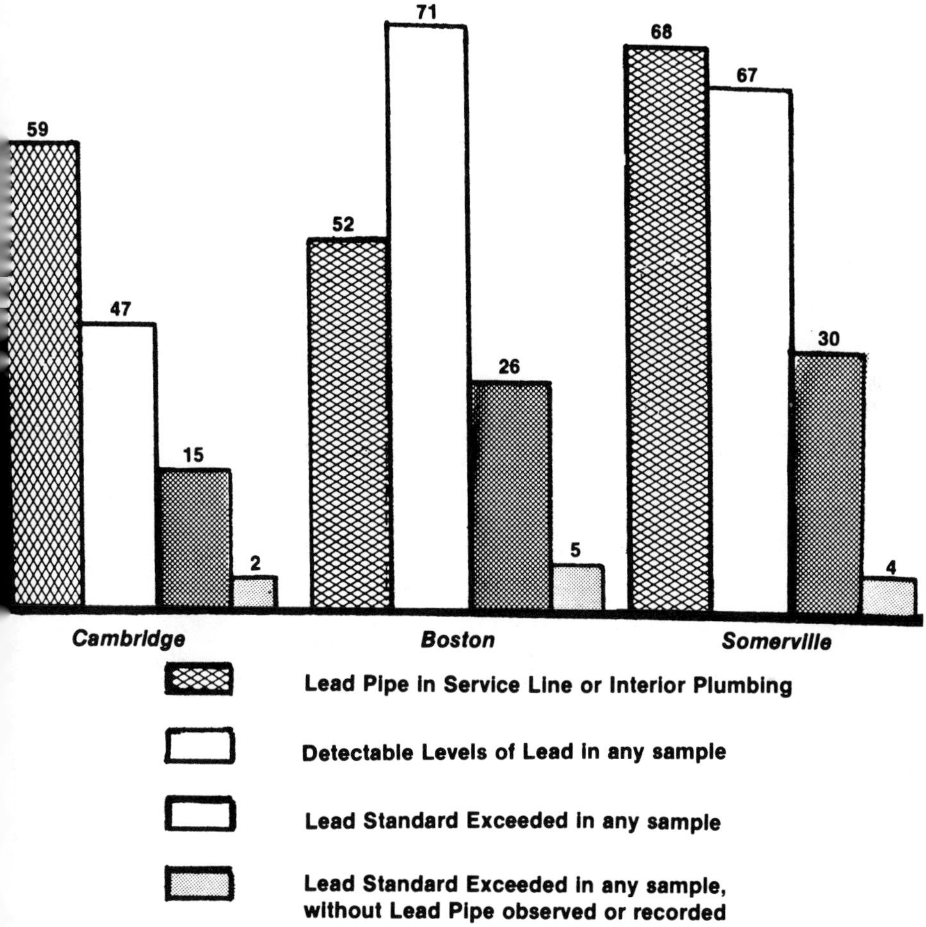

Fig. 1. Percent of households with lead pipe and lead in tap water in any sample taken.

ences in the lower limits of detection. Variations in the extent of use of lead pipe or lead materials, such as solder in the service lines or in interior plumbing, may also be a factor. The percentage of observed or recorded lead pipe in use in Boston (52 percent) compared to the pickup of lead (71 percent) may indicate the presence of hidden or unrecorded sections of lead pipe. Another possible explanation is that lead is being leached from solder used to join copper pipes. This would be consistent with Dangle's observations of pickup of lead in the distribution system in Seattle, a city where lead pipe is virtually unknown (Dangle, 1975). Few households without known lead pipe have water with lead values exceeding the standard.

Paint: Multiple measurements were made in 93 households. Maximum measured lead level in a household was used for analytic purposes. Flaking paint or broken plaster was observed in only 7 households. Teeth marks on window sills were noted in only 2 residences.

Dust Samples: The highest dust sample contains nearly 500 times as much lead as the lowest sample (Table 2).

Correlations Between Environmental Variables: Table 3 presents Pearson correlation coefficients for individual household lead levels in standing grab water (STH_2O), running grap water ($RUNH_2O$), dust and paint, and for traffic density measured by the 50 (TD50) and 100 (TD100) meter scales. Correlations were not determined for composite and early morning samples because these were obtained on a highly selective basis. For determination of these relationships, water lead levels were grouped beginning at or below detectable limits to 20 μg per liter, and thereafter by increments of 10 μg per liter, 20-30 μg per liter, etc. Significant correlations were found only between standing and running grab water samples, between the two measures of traffic density (TD50 and TD100), and between traffic density (TD50) and dust lead levels.

TABLE 3. Pearson Correlation Coefficients, Number of Cases of Significance Level. Lead Levels of Environmental Samples

	STH_2O	$RUNH_2O$	TD50	TD100	DUST
$RUNH_2O$	0.4637 (314) p=0.001				
TD50	-0.0394 (314) p=0.243	0.0393 (319) p=0.242			
TD100	0.0286 (314) p=0.307	-0.0003 (319) p=0.498	0.5542 (320) p=0.001		
DUST	-0.0073 (207) p=0.459	-0.0280 (211) p=0.343	0.1393 (212) p=0.021	0.0531 (212) p=0.221	
PAINT	0.0220 (90) p=0.418	0.1417 (92) p=0.089	0.0198 (93) p=0.425	0.1298 (93) p=0.108	0.0826 (57) p=0.271

The relationships between the water samples would be expected, since the samples were obtained under similar conditions, the only variation being the duration of water contact with the household water distribution system. The relationship between the traffic density measures results from the use of two scales to express an unvarying relationship. The relationship between proximity to major highways on the 50 meter scale (TD50) and dust lead levels is consistent with other studies suggesting that lead in gasoline contributes to elevation of local lead levels in soil and dust. Our failure to find a significant relationship between dust lead and proximity to heavy traffic on the 100 meter scale (TD100) may indicate that particulate lead is deposited locally only very near to the emission source. This relationship may be obscured using the larger scale (TD100).

Blood Samples: Blood samples were obtained from 793 individuals. The reliability coefficient for 43 duplicate samples, as determined using a Pearson product correlation, is 0.903. A frequency distribution of blood lead levels is shown in Table 4. For analytic purposes, blood leads were logarithmically transformed to obtain a normal distribution of value. The empirical cumulative distribution function of log blood lead was well-fitted by a normal curve with a mean of 3.0802 and a standard deviation of .3793 (this translates to a geometric mean of 21.76 μg per deciliter and a geometric standard deviation of 1.46 μg per deciliter). A Kolmogorov-Smirnov goodness of fit test supported the appropriateness of this transformation. To take into account the log normalcy of the measurements all means and standard deviations for blood lead, used for analyses, are geometric.

Table 5 shows the distribution of blood lead by age and sex. The highest blood leads were observed in the 2-6 year old group in both sexes. Males had higher blood lead than females except in the group under 2 years of age. The increased mean blood lead level in the 2-6 year old group led us to question whether this

TABLE 4. Frequency Distribution of Blood Lead

Blood Lead (μg/dl)	*Number of Individuals*	*%*
<10	*24*	*3.0*
11-20	*319*	*40.2*
21-30	*308*	*38.8*
31-40	*110*	*13.9*
41-50	*19*	*2.4*
51-60	*7*	*0.9*
>60	*6*	*0.8*
	793	*100.0*

TABLE 5. Geometric Mean (G.M.) Blood Lead Levels and Geometric Standard Deviations (G.S.D.) by Sex and Age

Age Group (Years)	Total f	Total G.M. (G.S.D.) μg/dl	Males f	Males G.M. (G.S.D.) μg/dl	Females f	Females G.M. (G.S.D.) μg/dl
0-2	50	21.84 (1.48)	23	20.20 (1.56)	27	23.35 (1.40)
2-6	168	26.79 (1.42)	84	26.99 (1.41)	83	26.70 (1.42)
6-20	182	22.18 (1.40)	88	22.67 (1.35)	94	21.73 (1.44)
20-50	302	19.38 (1.45)	53	23.04 (1.34)	249	18.68 (1.46)
50+	91	20.92 (1.45)	35	24.47 (1.50)	56	18.97 (1.36)
All Ages	793	21.76 (1.46)	283	23.95 (1.42)	509	20.64 (1.47)

group was over represented in referrals from the neighborhood health centers. We examined the blood lead results by age group and by referral pattern. Table 6 shows that the members of households referred from neighborhood health centers had higher blood lead levels in all age groups, but the peaking in the 2-6 year old group occurred in both the referred and the non-referred sample. In aggregate and in all subsamples, the 2-6 year olds displayed higher blood leads than all other age groups. They were significantly higher ($p < 0.05$) than all older groups, except for 50+ year old males, where they were higher, but not significantly so.

Pica: Children who, by parent report, displayed pica had significantly higher blood lead levels than those who did not (27.55 μg/dl as opposed to 23.51 μg/dl, $t = 3.78$, $p < .01$). Further attempts at elucidating the relationship by considering other variables gave rise to no consistent or significant results. The ineffectiveness of these latter analyses may be the result of the

TABLE 6. Geometric Mean Blood Lead Levels (G.M.) and Geometric Standard Deviations (G.S.D.) by Age and Referral Pattern

Age Group (Years)	Referred f	Referred BlPb GM μg/dl	Referred G.S.D.	Non-Referred f	Non-Referred BlPb GM μg/dl	Non-Referred G.S.D.
0-2	15	29.82	(1.35)	35	19.12	(1.42)
2-6	59	32.85	(1.30)	109	23.99	(1.41)
6-20	33	26.52	(1.38)	149	21.33	(1.38)
20-50	43	21.91	(1.34)	259	18.99	(1.47)
50+	--			91	20.92	(1.45)
All Ages	150	27.63	(1.39)	643	20.58	(1.45)

generality of our pica measure (i.e., not specifying the objects mouthed), coupled with the great variability in environmental lead sources and the possibility that self-report is not a reliable index of pica.

Relationship of Blood Lead to Environmental Factors

The summary data in Table 2 show that each type sample was not taken for each household. Complete data are available only for running and standing grab water samples, traffic density, and blood lead levels of residents. Dust and paint data were not collected from all households. The number of individuals for whom paint data were available was prohibitively small for inclusion in the regression analysis. However, correlational analysis revealed no significant relationship between paint and blood lead (r = .00; n.s.).

In an effort to avoid making assumptions regarding the comparability of water lead distributions below the different detectable levels in the two communities, we analyzed data in Cambridge separately from that of Boston and Somerville. There were no significant differences in mean blood leads, dust, and paint lead levels, sex and age distributions between the participants in Boston/Somerville and Cambridge (Table 7). There were differences in educational achievement of head of household, total household income, and the lead content of water.

Statistical Analysis

The statistical analysis was designed to isolate the effect of water lead on blood lead, independent of the effects of age, sex, and social-economic status, and to take into account other measured environmental factors. By adapting a method of analysis, suggested by Cohen and Cohen (1975), for the treatment of missing values in quantitative data in multiple regression analysis, we were able to make use of all the information gathered without making distributional assumptions and without dropping subjects or variables. The data for lead levels in water samples were incorporated in two aspects--one reflecting the above/below detectable level dichotomy and the other consisting of lead values on the standard metric. Measurements below detectable levels were assigned an imputed value for this second aspect.

In regression analyses concerned with the estimates of regression coefficients and the hierarchical partitioning of increments in explained variance, consideration was given to a suitable imputed value for the below detectable water leads in the full scale aspect. The hierarchical decomposition of explained blood lead variation will be invariant for different imputed values, provided the increment is considered after the missing value dichotomy (above/below detectable) has been entered into the regression.

TABLE 7. *Community Comparisons*

	Boston/Somerville			Cambridge				
Variable	x	s.d.	n	x	s.d.	n	t	p
Blood Lead μg/d	21.93	1.47	524	21.47	1.44	231	.71	n.s.
Water Lead mg/l STH_2O	.034	.093	524	.019	.013	331	3.71*	.000
Dust Lead mg/g	2.28	6.35	324	3.46	11.50	174	-1.25*	n.s.
Paint Lead mg/cm^3	14.97	11.37	203	14.32	10.63	67	.41	n.s.

Sex

	Males	Females	
Boston/Somerville	194 (37.0%)	330 (63.0%)	524
Cambridge	74 (32.0%)	157 (68.0%)	231

$x^2_{(3)} = 1.53$ (n.s.)

Age

	0-2 yrs.	2-6 yrs.	6-20 yrs.	20-50 yrs.	50+ yrs.	
Boston/Som.	35(6.7%)	115(21.9%)	127(24.2%)	196(37.4%)	51(9.7%)	524
Cambridge	15(6.5%)	48(20.8%)	50(21.6%)	88(38.1%)	30(13.0%)	231
						755

$x^2_{(4)} = 2.17$ (n.s.)

Educational Level of Head of Household

	0-8 yrs.	9-12 yrs.	13-16 yrs.	17+ yrs.	
Boston/Som.	15 (2.9%)	347 (66.2%)	97 (18.5%)	65 (12.4%)	524
Cambridge	1 (0.4%)	113 (48.9%)	55 (23.8%)	62 (26.8%)	231
					755

$x^2_{(3)} = 34.44$ ($p < .01$)

Income

	$0-5000	5000-10,000	10,000-20,000	20,000+	
Boston/Som.	75 (14.6%)	194 (37.7%)	240 (46.7%)	5 (1.0%)	514
Cambridge	33 (14.7%	52 (23.2%)	96 (42.9%)	43 (19.2%)	224
					738

$x^2_{(3)} = 90.05$ ($p << .01$)

*using separate variance estimate

The estimated regression coefficients are interpretable since the imputed values are the midpoints of the ranges of the below detectable values (6.5 μg/liter of Boston and Somerville and 10.5 μg/liter for Cambridge).

To determine whether water lead accounted for observed variance in log blood lead levels, we applied a statistical model which contains an above/below detectable dichotomy, as well as linear, quadratic, and cubic aspects of the water lead levels for each of the four water sampling techniques. To ensure comparability of the samples, the data were grouped so that comparisons were made within household groups arranged thus:

households where two types of samples were taken

households where three types of samples were taken

households where four types of samples were taken

These do not represent separate samples, but are hierarchically constituted.

These analyses show a high degree of consistency in the relationship between log blood lead levels and the lead levels observed in standing grab samples (Table 8). In all groupings of the Boston-Somerville water samples, observed lead in the standing grab water samples shows a significant relationship ($p < .01$) with log blood lead of the residents of the household. The running grab water samples show a less consistent relationship with log blood lead and the proportion of variance explained is not so great as that explained by the standing grab water sample. The strength of relationship of the standing grab water sample to blood lead is repeated when the same analyses were performed on

TABLE 8. Proportion of Variation in Log Blood Lead Explained by Water Lead, by City of Residence and Sampling Method

# Samples Taken per Household		STH_2O	$RUNH_2O$	EMH_2O	$COMH_2O$	n
4	Bos/Som.	.321**	.196	.126	.166	37
	Camb.†	--	--	--	--	11
3	Bos/Som.	.065**	.026	.062**	--	277
	Camb.	.102	.072	.130*	--	77
2	Bos/Som.	.044**	.016	--	--	551
	Camb.	.076**	.045*	--	--	242

*p .05
**p .01
†11 cases in this subsample were considered too few for analysis.

the Cambridge data. The lead in standing grab water samples explains 4.4% and 7.6% of the variation in log blood lead in the Boston/Somerville and Cambridge samples, respectively. Standing grab water samples, therefore, were used in the regression analysis exclusive of the other type water samples. Table 9 explicates the various components of this model for a single entry in Table 8, specifically, the 4.4% representing Boston/Somerville, standing grab sample, homes with two samples taken.

Analysis of the contribution of other potential influences on blood lead was performed using an additive model with hierarchical inclusion of the demographic variables, age, sex, and educational level of the head of household prior to consideration of environment variables. When income was used as a demographic variable and entered into the equation after educational level, it accounted for an insignificant proportion of variance in both Boston/Somerville and Cambridge (0.5% and 0.1% respectively). For this reason and to not lose cases with missing data, income was not entered into the final model. By entering the demographic variables into the model prior to environmental variables, we were able to isolate residual variation in blood lead to analyze its relationship with specific environmental sources. Thus, we removed variations in blood lead which could be explained by demographic differences, and focused on the added explanatory power of the measured environmental variables.

Age, sex, and educational level of the head of household each explain a statistically significant proportion of variance in the dependent variable, log blood lead (Table 10). The appropriateness of a covariance model was confirmed by the findings of no significant interactions between the background variables (as covariates) and the variables of environmental exposure. Water lead explains 3.8 percent of the variation in blood lead, even after the variation associated with the demographic factors has been considered. Dust and traffic density each contributes less explanatory power than water lead. The total explanatory power of the

TABLE 9. Decomposition of Proportions of Variance Explained Using a Hierarchical Model Regressing Ln(BlPb) on STH_2O Pb (n = 551)

Independent Variable	R^2	*F*	*df*	I_{R^2}	*F*	*df*
Above/Below Detectable	.002	.954	1.549			
+ Linear aspect	.020	5.722**	2.548	.019	10.473**	1.548
+ Quadratic aspect	.036	6.752**	3.547	.015	8.652**	1.547
+ Cubic aspect	.044	6.273**	4.546	.008	4.700*	1.546

I_{R^2} = increment in R^2

* $p < .05$

**$p < .01$

model is 17.8 percent of the variation in log blood lead. An identical model in Cambridge showed water contributing a significant 5.6 percent.

Table 11 shows the explained variation and significance level of each environmental variable, when considered independently for entry into the regression equation after the demographic variables have been entered. Water lead was the best environmental predictor of blood lead in both communities.

Failure to control for demographic variables tends to provide over-estimates of water's contribution to blood lead. When these are controlled for, the proportion of variance accounted for by water lead decreases (Table 12).

TABLE 10. Hierarchical Decomposition of Proportion of Variance in Ln (Blood Pb) Explained by Model

Boston/Somerville (n = 524)

Variables (in order of inclusion	I_{R^2}	*F*	*df*	*p*
Age	*7.2%*	*14.895*	*3.510*	*<.01*
Sex	*1.6%*	*9.930*	*1.510*	*<.01*
Education	*2.9%*	*8.999*	*2.510*	*<.01*
Water Pb	*3.8%*	*5.896*	*4.510*	*<.01*
Dust Pb	*2.2%*	*6.827*	*2.510*	*<.01*
TD 50	*0.2%*	*1.241*	*1.510*	*n.s.*
	17.8%	*8.510*	*13.510*	*<.01*

TABLE 11. Added Proportion of Explained Variation in Ln (Blood Pb) Following Adjustment for Age, Sex, and Educational Level

	Boston/Somerville				*Cambridge*			
	I_{R^2}	*F*	*df*	*p*	I_{R^2}	*F*	*df*	*p*
Water Pb	*3.8%*	*5.831*	*4.510*	*<.01*	*5.6%*	*4.049*	*4.217*	*<.01*
Dust Pb	*2.1%*	*6.417*	*2.510*	*<.01*	*2.2%*	*3.140*	*2.217*	*<.05*
Traffic Density	*0.3%*	*2.067*	*1.510*	*n.s.*	*1.2%*	*3.352*	*1.217*	*n.s.*

TABLE 12. Proportion of Variance in Ln (Blood Pb) Accounted for by Water Lead

	Boston/Somerville (n = 524)[a]	Cambridge (n = 231)*
Considered alone	4.1%	7.1%
Controlling for Demographic variables	3.8%	5.6%

[a]*This constitutes a reduced sample as compared with the largest sample considered in the water sampling section (see Table 7 - STH_2O, "Households where 2 types of samples taken"). This is due to the exclusion of 27 cases in Boston/Somerville and 11 cases in Cambridge which were missing data on demographic variables. This also explains the different R^2 for water lead considered alone.*

THE REGRESSION EQUATION

Analysis of the Boston/Somerville sample gave rise to the following regression equation:

$$\begin{aligned} Y = {} & .211A_1 + .090A_2 + .011A_3 + .101S \\ & + .180E_1 + .093E_2 + .027W_d \\ & + 2.729W - 4.699W^2 + 2.116W^3 \\ & + .040D_d + .011D + .042T + 2.462 \end{aligned}$$

where,

- Y - ln[Blood Lead(μg/dl)]
- A_1-A_3 - dummy variables for age: 0-6 years, 6-20 years, 20-50 years (the reference category being 50+ years)
- S - dummy variable for sex (male = 1, female = 0)
- E_1, E_2 - dummy variables for educational level of the head of household: 0-12 years, 13-16 years (17+ years - reference group)
- W_d - lead level of water - above/below detectable dichtomy (above = 0, below = 1)
- W - lead level of water as assessed from standing sample in units of mg/l - values below detectable level assigned the midpoint of the below detectable range
- D_d - lead level of dust - missing value dichotomy (missing = 1, not missing = 0)
- D - lead level of dust - in units of mg/gram - missing values imputed using mean dust lead of non-missing data
- T - traffic density (50 meter scale)

(R^2 = .178; $F_{13,510}$ = 8.510 ($p < .01$); standard error of prediction = .354)

The suitability of the regression equation for prediction was determined by cross validating the Boston/Somerville derived equation on Cambridge data used as a calibration sample. The proportion of variance accounted for in the Cambridge sample was 15.1%. This was expectedly smaller than the R^2 for the screening sample, both for statistical as well as practical considerations. On the practical side, there is the greater suppression of lead signal in Cambridge. This increased the lowest observable lead level and obscured variation in a region of the scale with a sizeable number of cases.

Cigarette Smoking

Attention was given to smoking as a possible confounding variable for those over age 20 years. Analysis was limited to individuals aged 20 years and over who were non-smokers or smoked cigarettes (pipe and cigar smokers were omitted). The general linear model was used, regressing log blood lead first on demographic variables, followed by environmental sources (Table 13). The age dichotomy showed no significant difference between blood lead levels in the 20-50 year old group and the 50+ years group in either the Boston/Somerville or Cambridge samples. Males had significantly higher blood lead levels in both samples.

When the number of cigarettes smoked was entered after the demographic variables, it contributed a small but significant proportion of explained variation in the Boston/Somerville subsample, but no such contribution in Cambridge. It may be further noted that in these sub-samples, water lead, after adjusting for background variables and cigarette smoking, contributes 5.5% and 12.7% to the explained variation in log blood lead in Boston/Somerville and Cambridge. When entered after both demographic and other environmental sources, number of cigarettes smoked contributed 2.2% and 0.4% to explained variation in log blood lead in Boston/Somerville and Cambridge, respectively.

Children

We used the regression equation to determine expected blood lead levels for children, aged 6 years or under at different tap water lead levels (Table 14). Mean values were substituted for all variables, except age and water lead levels. For the above/below detectable dichotomy (W_d), the mean (.437) represents the proportion of individuals in our sample residing in homes with water lead levels below the detectable level. The mean for the dust missing value dichotomy (.3817) represents the proportion of individuals in our sample for whom dust lead data are absent. The results are expressed as the predicted geometric mean blood lead, as well as in the expected percentage of that age group whose blood lead would be above a critical level, for various water lead levels.

TABLE 13. Explained Variation in Ln (Blood Pb) - A Model Incorporating Number of Cigarettes Smoked

	R^2	F	df	p
A. Boston/Somerville (n = 244)				
Variable (in order of inclusion)				
Age	0.4%	<1	1,231	n.s.
Sex	6.8%	20.09	1,231	<.01
Education	2.8%	4.14	2,231	<.05
# Cigarettes Smoked	2.1%	6.20	1,231	n.s.
Water Pb	5.5%	4.06	4,231	<.01
Dust Pb	4.1%	6.06	2.231	<.01
Traffic Density	0.0%	<1	1,231	n.s.
Total	21.8%	5.366	12,231	<.01
B. Cambridge (n = 116)				
Variable (in order of inclusion)				
Age	0.9%	1.25	1,104	n.s.
Sex	3.5%	4.87	1,104	<.05
Education	3.2%	2.23	2,104	n.s.
# Cigarettes Smoked	0.3%	<1	1,104	n.s.
Water Pb	12.7%	5.89	3,104	<.01
Dust Pb	7.0%	4.87	2,104	<.05
Traffic Density	0.5%	<1	1,104	n.s.
Total	25.3%	3.20	11,104	<.01

TABLE 14. Predicted Geometric Mean Blood Leads, and Predicted Percentage of Individuals at Risk for Various Blood Lead Levels, and Various Water Lead Levels (Age 0-6 years)

Water Pb	G.M. Pred. Bl.Pb	Critical Blood Pb's (μg/dl)							
(μg/l)	(μg/dl)	15	20	25	30	35	40	45	50
15	24.288	91.3%	70.9%	46.8%	27.4%	15.2%	7.9%	4.1%	2.1%
20	24.606	91.9	71.9	48.0	28.8	16.1	8.5	4.4	2.3
25	24.903	92.4	73.2	49.6	29.8	16.9	9.0	4.7	2.4
30	25.229	92.5	74.5	51.2	31.2	17.9	9.7	5.1	2.7
35	25.534	93.3	75.5	52.4	32.3	18.7	10.2	5.5	2.9
40	25.842	93.8	76.4	53.6	33.7	19.5	10.8	5.8	3.1
45	26.128	94.2	77.3	54.8	34.8	20.8	11.4	6.2	3.3
50	26.443	94.5	78.5	56.4	35.9	21.5	12.1	6.7	3.6
75	27.910	96.0	82.6	62.2	42.1	26.1	15.5	8.9	5.0
100	29.312	97.1	86.0	67.4	47.4	30.9	18.9	11.3	6.6

Discussion

This study of the relative significance of several household sources of lead indicates that lead in tap water and, to a lesser degree, lead in household dust explain a significant proportion of the observed variation in blood lead levels of the residents. We are unable to confirm a relationship between blood lead level and either the maximum paint level measured or the proximity of households to major highways, although we did find a significant relationship between distance from traffic as measured on the 50 meter scale with dust lead levels. The strength of the observed association between blood lead and the demographic variables, age, sex, and education of head of household, indicate that these factors should be taken into account in any analysis of the effect of environmental sources of lead. Age exerts the greatest effect on blood lead of any variable measured, reflecting the significant increase observed in blood lead of young children.

Barltrop *et al.* (1975) investigated the relationship between blood lead levels of children and their mothers with the lead content of house dust and soil in a rural district with varying soil lead levels but with minimal atmospheric pollution. Despite a general increase in lead exposure for children in high soil areas, they considered soil and dust to be relatively minor sources, due to the small blood lead increments observed.

Household dust as a source of lead poisoning has not been widely studied. Sayre *et al.* (1974) concluded that ingestion of lead in dust might explain above normal levels in urban children. Our study indicates that it is of low relative significance in contributing to blood lead levels in our geographic area.

The Silver Valley Lead Study identified five variables (air and soil lead, age, dustiness of the home, and occupational status) associated with blood lead of children in an area near a lead smelter (Yankel *et al.*, 1977). Several other studies have shown elevated blood lead levels in children of workers occupationally exposed to particulate lead (Roberts *et al.*, 1974; Landrigan *et al.*, 1975; HEW, 1976, HEW, 1977; Baker *et al.*, 1977). We did not have data to evaluate the relationship between occupation of any member of the household and blood lead levels of individual members. The observed significant relationship between blood lead and traffic density on the 50 meter scale but not the 100 meter scale may indicate that some proportion of lead in household dust is derived from nearby automobile emissions.

The route of absorption of dust, i.e., by ingestion or inhalation, is not clear, with either one or the other route being implicated by observations or assumptions by various authors. The origin of the lead in dust has been generally ascribed to fall-out from leaded gasoline, to paint chips, or an industrial exposure. In a study of lead absorption near an ore smelter, which demonstrated a significant correlation between lead in household dust and blood lead, Landrigan *et al.* (1975) review evidence that pulmonary uptake may be an additional route of lead absorption, particularly when particle size is small. Our study does not clarify the origin of lead in dust, nor its route of absorption.

Many studies of lead in drinking water have been conducted in the United Kingdom during the past decade. The presence of lead in drinking water has been demonstrated in association with lead piping in the water supply system (Beattie *et al.*, 1972). Moore (1973) has shown that the principle factors affecting the rate of solution of lead in water are pH, temperature, and hardness of the water. Addis and Moore (1974), Covell (1975), and Moore (1975) have reported significantly higher blood lead levels in people living in households having lead water pipes, and that their blood lead levels were related to water lead levels. Moore *et al.* (1977) have reported a curvilinear relationship between lead content of household drinking water and blood lead in Scotland. Our studies confirm that relationship.

Consideration of a Regulatory Standard for Lead in Drinking Water

Taylor (1977) has reviewed the history of legislation in the United States aimed at the prevention of diseases associated with drinking water. The first national standards regulating drinking water in 1914 prescribed only bacterial limits, but in 1925 limits were set for lead, copper and zinc. The 1925 standard specifying that lead should not exceed 100 µg per liter was continued until the 1962 Public Health Service Standards lowered the acceptable maximum concentration of lead in finished water to 50 µg per liter. This same level was adopted in the Interim Drinking Water Standards of the Environmental Protection Agency (EPA), which went into effect in June, 1977 (EPA, 1975). The National Academy of Sciences Report, Drinking Water and Health, provides a review for the United States of the adverse relationship between public health and certain constituents of drinking water (NAS, 1977). The inorganic contaminant cited with the greatest potential for toxicity was lead. Because of the additive effects of sources of lead in food, water and air, the Report states that "although further studies will be necessary to arrive at a reasonable limit, it is suggested that the limit be lowered" (below 50 µg per liter).

The World Health Organization (WHO) standard of 100 µg/l was established in 1971 (WHO, 1978). In 1972, the WHO Regional Office for Europe and the Institute of Occupational Health, Helsinki, convened a Working Group on the Hazards to Health of Persistent Substances in Water. This Working Group recommended that the WHO standard be dropped to the 50 µg/l standard (Hernberg, 1973).

These efforts at establishing a standard were apparently made in the absence of data showing a quantitative relationship between lead in water and lead in blood. An approach toward a more biologic basis for standards has recently been made both in the United States and under WHO auspices, by establishing an "acceptable" blood lead level for a population. A Working Group meeting in London in September, 1977, sponsored by the WHO Regional Office for Europe and the Government of the United Kingdom, suggested a maximum blood lead value of 20 mg lead per dl for 50% of the population (WHO, 1978). It was agreed that blood lead could be used as a measure of lead absorption into the non-skeletal or soft tissues of the body and a significant association between blood lead and the lead content of drinking water was recognized. The Working Group reaffirmed that the upper limit should be 50 µg per liter, measures be taken to reduce the exposure of consumers.

The United States EPA in its National Primary and Secondary Ambient Air Quality Standards has adopted this approach regarding regulation of lead in ambient air, and further has attempted to quantify the contribution of air lead to blood lead (EPA, 1978). In these standards, EPA has determined that young children, (age 1-5 years) should be regarded as a group within the general population that is particularly sensitive to lead exposure, and has adopted 30 µg lead/dl as the maximum safe blood level for this age group. This is in accord with the statement of the Center for Disease Control of the United States Department of Health, Education, and Welfare on increased lead absorption and lead poisoning of children, which defines "undue or increased lead absorption" at blood lead levels as low as 30 µg/dl(CDC, 1977). EPA regards the number of children predicted to be below 30 µg lead/dl as the critical health consideration, and estimates that at a population geometric mean of 15 µg lead/dl, 99.5 percent of children will be below 30 µg lead/dl.

In our consideration of establishment of a standard for lead in drinking water, we suggest, first, that the type of tap water sample used should be a standing grab sample. Analysis of our data indicates, in two different water supply systems, a stronger and more consistent relationship between household standing grab water samples and log blood lead levels of residents than between running grab water samples and log blood lead levels. Further studies may be required to demonstrate the applicability of this relationship in other locations.

The WHO suggestion for use of the running grab sample may have been made on an assumption that it was more reproducible than any other type sample. However, it seems reasonable to accept a demonstrated relationship with blood lead as a basis for a standard in the absence of contrary data. Further, we agree with an approach which attempts to quantify the relationship of a population blood lead level to an environmental exposure. We recognize the validity of EPA's statement that young children are at special risk, and our analyses show that children in the 2-6 year old age group have higher blood leads than other age groups.

To incorporate these concepts into a biologically based consideration of a standard for lead in water, we developed Table 14.

The following discussion of the efforts of WHO and EPA to set a standard is not intended to be definitive. Rather, we are proposing a method by which such efforts may be judged.

Using results from our sample in Table 14 as an example, we can evaluate the feasibility of regulatory action to achieve population goals. The recommendations of the WHO Working Group in London in 1977 for a goal of 50% of the population to have a blood lead less than 20 μg per dl was linked to the establishment of a water lead standard of 50 μg/l. Table 14 indicates that for our sample a water lead level of 50 μg per l is associated with 78.5% of 0-6 year olds exceeding the critical value of 20 μg per dl blood lead.

However, the Working Group did not consider a specific age group. If we use our equation to predict a geometric mean blood lead for our sample, using the overall age distribution in our sample, we obtain a value of 23.3 μg per dl. At this geometric mean blood lead, 66.6% of the population would have a blood lead level above 20 μg per dl. Although closer to the 50% figure, it is considerably higher.

Based on our analysis the 50% at risk figure cannot be achieved through regulation of water lead alone in the cities we surveyed, for even with 0 μg lead per l of tap water, our equation predicts a geometric mean blood lead level of 23.3 μg/dl for 0-6 year old group and 20.6 μg/dl for the general population. If a critical blood lead level of 25.0 μg/dl were used, the 50% at risk figure would be achieved for children at a water lead standard of 26.5 μg/l and for the general population at 82.5 μg/l. This wide disparity reflects the low contribution of water lead to blood lead and the peak in blood lead which we observed in the younger age group.

The proposed 50 μg per l water standard would place 35.9% of children under age 6 years in excess of 30 μg per l blood lead. At 25 μg per l, 29.8% of children would have blood leads above 30 μg per liter. Thus, by applying to our sample the EPA goal that 99.5% of children have a blood lead below 30 g per dl, we find that this cannot be achieved at the current United States water standard of 50 μ per liter, nor at any level of water lead down to 15 μg per liter, which is below detectable limits in one of our studied water supply systems. This is, no doubt, due to the multiple sources of lead in other substances which contribute to blood lead levels. However, we have demonstrated that with rising tap water lead levels, the percentage of children at risk of blood lead levels above 30 μg per dl also rises. This should indicate the necessity to keep the exposure to lead in tap water to as low a level as is achievable.

Mahaffey (1977) has reviewed the available data on quantitative relationships between lead ingestion and development of toxicity. She recommends that the maximal daily intake from all sources, for infants below age 6 months, should be as low as possible and less than 100 μg per day, and that intake should be no more than 150 μg lead per day for children between 6 months and

2 years. Based on an assumed consumption of one liter of water daily for infants and children, even at the current United States standard of 50 µg per liter a child could ingest 50 µg of lead per day from water alone. This would add a substantial quantity, considering the many known sources in food, dust and air.

Detectable levels of lead were found in 62.7% of all households surveyed. Field observations and water department records indicated that 59.8% of the households had lead pipe either in the service line and/or in interior plumbing. The relationship of lead in drinking water with lead pipe in the distribution system seems clear in these two corrosive water supply systems. Further studies are necessary to determine whether corrosion control will be effective in reducing lead content to an acceptable level when lead pipes are used to convey soft water of low pH for household use. If lead cannot be reduced in water by corrosion control, consideration of removal of lead pipes will be necessary. New use of lead materials in drinking water systems should be restricted to the maximum extent possible.

SUMMARY

We have demonstrated the presence of lead in household tap water in association with the use of lead in the water distribution systems of three cities in Massachusetts. These cities use two different natural sources of water, both of low pH and low mineral content. A statistical method is presented for evaluation of the effect of measured environmental factors on blood lead, independent of the effects of demographic factors. Failure to control for demographic factors provides overestimates of water's contribution to blood lead. After consideration of demographic factors, lead in drinking water explains 3.8 percent and 5.6 percent of the variation in log blood lead in Boston/Somerville and Cambridge, respectively. Dust, smoking habits, age, sex, and socioeconomic factors each explain varying proportions of blood lead in the two samples analyzed. Among factors measured, age explains the greatest variation in blood lead. Children, age 2 to 6 years of age, have higher blood lead levels than other age groups. Consideration is given to various regulatory levels for drinking water lead levels, suggested by two World Health Organization Working Groups, and by the National Academy of Sciences, and the United States Environmental Protection Agency. In our study area, it is not possible to protect all children from blood lead levels above 30 µg per deciliter by regulation of water lead content. However, we conclude that drinking water lead levels should be as low as possible, in view of the multiple environmental sources of body lead burden.

ACKNOWLEDGMENT

This study was supported in part by a grant from the U.S. Environmental Protection Agency (R-802794) and by the International Lead Zinc Research Organization.

REFERENCES

Addis, G., and Moore, M. R. (1974). Lead level in the water of suburban Glasgow. *Nature 252,* 120-121.

Baker, E. L., Jr., Folland, D. S., Taylor, T. F., *et al.* (1977). Lead poisoning in children of lead workers home contamination with industrial dust. *N. Eng. J. Med. 296,* 260-261.

Barltrop, D., Strehlow, C. D., Thornton, L., *et al.* (1975). Absorption of lead from dust and soil. *Postgrad. Med. J. 51,* 801-804.

Beattie, A. D., *et al.* (1972). Environmental lead pollution in an urban soft water area. *Brit. Med. J. 1,* 491-493.

Caldwell, J. S., *et al.* (1974). Evaluation of the Atomic Absorption Graphite Furnace for the Determination of Metals. In, *Proc. Second Annual Water Qual. Technol. Conf. of the Amer. Waterworks Assoc.,* Dallas, Texas.

CDC (1975). Increased Lead Absorption and Lead Poisoning in Young Children: A Statement by the Center for Disease Control. U.S. Department of Health, Education, and Welfare.

Cohen, J., and Cohen, P. (1975). Missing Data, pp. 265-290. In, *Applied Multiple Regression/Correlation Analysis for the Behavioral Sciences.* Lawrence Erlbaum Associates, Inc.

Covel, B. (1975). Lead content in household water in Edinburgh. *Health Bull. 33,* 114-116.

Dangel, R. A. (1975). Study of Corrosion Products in the Seattle Water Department Tolt Distribution System. Environmental Protection Agency Technical Series EPA-670-2-75-036, U.S. EPA, Cincinnati, Ohio.

Donaldson, W. (1924). The action of water on service pipes. *J. Amer. Waterworks Assoc. 11,* 649.

Durfor, C. M., and Becker, E. (1964). Public water supplies of the 100 largest cities in the U.S., 1962. U.S. Geological Survey Water Supply Paper 1812.

EPA (1975). National Interim Drinking Water Regulations. U.S. Environmental Protection Agency Water Programs. Fed. Reg. 40:59570.

EPA (1978). National Primary and Secondary Ambient Air Quality Standards. Fed. Reg. 43:46249.

Frontinus, S. J. (1973). *The Water Supply of the City of Rome,* AD 97 (translated by C. Herschel), J. J. Matera and R. M. Babcock, eds. New England Waterworks Association, Boston.

Hernberg, S. (1973). Health hazards of persistent substances in water. *WHO Chron. 27,* 192-193.
HEW (1976). Lead poisoning-Tennessee. *CDC Morbid Mortal Weekly Rept. 25,* 85.
HEW (1977). Increased lead absorption in children of lead workers-Vermont. *CDC Morbid Mortal Weekly Rept. 26,* 61-62.
Karelekas, P. G., Jr., Craun, G. F., Hammonds, A. F., *et al.* (1976). Lead and other trace metals in drinking water in the Boston Metropolitan Area. *J. N. Eng. Waterworks Assoc. 90,* 150-172.
Landrigan, P. J., Gehlbach, S., and Rosenbloom, B. F. (1975). Epidemic lead absorption near an ore smelter: the role of particulate lead. *N. Eng. J. Med. 292,* 123-129.
Mahaffey, K. R. (1977). The relation between quantities of lead ingested and health effects of lead in humans. *Pediat. 59,* 448-456.
Massachusetts State Board of Health (1899). 30th Ann. Rept. Wright and Potter Printing Co. State Printers, Boston.
Matson, W. R., Griffin, R. M., and Schreiber, B. B. (1971). Rapid Subnanogram Simultaneous Analysis of Ln, Cd, Pb, Cu, Bi, and Tl, pp. 396-406. In, D. Hemphill, ed., *Trace Substances in Environmental Health-IV.* University of Missouri, Columbia, Missouri.
Moore, M. R. (1973). Plumbosolvency of waters. *Nature 243,* 222-225.
Moore, M. R. (1975). Lead in Drinking Water and Its Significance to Health, Drinking Water Quality, and Public Health. Water Research Center.
Moore, R. M., Meredith, P. A., Campbell, B. C., *et al.* (1977). Contribution of lead in drinking water to blood lead. *Lancet 661,* Sept. 24.
NAS (1977). Drinking Water and Health: A Report of the Safe Drinking Water Committee. National Research Council, National Academy of Sciences, Washington, D.C.
Roberts, T. M., Hutchinson, T. C., Paciga, J., *et al.* (1974). Lead contamination around secondary smelters: estimation of dispersaL and accumulation by humans. *Science 196,* 1120-1122.
Sayre, J. W., Charney, E., Vostal, J., *et al.* (1974). House and hand dust as a potential source of childhood lead exposure. *Amer. J. Dis. Child. 127,* 167-170.
Taylor, F. B. (1977). Drinking water standards--Principles and history, 1914-1976. *J. N. Eng. Waterworks Assoc. 91,* 237-259.
Weston, R. S. (1920). Lead poisoning by water and its prevention. *N. Eng. Waterworks Assoc. 34,* 230-263.
WHO (1971). *International Standards for Drinking Water,* 3rd Ed. World Health Organization, Geneva.
WHO (1978). Health hazards from drinking water: standards for lead and nitrates. *WHO Chron. 32,* 132. (level for lead corrected in *Corrigendum 32,* 132, 1978).
Yankel, A. J., von Lindern, I. H., and Walter, S. D. (1977). The Silver Valley lead study: the relationship between childhood

blood lead levels and environmental exposure. *J. Air Pollut. Cont. Assoc. 27,* 763-767.

DISCUSSION OF PAPER BY DR. DOROTHY WORTH

Cooper: Does season have any affect on this at all? Is there any difference between summer and winter samples?

Dr. Dorothy Worth, Tufts Medical School: Mineral content and temperature, both of which vary seasonally, affect the solution rate. Our data were collected over an entire year, but we do not have seasonally different samples for each house. We have not accounted for seasonal variation in our analysis.

Chamberlain: I just wanted to comment that this ratio of about 1.2 between the young child's blood lead and its mother's blood lead keeps on cropping up. It appears in at least three surveys in Britain, Dr. Barltrop's survey in Derbyshire, and one by Lansdown, *et al.*, in the area of a smelter. All the time, the findings for this ratio indicate that the young child, 2 to 6 years of age, has about a 20 percent higher blood lead than its mother.

Worth: I was quite surprised when I first saw it, because I was not aware at that time of the other observations that this was occurring. I think a lot of the data were collected at about the same time. We are not making any implications that the child is absorbing more from water. I have no explanation for this. However, I think this requires an explanation: that of people who live in the same household, the child has a higher blood lead than the mother or other adults. I think you can see that it was not just the mother.

Theodore J. Kneip, New York University Medical Center: First, have you tried to rank the other variables in order? You've dealt with what I gather was accounting for some 4 to 6 percent of the variation in blood lead. However, if you were going to choose that variable that had the greatest effect, which would be first?

Worth: Age. If you take the non-environmental variables, age had the greatest effect.

Kneip: You're not going to control age. So if you were going to deal with those variables which you might be able to have an affect on, in what order would you proceed? Would you attack water first, where it accounts for only 6 percent of the variability in blood lead.

Worth: Well, that's another whole question: what you should control. Obviously, there are things in the household that we did evaluate. We did measure lead in paint on multiple surfaces. We were unable to find any correlation between the maximum paint lead level measured in a household and the blood lead of the residents. We don't know whether our dust method was a good dust method or not. Yesterday, I commented on Dr. Sayre's elegant method for collecting and analyzing dust and pica measures. In all of our analyses of the measured environmental variables, water had the greatest effect. Our model incorporated about several demographic variables. We don't understand why poorer people, less educated people, children in that age group, and males have different blood leads than others. But in those variables that we did measure, that small 8 or 9 percent, water showed the strongest relationship with blood lead, to our surprise.

CORRELATION OF RENAL EFFECTS WITH COMMON INDICES OF LEAD EXPOSURE*

P. B. Hammond

A study is reported concerning the interrelationship between indices of lead exposure in smelter workers and lead effects on the kidney and on the hematopoietic system. The indices of lead exposure were: 1) concentration of lead in the blood (PbB) at the time of the study of lead effects, 2) average PbB over the full period of employment, 3) concentration of erythrocytic porphyrin in the blood (FEP) at the time of the study, and 4) urinary excretion of aminolevulinic acid (ALAU). The hematopoietic effect measured was the concentration of hemoglobin in the peripheral blood. The renal effects measured included 1) plasma urea concentration (BUN), 2) plasma creatinine concentration, and 3) plasma creatinine clearance by the kidneys. Further, the interrelationships between PbB and FEP and between PbB and ALAU were determined.

Over the observed range of PbB (25-101 μg/dl) there was a decrement in hemoglobin with increasing lead exposure. A similar relationship was found between FEP and hemoglobin. BUN was positively correlated with concurrently-determined PbB. This effect persisted even when appropriate adjustments were made to take into account the effects of age on renal function.

In contrast to the significant interaction between PbB and BUN, the long-term history of lead exposure, as reflected in the product of PbB times years of exposure, did not correlate well with BUN. FEP did not correlate with BUN whereas ALAU did.

The workers also were subdivided into high and low exposure groups. The high exposure group consisted of men whose PbB had, on frequent occasion, equalled or exceeded 80 μg Pb/dl. The low exposure group consisted of men whose PbB had never equalled or exceeded 80 μg Pb/dl. When segregated in this manner, only the high exposure group differed significantly from the control group as to BUN. It is concluded from these data that the intensity of

**The paper is published in full in the Journal of Occupational Medicine, 22: 475, 1980. Reference should be made to the full paper for details of the data discussed herein.*

ISBN 0-12-460520-6

lead exposure is of greater importance than the duration of exposure as a determinant of adverse renal effects among occupationally exposed subjects.

DISCUSSION OF PAPER BY DR. PAUL B. HAMMOND

Williams: On your hemoglobin-blood lead regression, I think you pointed out that there was no significant relationship between hemoglobin and blood lead for the lead workers. Therefore, the apparent significant regression may have been due entirely to the controls being from a different population from the lead workers, perhaps having a different dietary status, and to attribute any change to blood lead, I think, is not quite fair. In addition, some of your regression lines appear to be linear where those reported by Ted King this morning were curvilinear. When you included the controls in the regression analyses, I think it was plasma ALA, the relationship looked curvilinear. Therefore, if you had included the controls in all of the analyses I think you may have gotten a curvilinear result. Thirdly, you mentioned chelation therapy. Perhaps your BUN regression line had a positive slope, not from blood lead, but from chelates. And fourthly, we heard this morning about the possibility of bioactive fractions of the blood lead causing different effects. I'd just like to make the point that there might be one bioactive fraction for colic, one for renal damage, one for hemoglobin, one for any other effect you care to mention.

Hammond: I'll take the last question first. I don't suggest that there be a separate index for each target organ. I simply say that, at least for the particular organs examined, it appears that overall the concentration of lead in the blood reflected, with greater fidelity, the renal status and the hemoglobin status if you pool consideration of all of these. The second point that you made was in regard to not including the controls in the regression of ALAU against blood lead. Well, you're quite right. One could have done it that way, and perhaps you'd get a different line. I am sure you would, in fact. Thirdly, you question the validity of concluding that there was an effect of lead exposure on hemoglobin on the basis of pooling the controls with the workers in 1976. I would certainly agree that there is a problem as to whether these are the right controls. In other words, ideally, you want the same men doing the same kind of thing, eating the same kind of food but wearing a respirator or something. However, I did show that in 1977 there was an interaction. I think it is debatable as to how the 1976 data should have been handled. But, nevertheless, the overall picture is not only suggestive of an interaction, but also is consistent, I think, with other people's observations, such as Tola's. In fact, it is not inconsistent with your own work, wherein you only saw anemia when the concentration of lead in the blood reached 110 μg/dl. And I would remind you that the regression line runs from 16 μg/dl to 14 μg/dl for hemoglobin and over the blood lead range of 0 to 120 μg/dl. So, I don't think that is inconsistent with your own observation.

Williams: I do think that your BUN increase may be due to chelation and not to increases in blood lead.

Hammond: There is an impression that versenate is nephrotoxic and that impression, I think, is erroneous. It's based on work by Harry Foreman, done many years ago, and which since has been pretty widely discredited, particularly by the work of Aronson of Cornell University.

Gonick: One of the conclusions that you are deriving is that the elevation in BUN that you're correlating as an end effect on the kidney reflects kidney damage. That is not necessarily so. What you may be seeing, indeed, is a degree of volume depletion in these individuals, which is a mechanism suggested by the Australian workers as they attempted to explain why hyperuricemia was so common in their lead intoxicated individuals. They related it to an effect of lead on the renin angiotensin/aldosterone system such that people who were "leaded" were incapable of stimulating renin appropriately in response to any mild salt depletion. So they may always have mild depletion which could increase the BUN without implying true renal damage.

Hammond: Well, that may be a reasonable interpretation of the data. I wasn't aware of the evidence you cited. However, I should point out that other workers, including Lilis and Albahary, have demonstrated a progression of events going from elevated BUN to elevated serum creatinine to renal failure. I don't know whether the elevated BUN is really the first stage of a progression which, eventually, clearly is renal failure.

KIDNEY FUNCTION IN LEAD WORKERS

C. H. Hine, M.D., Ph.D.
H. A. Lewis, M.D., Sc.D.
Judy Northrup, M.D.
Shirley Hall, M.D.
J. W. Embree, Ph.D.

KIDNEY DISEASE

Renal disease is frequently encountered in the adult male population. It is estimated that there are approximately eight million persons (4% of the population) in the United States who have identifiable kidney disease. The mortality rate due to end-stage kidney failure is about 50,000 per year. There are no estimates of the numbers who might have elevated glomerular filtration rates. Kidney disease does not rank in the first 10 principal causes of death among males in the employable age group over 18. Therefore, it is to be expected that kidney disease of all types, including those which have manifestations of defects in glomerular filtration rate and in tubular absorption, will be found to occur with some frequency among lead workers in the lead industries.

KIDNEY RESERVE

The kidney, like other organs of the body, has a considerable reserve. At any particular time, only about half of the approximately one million nephron units are functioning. Therefore, the normal kidney can function entirely satisfactorily with less than 60% of its units working. For example, the removal of one kidney as is done in kidney transplants, with a resulting reduction of 50% of pre-surgical function does not impair the renal function of the donor. As with other organs, it is possible to determine the range and capacity for function under different environmental influences, physiologic states, and in the presence of disease.[1]

ISBN 0-12-460520-6

SUSCEPTIBILITY TO DAMAGE

Because of the kidney's rich blood supply in relation to its mass, this organ is particularly liable to damage from chemicals. Drugs and chemicals may be concentrated in renal tissue during secretion or absorption, or by the osmotic concentration of fluids in the renal medulla, leading to impairment. Damage to the kidney from chemical substances may be either acute or chronic and may predominantly affect glomerular or tubular function or alternatively cause overall renal damage, with anuria. Usually, it is temporary and reversible. A great number of chemical types may produce kidney damage, among which are metals, solvents, diagnostic media, and therapeutic agents.

METAL NEPHROPATHY

The characteristic patterns of untoward effect on the kidney produced by metals falls into three categories, namely: acute metal nephropathy, chronic metal nephropathy, and the nephrotic syndrome.

In acute metal nephropathy, the portion of the kidney most generally affected is the proximal convoluted tubule, with a consequent impairment of function. Clinical manifestations are renal glycosuria, renal aminoaciduria, excessive loss of phosphate and urate, a defective excretion of hydrion, excessive urinary loss of potassium, and impaired urine concentrating ability. The clinical picture is generally that of potentially reversible acute renal failure. If damage is not too severe, regeneration of tubular cells occurs, with complete recovery. If nephrosis is extensive, damage and poorly functioning tubules remain. Glomerular changes usually do not occur.

Irreversible chronic renal disease due to initial renal damage caused by metals is termed "chronic metal nephropathy." This either occurs following an acute nephropathy or results (rarely) from long-continued contact which does not induce clinically apparent acute manifestations.

The nephrotic syndrome is most difficult to establish as being due to metals in the individual case and depends essentially on careful epidemiological studies. It may be considered as a sensitivity reaction to a metal-protein complex, manifesting itself as an alteration in the permeability of glomerular capillaries to protein.

ACUTE LEAD NEPHROPATHY

Acute lead nephropathy almost entirely occurs in the age group of one to three. It is generally due to repeated ingestion of lead paints. The syndrome rarely occurs in the adult. When it does, it is due to ingestion of lead-contaminated food and drink.

Structural changes seen most frequently in acute lead poisoning are atrophy of the epithelial cells of the proximal convoluted tubule, signs of regeneration and increase in the inter-tubular connective tissue with thickening of tubular basement membranes, and some round cell infiltration. A striking characteristic finding is the presence of eosinophilic intranuclear inclusion bodies in some of the cells of the proximal convoluted tubules. Glomerular changes are not commonly seen in acute lead poisoning.

CHRONIC LEAD NEPHROPATHY

Goyer and Rhyne (1973), in their review of the world literature, indicate that the occurrence of a chronic form of renal disease in man due to lead is controversial. End-stage renal disease was reported in the older medical literature in men who had many years of excessive exposure. Points which differentiate chronic lead nephropathy from other types of chronic renal failure include the exceptionally slow progress of renal disease (40% of a "lead nephrotic" group were alive 15 years after first presenting features in comparison with 5% of a control group with chronic glomerular nephritis), only moderate elevation of BP, a mild proteinuria, and the occurrence of gout in over 50% of patients.

Chronic intoxication of animals from lead produced in laboratory experiments may be divided into three stages: Stage 1: reversible tubular effects characterized by development of intranuclear inclusion bodies and generalized aminoaciduria; Stage 2: chronic irreversible nephropathy, with varying tubular changes and progressive interstitial fibrosis; and Stage 3: renal failure and cancer in which adenocarcinoma, interstitial scarring and sclerotic glomeruli appear (Goyer & Rhyne, 1973).

In man, the progression (if it occurs) from acute to chronic kidney effects is less well-documented, due primarily to the paucity of renal biopsy material. While the animal model leads to an understanding of the possible chronic effects in man, there are unresolved differences. These include the particular renal content of lead in soft tissues that is capable of producing chronic nephropathy, the nonspecificity of the lesion in the absence of the demonstration of intranuclear inclusion bodies, and the absence of any epidemiologic evidence that lead increases malignancies of the kidney in man.

There are differences in opinion as to the occurrence and

TABLE 1. Nephrotoxicity from Lead

Year	Observer	Report
1949	Lane	Chronic interstitial nephritis
1956	Clarkson	Aminoaciduria
1963	Lane	No problems with reduced exposure
1964	Richet	Chronic renal failure
1967	Lilis	Renal insufficiency
1970	Malcolm	No interstitial nephritis
1975	Wedeen	10% of work force
1978	Ramirez	Correlation of creatinine and lead
1978	Cooper	No ↑ in SMR from renal disease

frequency of renal involvement among humans exposed to lead in work situations (Table 1).

Richet *et al.* (1946) viewed chronic renal failure as an inevitable sequellae of long term lead exposure. Lilis *et al.* (1967) observed that approximately 17% of those exposed chronically to lead would develop renal problems. Lane *et al.* (1963), Malcolm (1970), Ramirez *et al.* (1978), and Cooper (1978) found neither interstitial nephritis, clinical evidence of kidney failure, or increased mortality due to renal disease among lead workers.

Cramer *et al.* (1974) studied renal ultrastructure and renal function in persons with lead exposure from less than a year to more than 30. Renal function tests were normal, except for a reduced GFR in one worker. Ultrastructural changes were localized to the proximal tubules. Mitochondrial changes were found in all persons. There was a decrease of the numbers of inclusion bodies formed in persons designated as having lead nephropathy. Cramer concluded that, depending on the length of exposure, there may be different stages in the response of the human kidney to chronic lead exposure:

Phase 1: Characterized by relatively high urinary output of lead, no impairment of renal functions, and formation of nuclear inclusion bodies in proximal renal tubular cells.

Phase 2: Occurring after 4 or more years of exposure, without gross impairment of renal function, but with morphological changes, including a moderate degree of interstitial fibrosis and lessening of the numbers or disappearance of nuclear inclusion bodies.

Phase 3: Frank renal failure, probably occurring infrequently. In the past, a number of cases of chronic interstitial nephritis were observed by at least one investigator (Lane, 1949) (Table 2).

Lead intoxication was reported to have an effect on the renin-aldosterone response to sodium deprivation. The plasma renin ac-

TABLE 2. Three Phases of Lead-Induced Kidney Changes

I.	*Inclusion bodies* ↑ *Urinary lead excretion* ↑ *Normal kidney function*
II.	*Inclusion bodies* ↓ *Urinary lead excretion* ↓ *Interstitial fibrosis* ↑ *Normal kidney function*
III.	*Interstitial fibrosis* ↑↑ *Reduced glomerular filtration rate* *Abnormal kidney function*

tivity (PRA) and aldosterone secretory rate (ASR) did not increase to the normal range in leaded persons, as would be expected normally with the test procedure used.

Since it has been the practice among investigators to publish positive rather than negative data, there are few reports in scientific literature which point to the lack of effect of lead on the kidney. However, there are reports which fail to confirm the kidney damage in adults who experienced lead poisoning when they were children (Tepper, 1963) and which document the lack of effect among workers with some elevation in their lead body burdens (Robinson, 1976).

The magnitude of the problem was viewed in an entirely different light by Wedeen in his publication in the American Journal of Medicine (1975) and through his presentation to an OSHA hearing on the effects of lead. As a result of his studies, Wedeen concluded that blood lead levels could not be used to predict the presence of significant kidney damage due to lead and that other clinical tests should be used to determine whether the person was at risk of kidney damage because of his lead exposure. The inadequacy of the blood lead level as a predictor of possible renal involvement by lead was related, in part, to the fact that few of the workers were still exposed to lead at the time they were examined in his study.

Wedeen made a recommendation that the sub-clinical renal effects of lead should be detected and prevented, since this represents a material loss of functional capacity which has serious adverse health implications. He concluded that persons may have asymptomatic or pre-clinical renal failure with a 30-50% loss of kidney function, without being symptomatic. None of the workers he examined had symptoms of renal disease, but their loss of kidney function was ominous in terms of the life expectancy. He defined renal disease by a glomerular filtration rate of less than 90 ml/min. None of the men he examined had symptoms of kidney disease, although four were reported to have symptoms and signs of lead poisoning. His observations in a group of 69 persons are summarized in Table 3.

TABLE 3. Wedeen's Observations on 69 Lead-Exposed Workers

Observation	*No. Involved*
Persons with symptoms of lead poisoning	*4*
Persons with symptoms of kidney disease	*0*
Cases with laboratory evidence of elevated BUN, creatinine and uric acid	*1*
Employees with increased body burden of lead	*41*
Employees with decreased GFR	*19*
Persons with EM evidence of structural changes	*10*
Persons with intranuclear occlusion bodies	*0*
Persons designated as having nephropathy	*13*

Wedeen concluded that occupational lead nephropathy had not previously been detected within the United States in any numbers. He concluded that lead nephropathy has been overlooked because:

1. When lead nephropathy becomes symptomatic, the patient leaves his job and is lost to follow-up.
2. When lead nephropathy is advanced, hypertension develops and is considered the cause of renal failure.
3. Routine clinical tests do not indicate the presence of kidney disease until lead nephropathy is advanced. Blood tests such as the blood-urea-nitrogen and serum creatinine are increased only when two-thirds of kidney function is lost. The urinalysis which is routinely used to screen for kidney disease is normal in early lead nephropathy. Thus, moderate renal disease is not routinely detected in the physician's office.
4. Reliance on blood lead concentrations to make the diagnosis of lead poisoning is not sufficient to detect possible renal involvement, as the concentration of lead in tissues which produced kidney injury is not known.

Wedeen concluded that at least 10% of American lead workers might have occupational lead nephropathy, and defined this as a minimal estimate. Since NIOSH, National Institute for Occupational Safety and Health (1978), estimated about one million workers were exposed to lead, according to Wedeen's estimate, there were 100,000 cases of preventable renal disease due to occupational exposure to lead.

RENAL FUNCTION TESTS

Renal function tests are generally designed to assess three parameters, i.e., glomerular filtration (GFR), tubular function, and renal plasma flow. The determination of the GFR gives the greatest amount of information and is relatively easy to perform. The test is usually conducted for 24 hours, but shorter periods of time can be utilized.

Measurement of the glomerular filtration rates gives an estimate of the functional impairment of the kidneys, whether due to renal disease or to some other cause, e.g., dehydration or blood loss, both of which reduce the glomerular filtration rate and impair renal function as a result of low arterial pressure and a reduced renal blood flow.

If any substance were present in the glomerular filtrate and it was neither reabsorbed (actively or passively) from, nor secreted into, the renal tubules, then the rate of its excretion in the urine would be the same as its rate of entry into the glomerular filtrate: $GFR = UV/P$, where U is its urinary concentration (mg/ml), V is the volume of urine per minute, and P is its concentration in the plasma.

Of the substances which have been studied, the one whose clearance is believed to give the best estimate of the glomerular filtration rate is the fructose polysaccharide inulin. However, since it is not present normally in plasma, it is necessary to infuse it intravenously at such a rate that its concentration in the plasma remains constant over the period during which its rate of appearance in the urine is measured. This makes the procedure rather complicated. For a number of species, including man, the creatinine clearance gives an estimate of glomerular filtration which is very nearly as good as that derived from inulin. Creatinine has the advantage of being naturally present in the plasma, since it is produced metabolically (Harvey, 1974).

Creatinine clearance was proposed as an estimation of the glomerular filtration rate in 1926 by Rehberg. However, in this procedure, the plasma levels were augmented with exogenous creatinine. In 1938, the endogenous creatinine test was introduced by Miller and Winkler. Although not ideal for measuring glomerular filtration, the endogenous creatinine clearance test has many practical advantages. The plasma concentration of this endogenous product of metabolism is relatively constant and daily excretions are not influenced markedly by urine flow rate, exercise, or diet.

Creatinine clearance rates show a lower GFR for the average individual than values obtained by the use of either inulin or iothalanate. As an estimate of all ranges of glomerular filtration rate, the creatinine clearance technique is found to compare favorably in accuracy and reliability with inulin clearance. GFR's determined by this method of measurement of creatinine clearance give a wider range of normal values than does that utilizing inulin alone. Some values cited as normal for adult males appear in Table 4.

TABLE 4. Normal Ranges of Creatinine Clearances

ML/MIN.	
75 - 125	*Cantrow and Trumper*
123 ± 16	*Davidshon and Henry*
84 - 162	*Henry, Cannon and Winkelman*
105 ± 20	*Tietz*

There is a decrease in the GFR with age (Davis and Shock, 1950). The measurement of inulin clearance made under basal conditions, in ambulatory, afebrile males free from history of clinical evidence of renal disease, hypertension, or heart disease, indicated that there was a linear decrease in clearance beyond the age of 30 years. The average inulin clearance dropped from 122.8 to 65.3 ml per minute per 1.73 M^2 between the ages of 20 and 90 (46%).

In some forms of renal disease, renal function is impaired as a result of loss of entire functional nephrons; in others, partial impairment of each nephron results. For both of these reasons, an estimate of the rate of formation of glomerular filtrate gives some idea of maximum capability of the kidneys for providing compensation for alteration in the internal environment, and comparison with the value for a normal but otherwise similar subject of the same species gives some idea of the degree of impairment of renal function which may exist.

PROCEDURES AND METHODS

The present study was designed to investigate certain questions relative to the possible effects on the kidneys of lead in an adult population working with lead. The questions considered were:

1. Is there a correlation between glomerular filtration rate and the body burden of lead?
2. Do persons who have symptoms compatible with lead intoxication also have decreases in their GFR?
3. Can measurement of the GFR serve as a useful indicator of an undesirable body burden of lead?
4. Will persons who have had past documented episodes of signs and symptoms of lead intoxication sufficient in extent to require chelation therapy have decreased glomerular filtration rates?

5. What is the magnitude of the effect of the aging process in terms of decline of glomerular filtration rate?

6. Are values obtained through the short-term clearance measurement of GFR congruent with those values obtained through the standard 24-hour clearance test?

7. What is the reproducibility and the variation in GFR values when the test is repeated during either the short-term or 24-hour test?

8. What is the degree of congruity of glomerular filtration rates with other possible subtle correlates of increased lead body burdens such as the erythrocytic protoporphyrin (FEP), latency time at the synapse, and nerve conduction velocity?

9. What numbers of employees in the population exposed to lead have decreased glomerular filtration rates?

METHODS

Six groups were chosen for study based on their history of lead exposure. There were three lead-exposed groups and three control groups. The lead-exposed groups included 1) persons engaged in the manufacture of batteries, 2) lead smelter workers, and 3) workers in a lead refinery. Control groups consisted of 4) copper refinery workers, 5) prison volunteers, and 6) retirees from a cadmium refinery (Table 5).

Methods of Selection of Subjects

All persons on whom observations were made, with the exception of the volunteer group of prisoners, were or had been at risk because of their past work exposures, and were evaluated because of this. There was no pre-selection and no matching made as to age, job classification, economic status, or years of work.

TABLE 5. Characteristics of Groups

Site	Number
1. *Battery Manufacturers*	47
2. *Lead Smelter Workers*	30
3. *Lead Refinery Workers*	171
4. *Copper Refinery Workers*	51
5. *Prison Volunteers*	20
6. *Cadmium Refinery Workers*	35
Total	354

Lead Measurements

Lead was determined by atomic absorption spectroscopy. Measurements were made in one of three laboratories; all maintained adequate internal controls and participated in check sample evaluations. Stated accuracies were ±10% in the range of 60 μg/dl.

Hemoglobin Determinations

Hemoglobin was determined by the method of Crosby (Crosby and Furth, 1956) (cyanmethemoglobin) or indirectly by extrapolation based on the total iron content which was determined by atomic absorption. Accuracy was ±5%.

FEP Determinations

Measurements were made according to the modification of the method of Chisolm and Brown (1977). Stated accuracy is ±10%.

Creatinine Determinations

The method of Tietz (Jaffe reaction) was employed for both blood and urine. Accuracy is ±10%.

Nerve Latency and Conduction Measurements

Motor nerve conduction velocity studies were performed on the right median, radial, and deep peroneal nerves. Skin temperature was recorded at the wrist with an M-99 Electrotherm and a correction was made of 2 meters per second for each 1 degree centigrade less than 31°C.

All subjects were tested supine in the same room by the same examiner with a TECA JM electromyograph. Monopolar needles were used for electromyography. Surface electrodes and stimulation was used for the nerve conduction studies.

Symptoms

All the participants at one location were interviewed as to symptoms compatible with increased lead loading which were experienced in the past or which were concurrently present. This was followed by an interrogation as to whether particular symptoms were present (Table 6). All statements of positive effects were given equal weight and no quantitation was attempted as to their duration or severity, although these data were recorded.

TABLE 6. Symptoms Accompanying Lead Loading

Abdominal discomfort	*Insomnia*
Anxiety	*Metallic taste*
Arthralgia	*Myalgia*
Decreased appetite	*Nervousness*
Colic	*Personality change*
Constipation	*Decreased sexual drive*
Diarrhea	*Sick feeling*
Dizzy spells	*Tremor*
Fatigue	*Vomiting*
Headache	*Weakness*

Collection of Samples

Urine samples were obtained from subjects according to the prescription that they void completely and then collect, in a plastic container, all the urine excreted between that time and the end of the collection period. Priming doses of one pint of water were used in some subjects. The length of the collection period varied between one and 24 hours, although a two-to-four hour collection was used with most subjects. The volume of urine excreted was measured in a graduated cylinder and the time of collection recorded to the closest five minutes.

Standardization of Data to Body Surface Area

All glomerular filtration rate calculations were reduced to a normalized body area of 1.73 M^2, based on the prediction for body surface area derived from a normogram based on the height and weight of the individual.

Statistical Procedures

All data were transferred to punch cards by means of key punch. A special program was written using an SPSS in Fortran and statistical analyses were carried out with an IBM 370 148 computer. The means of each of the variables were compared for Groups 1-6. For each variable, the means of the group were determined, along with the standard error, standard deviation, F value, and 2-tail probability value using both pooled and separate variance estimates. The latter probability value was used to ascertain whether two means were significantly different; probabilities equal to or less than .05 were considered significant.

The individual values of each variable were plotted against the GFR value for the subject and the interceptance slopes for the least square regression ("best" line through those points) were calculated, along with the correlation coefficient and its probability of significance. The value obtained for significance of the correlation coefficient was used to determine whether any such trends were significant.

For each variable, the numbers and percentages of values outside the normal range were obtained. The chi-square test was used to determine whether differences among the groups were significant.

Validation of Abnormal GFR

Evaluation of the persistency of decreased filtration rates was done on a limited number of subjects by repeating the determination with one additional measurement.

RESULTS

1. Characteristics of the groups studied:

A. Battery Manufacturers

There were 47 subjects in the group, 45 of whom were males and two females. One-third of the group had received chelation therapy on one or more occasions in the past. Six of them were no longer employed in battery manufacturing, having sought jobs elsewhere at the advice of their physician or for other reasons. A number were engaged in Workmen's Compensation litigation alleging injury from past lead exposure. Their ages ranged from 23 to 57, with a median age of 35.4 years.

B. Lead Refinery Workers

There were 32 members of 140 persons employed at a lead refinery located in a rural setting. They were undergoing a multiphasic screen at the time of the evaluation. All were male; the age range was from 21 to 56, with a median age of 34.4. None had been restricted because of his particular job assignment, or because of elevated lead body burden.

C. Lead Smelter Workers

There were 171 persons included in this group out of a total work force of approximately 250. All of them were employed. None of them were on job restriction because of elevated body burden, although 20 (11.7%) had blood lead concentrations of

60 μg/dl or greater, as determined by recent biologic monitoring. The mean age was 38.9 years.

D. *Copper Refinery Workers*

There were 51 workers in this group, 49 male, two female. Their ages ranged from 18 to 60, with a median age of 34.9. Their occupational exposures included selenium, tellurium and silver fumes, and mineral acid gas. A symptom inventory was obtained and a complete medical examination was conducted on each person.

E. *Prison Volunteers*

Twenty males incarcerated in a correction facility volunteered for repeated measurement of GFR. None had a previous history of working with lead or ever having had a diagnosis of lead intoxication. They were divided into five groups according to 10-year increments, starting at age 20 and ending with age 59. None of them had any active or past kidney disease.

F. *Cadmium Refinery Retirees*

There were 35 male retirees whose ages ranged from 63 to 85. The mean age was 70.2. A number of them had diseases compatible with any older age group and included such medical conditions as hypertension, emphysema, and arteriosclerosis. All were given physical examinations during the evaluation of their medical status.

2. Glomerular filtration rate:

A. *Mean Values*

Glomerular filtration measurements were made on 274 persons. The mean values, their standard deviation, and the percentage of values outside the expected range are indicated in Table 7 for each of the groups. In determining the normal range, the values were arbitrarily chosen as 80-160 ml/min. This was chosen as encompassing the majority of rates reported by others in the literature.

Only 8% of the persons evaluated at the lead battery manufacturing plant fell below the normal range, while 26% were above the normal range. Only 3% of persons exposed at the lead smelter had less than expected flow rates, while 13% were above. The third lead-exposed group, those at the refinery, had 20% below and 7% above expected rates. In the control group at the copper

TABLE 7. Glomerular Filtration Rate Values

Group	Mean	Standard Deviation	% Outside Normal Range Below	Above
1	112	45	8	26
2	130	47	3	13
3	133	50	20	7
4	107	45	29	8
6	96	38	43	0

refinery, 29% of the persons fell below the expected filtration rate and 7% were above. Forty-three percent (43%) of the group of retired cadmium workers were below the expected normal, the age distribution of the group probably being responsible.

There was no significant difference between Groups 1 and 2, between Groups 1 and 3, or between Groups 2 and 3. There was no difference between Control Group 4 and Groups 1, 2, or 3. When the retirees (Group 6) were compared with Groups 2 and 3, there was a significant difference, but this disappeared when compared with Groups 1 and 4.

3. *Lead Burdens (Recent)*

A total of 242 determinations of recent blood lead concentrations was made. The mean values for Groups 1-3 were 42.7, 45.6, and 46.7 μg/dl, respectively. There was no significant difference between Groups 1 and 2, Groups 1 and 3, or Groups 2 and 3 (Table 8).

A comparison of the GFR with recent blood lead concentrations was made. There was no significant difference between Groups 1 and 2, between Groups 1 and 3, and between Groups 2 and 3. The comparison of GFR with recent lead burdens indicated no significant correlation ($r = .02$) of GFR with the concentration of blood lead in any of the groups.

4. *Lead Burdens (Averaged)*

A total of 207 determinations was made of lead values reflecting results of from 1 to 5 years repeated samplings. The mean value for Group 1 was 41.8 and for Group 3, 49.5 μg/dl. There was a significant difference between the groups. The comparison of GFR with average lead burdens of Groups 1 and 3 showed a slight negative correlation ($r = -.11$).

TABLE 8. Correlation of GFR with Certain Variables

Measurement	*Correlation*	*R Value*
Age	*Slight (-)*	*-.19*
Hemoglobin	*None*	*.04*
FEP	*None*	*.005*
Blood Lead (recent)	*None*	*.02*
Blood Lead (average)	*Slight (-)*	*-.11*
Creatinine	*Slight (-)*	*-.23*

5. *Age*

There were 271 examinees for whom the age was recorded. The mean values for Groups 1-6 were 35.4, 34.4, 38.9, 34.9, and 70.2. There was no significant difference between Group 1 when compared with Groups 2, 3, and 4, but there was when compared with Group 6. There was no significant difference between Groups 2, 3, and 4, but there was between the Groups 3 and 6. There was a slight negative correlation (r = -.19) of GFR with age among the workers.

6. *Blood Creatinine*

A total of 274 determinations of blood creatinine was made. The mean values for Groups 1, 2, 4, and 6 were 1.19, 1.14, 1.05, and 1.48 mg/dl, respectively. There was no significant difference between Groups 1, 2, 3, and 4, but there was between Group 6 and all other groups. There was a slight negative correlation (r = -.23) of GFR with the concentration of blood creatinine among the workers.

7. *Hemoglobin*

A total of 179 determinations of hemoglobin was made. The mean values for Groups 1, 2, and 3 were 16.4, 15.4, and 15.3 gm/dl, respectively. There was a significant difference between Group 1 when compared with Groups 2 and 3. There was no significant difference between Groups 2 and 3. There was no significant correlation (r = .04) of GFR with hemoglobin for the workers.

8. Erythrocytic Protoporphyrins

A total of 46 FEP determinations was made. This measurement was performed only on the battery workers. The mean value for this group was 75.3 μg/dl. There was no correlation of the determinant with the GFR (r = .005) for the workers.

9. Latency

A total of 45 determinations of latency of both the medial and peroneal nerves was made. This determination was made only in the group of lead battery workers. The mean value for the median nerve was 3.9 milliseconds and 4.9 milliseconds for the peroneal nerve. There was no significant correlation (r = .02) of the determination with the GFR for the medial nerve, and a slight positive correlation (r = .15) with the peroneal nerve. Eight subjects had increased medial nerve latency; of these, two also had decreased nerve conduction. Three subjects had increased peroneal latency; none of these had abnormal nerve conduction velocities.

10. Nerve Conduction

A total of 45 determinations of nerve conduction velocity (NCV) was made on each of three nerves--the medial, radial, and peroneal nerves. This determination was made only in the battery lead worker group. The mean values for the three nerves were 55.9, 61.2, and 49.9 meters/second, respectively. There were five subjects with slowing of conduction in the medial nerve, five with slowing in the radial nerve, and two with slowing in the peroneal nerve. Decreases in velocity were not considered to be medically significant unless there was decreased nerve conduction velocity in at least two of the three nerves. None of the subjects had slowing in two, but one subject had slowing in all three nerves.

There were no factors which would cause increased nerve conduction. Four persons who currently had complaints of weakness all had normal nerve conduction. Among five subjects who have had past complaints of weakness, only one had an abnormality in one nerve. There was a slight positive correlation between GFR and NCV in the medial nerve (r = .13), no significant correlation in the radial nerve (r = .02), and slight positive correlation with the peroneal nerve (r = .19) (Table 9).

11. Past Complaints

Of the employees who worked at the battery plant, 70.2% indicated past complaints compatible with increased lead loading. The mean number of complaints among those indicating their presence was 6.8. There was no correlation between these and the GFR (r = -.03).

TABLE 9. Correlation of GFR with Latency and Nerve Conduction Velocity

Measurement	*Correlation*	*R Value*
Latency Median Nerve	*None*	*.02*
Latency Peroneal Nerve	*Slight (+)*	*.15*
Conduction Velocity Median Nerve	*Slight (+)*	*.13*
Conduction Velocity Radial Nerve	*None*	*.02*
Conduction Velocity Peroneal Nerve	*Slight (+)*	*.19*

12. Present Complaints

Of employees who worked at the battery plant, 78.7% had present complaints compatible with increased lead loading. The mean number was 4.9. There was a slight negative correlation ($r = -.16$) between GFR and present complaints.

13. Validation

Thirty-one of the 274 values obtained showed filtration rates greater than 160 ml/min. These were not considered as indicating kidney dysfunction and further evaluation was not carried out. There were 70 values which, on first testing, were less than 90 ml/min. These were considered as suggesting impairment of function. As decreased rate could be due to technical difficulties, circadian rhythm, or temporary physiological imbalance, a second analysis was conducted on 16 of these. There were three who had decreased flow a second time (5.3%). However, when 18 subjects with normal expected rates were retested, 11 had abnormal rates.

DISCUSSION

The Council of Kidney and Cardiovascular Disease of the American Heart Association published criteria for the evaluation of the severity of renal disease (1971). They recommended placing patients in one class for each of three major categories, depending on the severity of the signs and symptoms, the severity of renal function impairment, and the level of performance.

Data obtained from this study would not warrant placing the persons studied in Class I insofar as severity of signs and symptoms was concerned. Further, utilizing the recommendation that the experiment classification should be based on measurement of glomerular filtration rate "as is commonly approximated by the creatinine clearance" is not possible since the majority of persons there had a reduction of more than 50% of the expected GFR, but only a few had serum creatinine levels elevated above 2.4 mg/ml. All employees would be Class I in the performance classification, since they were "capable of performing all the usual types of physical activity."

In their review on glomerular filtration, Renkin and Robinson (1974) indicated that at rest, the GFR remains remarkably stable on a daily basis. However, they indicated that factors which cause afferent arteriole vaso constriction may be associated with transient reduction. They predicted that at age 60, the GFR would be approximately 70% of the normal values observed in young adults, and that while reductions with advancing age have been attributed mainly to the loss of functioning nephrons, residual functional glomeruli will remain unchanged in size and the glomerular filtration rate per unit of glomerular surface area is constant. Loss of renal mass, on the other hand, is followed by compensatory hypertrophy of the remaining nephrons.

While measurement of the glomerular filtration rate is of value in a variety of clinical conditions and circumstances, it must be emphasized that underlying renal dysfunction cannot be excluded with confidence in the presence of a normal glomerular filtration rate, nor can it be established with finality by the demonstration of a reduction. Some renal disorders may be associated with a normal rate at least for a time and a substantial loss of renal mass may occur before reduction of the GFR is manifested clinically. Conversely, a reduction may occur in the absence of intrinsic renal disease as a consequence of hemodynamic alterations such as occur with contraction of the extracellular fluid volume.

Considering all these factors, the GFR measurement still adds an important dimension to the clinical evaluation of patients with progressive kidney disease. Unfortunately, at this time, it is not possible to show congruence of the 24-hour and shorter sampling periods for determination of GFR in the uncatheterized employee.

While none of the plasma creatinine values exceeded the upper limit of 1.6 mg/dl, this cannot be interpreted as assuring a lack of effect of lead body burden on GFR, since other studies of non-lead-exposed persons have indicated that the GFR may fall to as low as 40 ml/min without exceeding this creatinine value.

The duration of the sampling in order to obtain stable values reflecting the true GFR's has not yet been established. After plotting the volume, time of collection, and GFR's, we recommend that the samples should be collected over a period of at least 3 hours and that during this time, water intake should be increased. The test is simple to perform, but does require some training of the employee in order to obtain a true sample. Unlike the measurement of hemoglobin, FEP and pulmonary function which reflect rather subtle changes in impairment in function as with other kidney function tests, GFR may not become abnormal until there is loss of a considerable portion of organ function. Whether this loss is permanent in lead loading states has not been established firmly, since according to Wedeen, an improvement in function may occur after chelation. Because of the seriousness of loss of kidney function, it would seem prudent that in monitoring the lead worker, in addition to the usual measurements, that measurement of the GFR be included on at least an annual basis.

CONCLUSIONS

With reference to the stated aims and goals of the study, the results to date support the following conclusions:

1. The GFR of lead workers exposed in battery manufacture and the smelting and refining of lead is not significantly different from that of industrial workers not exposed to lead.
2. There was no positive correlation between persons with symptoms of increased lead loading and GFR.
3. There is a slight negative correlation of GFR with age in the non-lead retired population.
4. There is no correlation between hemoglobin, FEP, or recent blood lead concentration and GFR.
5. There may be a negative correlation between past lead loading and GFR, although this conclusion is preliminary, since these data have not as yet been corrected for the confounding factor of aging.
6. Increases in plasma creatinine concentration may be indicative of decreased GFR, although again, these data are not age-corrected.
7. There is no meaningful correlation between nerve latency or nerve conduction time with GFR.
8. The short-term creatinine clearance measurement, i.e., less than 24-hours, may be used as a screening device for measurement of kidney function in leaded employees, although if less than 4-hour collection periods are used, there is considerable daily variation.
9. Further work is required to define the factors causing variability in the procedure.

REFERENCES

American Heart Association, Council on the Kidney in Cardiovascular Disease, Criteria for the Evaluation of the Severity of Established Renal Disease (1971). *Ann. Intern. Med.* *75*, 251-252.

Berlyne, G. M., Varley, H., Nilwarangku, S., *et al.* (1964). The endogenous-creatinine clearance and glomerular-filtration rate. *The Lancet* 874-876, 10/24.

Cantarow, A., and Trumper, M. (1962). *Elimination of Non-Protein Nitrogenous Substances in Clinical Biochemistry.* Sixth Edition, W. B. Saunders Co., Philadelphia.

Chisolm, J. J., and Brown, D. H. (1977). Micro-Scale Photofluorometric Determination of Free Erythrocyte Porphyrin (Protoporphyrin IX). From: *Selected Methods of Clinical Chemistry, Volume 8,* Editor: Gerald R. Cooper, Center for Disease Control, Atlanta, Ga.

Clarkson, T., and Kench, J. (1956). Urinary excretion of amino acid by men absorbing heavy metals. *Biochem. J. 62,* 361-372.

Cooper, W., and Gaffey, W. (1975). Mortality of lead workers. *J. Occup. Med. 17,* 100/107.

Cramer, K., Goyer, R. A., Jagenburg, R., *et al.* (1974). Renal ultrastructure, renal function and parameters of lead toxicity in workers with different periods of lead exposure. *Brit. J. Indus. Med. 31,* 113-127.

Crosby, W. H., and Furth, F. W. (1956). A modification of the benzidine method for measurement of hemoglobin in plasma and urine. *J. Hematol.* Vol. XI, No. 4, 380-383.

Davies, D. F., and Shock, N. W. (1950). Age changes in glomerular filtration rate, effective renal plasma flow and tubular excretory capacity in adult males. *J. Clin. Invest. 29,* 496-507.

Davidsohn, I., and Henry, J. B. (1974). Renal function and its evaluation. In *Clinical Diagnosis by Laboratory Methods,* 15th Edition. W. B. Saunder Co., Philadelphia, Pa.

Dooland, P. D., Alpen, E. L., and Theil, G. B. (1962). Endogenous creatinine clearance. *Am. J. Med. 32,* 65-79.

Dunea, M. D., and Freedman, P. (1968). Renal clearance studies. *J. Am. Med. Assoc. 205* (3), 170-171.

Emmerson, B. T., ed. (1973). Chronic Lead Nephropathy. *Kidney International 4,* 1-5.

Goyer, R. A., and Rhyne, B. C. (1973). Pathological effects of lead. *Intern. Rev. of Exp. Pathol. 12,* 38-53.

Harvey, R. J. (1974). From: *The Kidneys and the Internal Environmental,* p. 55. John Wiley & Sons Publishers, New York.

Henderson, D. A. (1955). Chronic nephritis in Queensland. *Aust. Ann. Med. 4,* 163-177.

Henry, R. J., Cannon, D. C., and Winkelman, J. W. (1974). *Clinical Chemistry Principles and Techniques,* 2nd Edition. Harper & Row Publishers, Hagerstown, Md.

Lane, R. E. (1949). The care of the lead worker. *Brit. J. Indus. Med. 6,* 125-143.

Lilis, R., Dimitriu, C., Roventa, A. *et al.* (1967). Renal function in chronic lead poisoning. *Medicina del Lavoro 58,* 506-512.

Macadam, R. F. (1969). The early glomerular lesion in human and rabbit lead poisoning. *Brit. J. Exp. Pathol. 50,* 239-240.

Malcolm, D. (1970). The effects of lead on the kidney. *Transactions of the Society of Occ. Med. 20,* 50-53.

Malcolm, D. (1971). Prevention of long-term sequelae following the absorption of lead. *Arch. Environ. Health 23*, 292-298.
Morgan, J. D., Hartley, M. D., and Miller, R. E. (1966). Nephropathy in chronic lead poisoning. *Arch. Intern. Med. 118*, 17-29.
National Institute for Occupational Safety & Health (1978). Occupational Exposure to Inorganic Lead. Criteria for a Recommended Standard. Cincinatti, U.S. Dept. of Health, Education & Welfare.
Ramirez-Cervantes, B., Embree, J. W., Hine, C. H., Nelson, K. W., Varner, M. O., and Putnam, R. D. (1978). Health assessment of employees with different body burdens of lead. *J. Occ. Med.* Vol. 20, #9.
Renkin, E. M., and Robinson, R. R. (1974). Glomerular filtration. *New England J. Med. 290*. 785-792.
Richet, G., Albahary, C., Ardaillov, R., *et al.* (1964). La rein du saturnisme chronique. *Revue Francais d'Etudes Cliniques et Biologist 9*, 188-196.
Robinson, T. R. (1976). The health of long service tetraethyl lead workers. *J. Occ. Med. 18*, 31-40.
Sandstead, H. H., Michelakis, A. M., and Temple, T. E. (1970). Lead Intoxication - its effect on the renin-aldosterone response to sodium deprivation. *Arch. Environ. Health 20*, 356-363.
Tepper, L. B. (1963). Renal function subsequent to childhood plumbism. *Arch. Environ. Health 7*, 76-85.
Tietz, N. W., ed. Renal Function. In: *Fundamentals of Clinical Chemistry*. W. B. Saunders Co., Publishers (Philadelphia).
Wedeen, R. P., Maesaka, J. K., and Weiner, B. (1975). Occupational lead nephropathy. *Am. J. Med. 59*, 630-641.

DISCUSSION OF PAPER BY DR. CHARLES H. HINE

Saric: I have just a short comment about our clinical experience with glomerular filtration rates during the acute phase of lead poisoning. We almost regularly noticed diminished or decreased glomerular filtration rates in those subjects, but this was, as a rule, a transitory lesion. If we repeated the test after treatment or after a certain period of time after removal from exposure, we usually got normal results. Our interpretation was that this decreased glomerular filtration rate was due to spasm of afferent arterioles during the acute phase of intoxication.

Hine: Thank you, Dr. Saric, I think that confirms the observations in the literature. In general, and from discussions with Dr. Chisolm about what happened in children, this would probably be true. Of course, our concern is what happens subtly subclinically or in persons who have had episodes of real symptomatolgy and the blood lead concentrations are from 100 to 150 μg/dl for some months or years. Those are the persons of whom I have concern; those are the persons we're trying to identify as a possible group at increased risk. If Wedeen is correct, and you can find these people and chelate them and show increase in function, then I think it behooves us to search for, identify, and treat these people. I am not in the position at the present time to recommend chelation therapy for such a group because I am not sure that 1) it is really reversible and 2) chelation is the best thing to do in the face of what may be impaired kidney function, although I cannot base that on clinical experience, only on clinical judgement formed from reading the literature.

ERYTHROCYTE LEAD-BINDING PROTEIN: RELATIONSHIP TO BLOOD LEAD LEVELS AND TOXICITY

H. C. Gonick
S. R. V. Raghavan
University of California Center for Health Sciences
Los Angeles, California

B. D. Culver
University of California
Irvine, California

Although knowledge of the relationship between cadmium and its specific binding protein, metallothionein, has been critical in understanding both the transmembrane transport and toxicity of this heavy metal (Kimura *et al.*, 1974; Nordberg *et al.*, 1975), development of similar information concerning lead and its binding protein(s) has lagged far behind. The finding of a nuclear inclusion body within the proximal tubule of the kidney has served as a pathological hallmark of renal lead toxicity for many years (Galle and Morel-Maroger, 1965), but it has only recently been appreciated, chiefly through the work of Goyer and his associates (Goyer *et al.*, 1970; Moore and Goyer, 1974), that the inclusion body represents the binding of lead to a 27,500 molecular weight lead-binding protein. Goyer has suggested that this protein complex may serve as an important detoxification mechanism for lead, and may play a role similar to that of metallothionein for cadmium within the kidney (Moore and Goyer, 1974).

A circulating form of lead-binding protein was first recognized in 1977, when our laboratory described a 10,000 molecular weight lead-containing protein present in the erythrocytes from lead-exposed workers but not in erythrocytes from normal age-matched controls (Raghavan and Gonick, 1977). We have speculated that this circulating form of lead-binding protein also serves as a detoxification mechanism and that individual variation in the ability to synthesize the protein in response to a lead challenge may account for previously observed discrepancies between total blood lead and clinical and biochemical indices of lead toxicity. The observations presented in this report tend to support this hypothesis.

ISBN 0-12-460520-6

PATIENT SELECTION

In the initial study, five normal control individuals were selected from a group of male industrial workers not exposed to lead, ranging in age from 25 to 45 years. Three groups of lead-exposed workers (5 members in each group) were selected from a single lead smelting plant. The first group were workers with toxicity at high blood lead levels, who had been referred to one of us (HCG) for chelation treatment because of both symptomatic and biochemical evidence of toxicity. In all instances, total blood lead levels exceeded 80 µg% prior to referral and urine ALA levels were elevated.

A second group, whom we have characterized as "low lead" symptomatic (Table 1), were selected by one of us (BDC) because of their presentation at the plant with symptoms and urinary ALA levels strongly suggestive for toxicity, but with blood lead levels consistently well below the 80 µg% level. As a final control group, we included 5 workers from the same plant who were asymptomatic and whose blood lead levels were in the "safe" range, varying from 50 to 75 µg%.

In this initial study, we focussed our attention on measurements of lead in the RBC hemolysate and lead bound to total hemoglobin, to hemoglobin A_2, to the 10,000 molecular weight lead-binding protein, and free lead. The hemoglobin A_2 determinations were done because of previous observations by Bruenger *et al.* (1973) that there is normally a disproportionately high binding of lead to the hemoglobin A_2 component of hemoglobin, suggesting the possibility that an individual genetically incapable of producing a normal amount of this component might be at increased risk for developing lead toxicity at relatively low blood levels.

In a second study, which is still in progress, the study population consisted of eight normal controls, 14 asymptomatic lead workers, three "low lead" symptomatic workers, and one "high lead" symptomatic worker. Lead bound to a high molecular weight pre-hemoglobin fraction and to the erythrocyte membrane was measured

TABLE 1. Biochemical Data on "Low Lead" Symptomatic Workers

Pt.	*Blood Lead µg%*	*Hb gm%*	*Urine ALA mg%*	*Urine Copro-P µg/24 hr.*
1	*48*	*13.6*	*4.90*	*1253*
2	*43*	*13.6*	*2.55*	*294*
3	*52*	*15.4*	*2.35*	--
4	*54*	*14.1*	*0.90*	--
5	*45*	*14.4*	*1.95*	--
Normal values	*<40*	*14.5-16.5*	*<0.54*	*<160*

in this study, together with lead bound to hemoglobin and to the 10,000 molecular weight binding protein. The correlation between membrane Na-K-ATPase activity, as a measure of lead toxicity, and both total RBC lead and membrane-bound lead was then examined.

METHODS

Erythrocytes were separated from plasma by centrifuging heparinized blood at 300 g for 15 minutes. The RBC were washed, then hemolyzed by freezing and thawing. One ml of hemolysate and 2.0 ml of Tris buffer (0.05 M, pH 7.4) were applied to a 90 × 1.5 cm Sephadex G-75 column (V_0 = 75 ml) and eluted with Tris buffer at a rate of 12 ml/hour. Four fractions were collected. The first fraction (termed "high molecular weight fraction") extended from the void volume to the first appearance of hemoglobin (V_E/V_0 = 1.0-1.3). The second fraction consisted of hemoglobin, as detected by appearance of red color and maximum absorbance at 545 nm (V_E/V_0 = 1.3-1.5).

The third fraction contained the 10,000 molecular weight protein in an elution volume encompassed by a V_E/V_0 of 1.8-2.1. The fourth fraction (termed "free lead") extended from V_E/V_0 = 2.1-2.5, and presumably contained both inorganic lead and lead bound to amino acids or small peptides. Succeeding fractions contained no detectable lead. The lead was measured in lyophilized fractions by atomic absorption spectrophotometer with graphite furnace attachment and the lead content of each fraction was expressed as μg per 100 ml of original hemolysate.

In the initial study, hemoglobin A_2 was also separated by DEAE cellulose chromatography according to the method of Abraham *et al.* (1977). The total hemoglobin A_2 was measured by spectrophotometry and the lead content of this fraction was measured by atomic absorption spectrophotometry.

In the second study, the membrane fraction from erythrocytes was obtained by centrifugation at 20,000 g for 15 minutes. Membranes were washed twice with isotonic saline and dissolved in 0.1% Triton X-100. Lead content was measured by atomic absorption and expressed as μg per 100 ml of original hemolysate. Activity of Na-K-ATPase was measured in the membrane fraction by standard techniques (Hasan *et al.*, 1967) and expressed as μmoles Pi/mg protein/hr.

TABLE 2. Total Lead in RBC, Hemoglobin and 10,000 Mol. Wt. Fractions (μg%)

Pt.	*RBC*	*Hemoglobin*	*10,000 mol. wt. Fraction*	*Free*	*Residual*
			I. Normals		
1	39	23	2	10	4
2	47	20	2	16	9
3	49	22	3	13	11
4	54	29	3	12	10
5	42	20	2	11	9
			II. Asymptomatic Workers		
1	121	81	28	3	9
2	150	90	31	5	24
3	107	78	19	3	7
4	124	84	26	2	12
5	148	92	34	3	19
			III. Symptomatic Workers		
1	209	98	53	3	55
2	185	109	28	2	46
3	205	106	31	4	64
4	172	78	38	3	53
5	193	86	41	4	62
			IV. "Low Lead" Symptomatic Workers		
1	116	56	5	1	54
2	106	45	4	1	56
3	108	63	4	2	39
4	120	72	5	2	41
5	102	60	4	1	37

RESULTS

First Study: The quantitation of lead in the RBC hemolysate and distribution in the hemoglobin, 10,000 molecular weight, and "free" fractions for the control and the three lead worker sub-groups are presented in Table 2. "Residual" lead has been calculated as the difference between total lead in the RBC hemolysate and the sum of lead in the individual fractions. Our initial assumption was that the residual lead was distributed between the high molecular weight and membrane fractions. This assumption was later confirmed in the second study.

The differences between the four groups may be summarized as follows (Fig. 1):

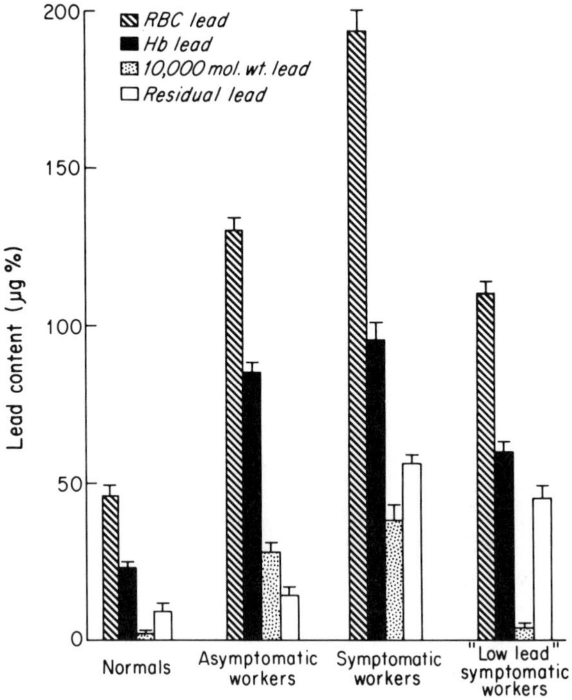

Fig. 1. Distribution of lead in human erythrocytes in normal controls and in three sub-groups of industrially-exposed workers.

1. Less than 5 µg% of lead was found in the 10,000 molecular weight fraction in the normal controls and in the "low lead" symptomatic workers.
2. There was significantly less lead in the hemoglobin fraction of the "low lead" symptomatic workers than in the asymptomatic group or the "high lead" symptomatic group (59±4 vs. 85±3 vs. 95±6 µg%,* $p < 0.05$).
3. "Residual" lead was always less than 24 µg% in the asymptomatic and normal groups but in excess of 39 µg% in the symptomatic groups, averaging 45 ± 4 µg% in the "low lead" symptomatic group and 56 ± 3 µg% in the "high lead" symptomatic group.

In Figure 2 we present the data relating the total RBC lead to the lead content of hemoglobin A_2 fractions in the four groups. The lead content of hemoglobin A_2 was remarkably constant in all groups but the "low lead" symptomatic workers, averaging 13 ± 2 µg% in the latter group and 20 to 23 µg% in the other three groups. The amount of hemoglobin A_2 as a percent of total hemoglobin was also slightly reduced in the "low lead" symptomatic workers

Mean ± S.E.

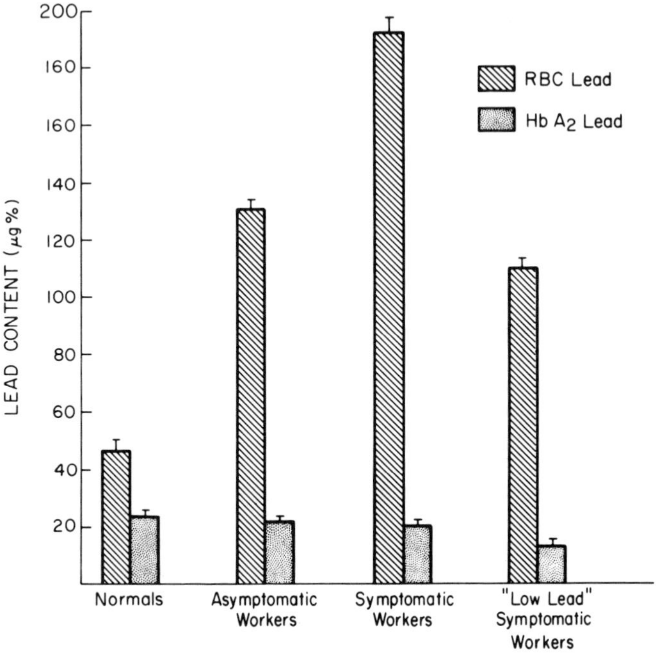

Fig. 2. Relationship of lead in hemoglobin A_2 fraction to total RBC lead in normal controls and in three sub-groups of industrially-exposed workers.

(1.5±0.2% in "low lead" workers vs. 2.1±0.1% in the remaining groups).

Second Study: In the second study, lead was measured directly in the high molecular weight fraction and the membrane fraction as well as in the hemoglobin and 10,000 molecular weight fractions. Results are shown in Table 3. Direct measurement of lead in the high molecular weight fraction was of interest because activity of the enzyme, ALA dehydrase, is found in this fraction (Raghavan and Gonick, unpublished) and lead located in this compartment might be anticipated to inhibit this important heme enzyme. Similarly, lead segregated in the membrane fraction was of interest because it might inhibit the membrane transport enzyme, Na-K-ATPase.

In the 26 individuals studied, total recovery of lead summated from the individual fractions averaged 107% of the original RBC hemolysate lead value, verifying the accuracy of the determinations and also the prediction that the calculated "residual" lead would be found in the high molecular weight and membrane fractions. In the normal control population, lead content of each of these fractions averaged 5 µg%, accounting for 12% of the total RBC hemoly-

TABLE 3. Lead Content of RBC and RBC Lead-Binding Fractions (μg%)

Pt.	RBC	Hemoglobin	10,000 mol. wt. Fraction	High mol. wt. Fraction	Membrane Fraction
			I. Normals		
1	60	34	2	6	5
2	43	23	2	5	5
3	52	26	3	5	5
4	42	23	3	5	4
5	46	22	2	6	5
6	29	19	2	4	4
7	32	19	2	4	4
8	31	20	2	4	4
			II. Asymptomatic Workers		
1	163	98	39	18	15
2	141	83	37	14	10
3	172	103	53	9	12
4	157	92	36	14	17
5	178	106	47	9	20
6	128	81	30	6	11
7	145	89	36	10	9
8	145	93	21	14	13
9	128	87	26	8	9
10	102	65	28	6	7
11	150	93	36	8	14
12	107	68	26	6	10
13	154	98	34	12	14
14	124	79	33	8	9
			III. Symptomatic Workers		
1	168	81	33	33	24
			IV. "Low Lead" Symptomatic Workers		
1	108	63	4	27	16
2	120	72	5	29	18
3	102	60	4	24	19

sate lead. In the asymptomatic lead workers, the lead in the high molecular weight fraction averaged 10 μg% or 7% of the total RBC hemolysate lead, and the lead in the membrane fraction averaged 12 μg% or 8% of the total RBC lead. In the four symptomatic lead workers, the lead in the high molecular weight fraction averaged 28 μg% or 22% of the total RBC lead. The lead in the membrane fraction averaged 19 μg% or 15% of the total RBC lead.

Membrane Na-K-ATPase activity averaged 0.34 μmoles Pi/mg protein/hr in the normal controls, 0.23 μmoles Pi/mg protein/hr in the asymptomatic workers, and 0.11 μmoles Pi/mg protein/hr in the symptomatic workers. Although the number of symptomatic workers is too small to permit valid statistical comparisons, the differences in enzyme activity are striking.

The validity of the approach that we have adopted may be assessed by examining the relationship between, on the one hand, total RBC lead and membrane Na-K-ATPase activity (Fig. 3), and on the other hand, membrane lead and membrane Na-K-ATPase activity (Fig. 4). Regression lines and correlation coefficients have not been indicated on these graphs because the data are incomplete. However, in the group studied to date, the correlation between total RBC lead and membrane Na-K-ATPase activity was not significant (r=0.039), whereas the correlation between membrane lead and membrane Na-K-ATPase activity was highly significant (r=0.876, $p < 0.001$).

The *in vitro* inhibition by lead of a highly purified membrane Na-K-ATPase preparation obtained from hog cerebral cortex is demonstrated in Figure 5. Fifty percent (50%) inhibition is seen at a lead concentration of 3×10^{-5} M, and essentially 100% inhibition at 6×10^{-5} M. A direct comparison between *in vivo* and *in vitro* re-

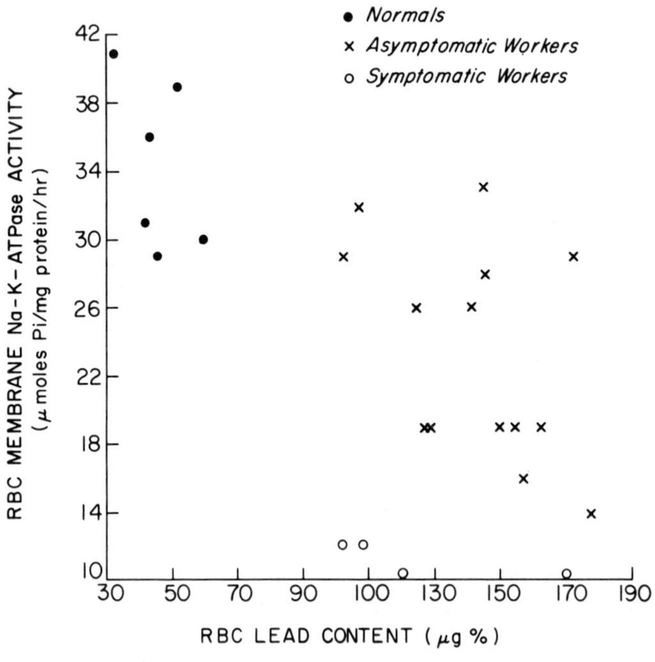

Fig. 3. Relationship between RBC membrane Na-K-ATPase activity and total RBC lead content.

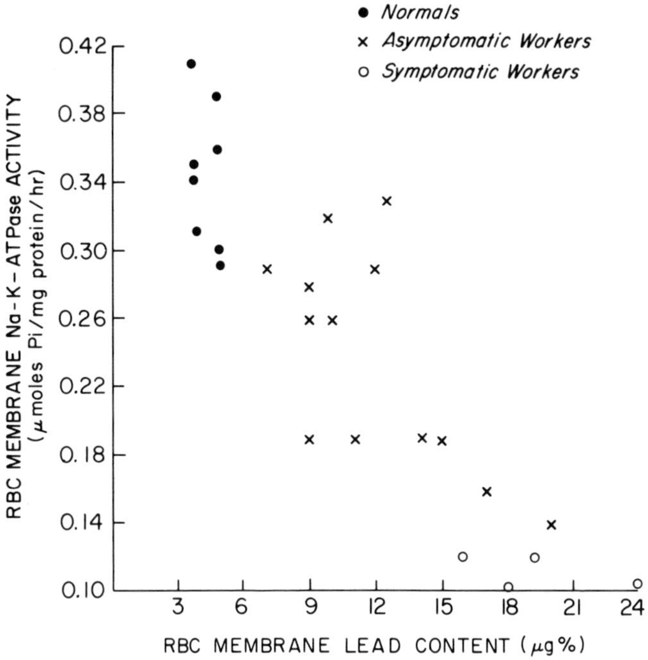

Fig. 4. Relationship between membrane Na-K-ATPase activity and RBC membrane lead content.

sults, however, will be possible only after Na-K-ATPase has been purified from RBC membrane.

DISCUSSION

The toxic manifestations of lead on the hematopoietic system are exhibited by defects in both heme (Goldberg, 1968) and globin (White and Harvey, 1972) synthesis, with resultant retardation of erythrocyte maturation, and by hemolysis of circulating mature erythrocytes (Hasan *et al.*, 1967a; Beck *et al.*, 1970). Deficient erythrocyte membrane Na-K-ATPase activity in lead poisoning, first demonstrated by Hasan *et al.* (1967), may cause affected cells to be lysed prematurely on an osmotic basis. Premature hemolysis may be related also to inhibition by lead of pyrimidine 5'-nucleotidase, an enzyme found in the cytosol of mature erythrocytes, as shown by Paglia *et al.* (1975). Treatment of lead-intoxicated individuals with EDTA rapidly reverses the hemolytic component, but not the effect of lead on heme biosynthesis (Beck *et al.*, 1970).

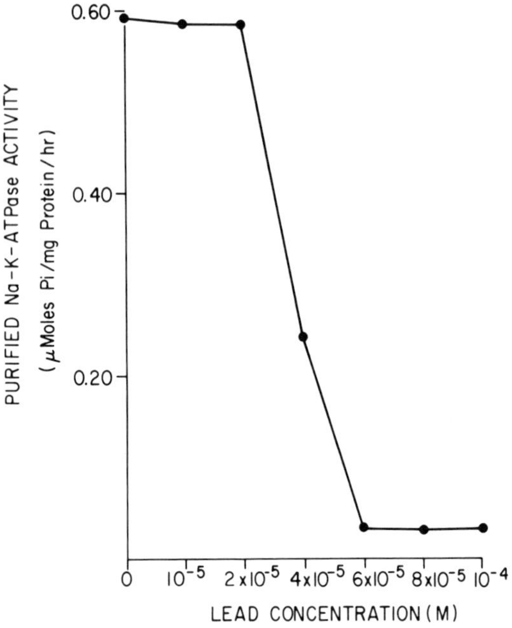

Fig. 5. In vitro inhibition of highly purified Na-K-ATPase by lead.

These observations suggest that there is a mobilizable fraction of lead in one or more compartments of the mature erythrocyte which contributes to hemolysis via an effect on susceptible enzymes.

In a previous communication (Raghavan and Gonick, 1977), we presented evidence for the presence of a 10,000 molecular weight lead-binding protein in erythrocytes from lead workers. As this lead-binding protein was not found in normal subjects, we have speculated that the synthesis of this protein is induced in response to lead exposure, and that the protein may serve to segregate lead in a non-toxic form. The results of the present study appear to support this hypothesis. We have demonstrated that there is a sub-group of industrial workers particularly susceptible to lead intoxication because of their inability to bind lead to this lead-binding protein ("low lead" symptomatic workers). This same group of workers also showed a diminution in lead bound to the hemoglobin A_2 fraction of hemoglobin, but the difference was not as striking as with the 10,000 molecular weight protein.

Reduced binding of lead to these fractions resulted in an accumulation of lead in the "residual" fraction, composed of lead in a high molecular weight fraction of the erythrocyte hemolysate and membrane-bound lead. These latter compartments contain enzymes (ALA dehydrase and pyrimidine 5'-nucleotidase in the cytosol and Na-K-ATPase in the membrane) which are particularly sensitive to inhibition by lead. Our demonstration that membrane Na-K-ATPase activity correlates well with membrane-bound lead, but not with total erythrocyte lead, is in keeping with the hypothesis that the primary determinant of lead toxicity in any cell is the distribution of lead between non-toxic binding proteins and compartments which contain susceptible enzyme systems.

There are several unanswered questions which can only be addressed by subsequent studies. At this point, we have directed our attention to the circulating erythrocyte as a source of the 10,000 molecular weight binding protein. It is unlikely that the protein originates in the mature erythrocyte or even in the nucleated erythrocyte precursors in the bone marrow. As an initial speculation, we would suggest that the 10,000 molecular weight protein, like metallothionein, is synthesized in the liver and then carried to other organs where it may serve either a transport or detoxifying function.

If the renal tubule proves to be one of the target organs, then the relationship between the 10,000 molecular weight protein and the 27,500 molecular weight lead-binding protein in the nuclear inclusion body will require definition. It will also be important to explore whether there is a genetic basis for failure to synthesize the 10,000 molecular weight protein in response to a lead challenge. HLA typing of "low lead" symptomatic workers may be helpful in this regard. Finally, age-related differences in ability to synthesize the 10,000 molecular weight protein need to be examined as a possible etiological factor in the known enhanced susceptibility of the young to central nervous system and kidney lead toxicity.

ACKNOWLEDGMENTS

This work was supported by USPHS grant No. R01 OH 00730 and by a grant-in-aid from the International Lead Zinc Research Organization. The authors gratefully acknowledge the cooperation of NL Industries and the secretarial assistance of Erica Brookes, DeAnna Mackey, and Ruby McCarty.

REFERENCES

Abraham, E. C., Reese, A., Stallings, M., *et al.* (1977). Separation of human hemoglobins by DEAE-cellulose chromatography using glycine-KCN-NaCl developers. *Hemoglobin 1*, 27-44.

Beck, P. D., Tschudy, D. P., Shipley, L. A., *et al.* (1970). Hematologic and biochemical studies in a case of lead poisoning. *Am. J. Med. 48,* 137-144.
Bruenger, F. W., Stevens, W., and Stover, B. J. (1973). The association of ^{210}Pb with constituents of erythrocytes. *Health Phys. 25,* 34-42.
Galle, P., and Morel-Maroger, L. (1965). Les lesions renales du saturnisme humain et experimental. *Nephron 2,* 273-286.
Goldberg, A. (1968). Lead poisoning and haem biosynthesis. *Sem. Hematol. 5,* 424-433.
Goyer, R. A., May, P., Cates, M. M., *et al.* (1970). Lead and protein content of isolated intranuclear inclusion bodies of lead-poisoned rats. *Lab. Invest. 22,* 245-251.
Hasan, J., Vihki, V., and Hernberg, S. (1967). Deficient red cell membrane Na-K-ATPase in lead poisoning. *Arch. Environ. Hlth. 14,* 313-318.
Hasan, J., Hernberg, S., Petsälä, P., *et al.* (1967a). Enhanced potassium loss in blood cells from men exposed to lead. *Arch. Environ. Hlth. 14,* 309-312.
Kimura, M., Otaki, N., Yoshiki, I., *et al.* (1974). The isolation of metallothionein and its protective role in cadmium poisoning. *Arch. Biochem. Biophys. 165,* 340-348.
Moore, J. F., and Goyer, R. A. (1974). Lead-induced inclusion bodies: Composition and probable role in lead metabolism. *Environ. Health Perspect. 7,* 121-127.
Nordberg, G. F., Goyer, R. A., and Nordberg, M. (1975). Comparative toxicity of cadmium metallothionein and cadmium chloride on mouse kidney. *Arch. Pathol. 99,* 192-197.
Paglia, D. E., Valentine, W. N., and Dahlgren, J. G. (1975). Effects of low level lead exposure on pyrimidine 5'-nucleotidase in the pathogenesis of lead-induced anemia. *J. Clin. Invest. 56,* 1164-1169.
Raghavan, S. R. V., and Gonic, H. C. (1977). Isolation of low molecular weight lead-binding protein from human erythrocytes. *Proc. Soc. Exp. Biol. Med. 155,* 164-167.
Raghavan, S. R. V., and Gonick, H. C. Unpublished observations.
White, J. M., and Harvey, D. R. (1972). Defective synthesis of α and β globin chains in lead poisoning. *Nature 236,* 71-73.

DISCUSSION OF PAPER BY DR. HARVEY C. GONICK

Lalitha Murphy, University of Cincinnati: Dr. Gonick, in your electrophoresis with polyacrylamide gel, did you use cytosol with the cell membrane or without the cell membrane?

Gonick: This was the hemolysate without the cell membrane. The cell membrane was separated out first.

Murphy: What method did you use--the radioactive technique or atomic absorption--to detect the lead?

Gonick: Lead was detected by the appearance of tracer, not by the appearance of lead by atomic absorption. The amounts present in the gel would have been insufficient to detect by almost any atomic absorption method that is available today.

Murphy: So you used radioactive lead.

Gonick: That's correct. It is certainly possible that lead was present as "cold" lead in that second band that we showed on the acrylamide gel.

Murphy: You stated that you filtered the cell membrane and then used the cytosol for electrophoresis. Did you count the cell membrane before discarding?

Gonick: Yes. The cell membrane contained about 16% of the total radioactivity.

Murphy: Have you actually measured the deficiency of the synthesis of this protein, or are you assuming that there is a deficiency because of the lower binding that you observed?

Gonick: I think that's really the critical question, and I thank you for posing it. The same question can be posed in the cadmium metallothionein story. In order to truly answer the question, a good method for quantitating the presence of the protein, rather than the binding of the metal to the protein, is necessary. We would hope that an immunoassay could be developed that would allow that to be done appropriately. In the absence of a good immunoassay, I don't think that one can answer that question accurately.

Joseph Graziano, Cornell University Medical College: Have you ever fractionated hemoglobin to see whether the lead is in fact associated with glycosylated hemoglobins?

Gonick: If I might, I would like to ask Dr. Raghavan, who is the biochemist of our group, to answer that question.

S. R. V. Raghavan, University of California, Los Angeles: No, we have not.

Graziano: Have you looked into the possibility that these patients who are symptomatic at relatively low lead levels have G6PD deficiency or are carriers of G6PD deficiency?

Gonick: Not in this particular group, but I have looked at that in other patients, and this did not appear to be the case.

David A. Lawrence, The Albany Medical College of Union University: Is there any lipid or nucleotide associated with your 10,000 MW binding protein?

Gonick: This is a question that Dr. William Valentine has posed because of his interest in the effect of lead on nucleotidase. Unfortunately, we have not had a chance to interrelate our work with his. I don't know the molecular weight of the enzyme that he has found to be inhibited by lead, but it's possible that these are closely interrelated.

Hammond: What was the basis for classifying patients as symptomatic versus asymptomatic?

Gonick: First was the severity of their complaints when they presented themselves to Dr. Culver, who is usually capable of distinguishing between complaints that are real and complaints that, perhaps, have other bases. Secondly, the symptomatic patients had persistently elevated urinary ALA levels and tended to be slightly anemic, as compared to age-matched normals, and when measured, had elevated urinary coproporphyrins.

Carl C. Smith, University of Cincinnati: Metallothionein is usually thought to have a molecular weight of about 6,600, and there is a chance that the early data were based on confirmation of the protein. Is your material likely to have the same differences, or is it really 10,000 molecular weight? Also, is there any reason to think that there's organ specificity or tissue specificity? Is it synthesized in location, or must it be? Could it be made in liver or some other place and delivered here? The synthesizing properties, of course, could be there but they are often pretty deficient in mature red cells.

Gonick: I appreciate your questions. Those are all questions which require solid answers. The molecular weight is merely an es timation, as it was initially for metallothionein. It is certainl possible the molecular weight of this compound differs by two or three thousand from what it will eventually prove to be. Accurate metallothionein, that, again, really requires both immunoassay techniques and purification for some finalized answers. We would speculate, as you have, that it's possibly made in the liver and transported to various organs. But we have absolutely zero data on anything other than the red blood cells at this time.

LEAD EFFECTS ON THE IMMUNE SYSTEM

David A. Lawrence

Albany Medical College of Union University
Albany, New York

INTRODUCTION

Lead (Pb) compounds are known to exert toxic effects on the renal, haemopoietic, embryonic, and central nervous systems (Goyer, 1971; Chisolm, 1971; Goyer and Rhyne, 1973; Jacquet *et al.*, 1975). The toxicity of Pb is a major concern, because a relatively high level of this heavy metal pollutant is in the environment (Kehoe, 1976). The average individual has been shown to be acutely or chronically exposed to only subclinical doses of Pb (Waldron and Stofen, 1974); however, the physiologic effects of subclinical exposure to Pb are not well-documented. In addition, most studies have attempted to assess the direct mechanisms of Pb toxicity and, therefore, have been pathologic investigations. A limited number of toxicity studies have examined the indirect effects of Pb-induced toxicity by evaluating the effects of Pb on the immune system (Vos, 1977). Since the immune system maintains the integrity of self and protects the host from pathogens, Pb alteration of the immune system could upset homeostatic mechanisms and natural and acquired resistance to invading organisms.

The immune system is composed of two effector branches known as humoral immunity (HI) and cell-mediated immunity (CMI), which result from direct, specific responses of B-lymphocytes and T-lymphocytes, respectively. Acute and chronic exposure of experimental animals to low levels of Pb has been shown to alter the immune system and enhance the host's susceptibility to bacterial infections (Selye *et al.*, 1966; Hemphill *et al.*, 1971; Cook *et al.*, 1975). Although Pb has been shown to enhance morbidity and mortality (Vos, 1977; Cook *et al.*, 1975), the mechanisms involved have not been delineated. Most of the studies to date have investigated only humoral immunity, the production of antibodies. The effects of Pb on T-lymphocyte functions are not known, and suppression of T-lymphocyte function as well as B-lymphocyte function dramatically increases mortality. T-lymphocytes are not only re-

ISBN 0-12-460520-6

sponsible for CMI, but they regulate HI by enhancing or suppressing B-lymphocyte proliferation and differentiation. A third cell type also is involved in immunity; T-lymphocyte activation (Rosenthal, 1978), and the effector functions (CMI and regulation of HI) of T-lymphocytes (Rosenthal, 1978; Treves, 1978) are controlled by macrophages. Therefore, the immune system is dependent on three main cell types: B-lymphocytes, T-lymphocytes, and macrophages. The ability of Pb to alter the activities of these sets of cells has not been assessed adequately.

In order to determine the effects of an immunologic potentiator or suppressive agent on the immune system, the system must be dissected into its functional components and the effects of the agent examined on each separately and combined. The effects of Pb on the immune system have been investigated only in *in vivo* studies; thus, assessment of Pb alteration of B-lymphocyte, T-lymphocyte, and macrophage function cannot be determined. *In vivo* studies can measure only the overall effect, that is, a suppressed or enhanced HI or CMI response, and direct or indirect effects of Pb on the immune system cannot be evaluated accurately. *In vivo*, Pb has been shown to reduce resistance to bacteria (Selye *et al.*, 1966; Hemphill *et al.*, 1971; Cook *et al.*, 1974; Cook *et al.*, 1975) and viruses (Gainer, 1974; Gainer, 1977) and to inhibit antibody production (Koller, 1973; Koller and Kovacic, 1974). Since these effects could be due to direct suppression of B-lymphocyte, T-lymphocyte, and/or macrophage functions, indirect effects on the pathogens or antigens employed, or indirectly via alteration of aspects of innate immunity such as neutrophil function, natural flora, or mucous barriers (Raffel, 1961), the present investigation was undertaken to determine the *in vivo* and *in vitro* effects of Pb on HI and CMI (Table 1).

In this study, the ability of Pb to alter immunologic activities was assessed by three different approaches, as shown in Figure 1: a) Pb was administered *in vivo* and immune function assessed *in vivo*; b) Pb was administered *in vivo*, lymphocytes removed and washed, and their activities assessed *in vitro*; or c) lymphocytes were exposed to Pb *in vitro* and the *in vitro* effects of Pb were determined. The results indicate that assessment of the Pb effects on immunity vary dependent on the system employed. The *in vivo* effects of Pb did not always correlate with the *in vitro* effects of Pb.

MATERIALS AND METHODS

Animals. Female CBA/J and C57Bl/6 mice (Jackson Laboratory, Bar Harbor, Maine) or male Nylar mice (Division of Laboratories and Research, New York State Department of Health, Albany, N.Y.) were used interchangeably throughout these studies. Mice were maintained on laboratory chow and acidified, chlorinated water,

TABLE 1. Investigation of the Immune System

Humoral Immune Response (B-lymphocyte Activities)	*Cell-Mediated Immune Response (T-lymphocyte Activities)*
1. Enumeration of Antibody-producing Cells	*1. Mixed lymphocyte culture (MLC) Response*
2. Mitogen responsiveness: a) 2-mercaptoethanol b) Lipopolysaccharide	*2. Resistance to Listeria*
	3. Mitogen responsiveness: a) Concanavalin A b) Phytohemagglutinin

pH 3.0, ad libitum, prior to use in the studies. During the individual experimental studies, the above described water was replaced with distilled, Millipore filtered water (2μMHO;21°C) with various concentrations of lead (Pb) acetate (0.08 mM-50 mM); 50 mM Pb acetate was employed only in the initial 1 week study, because the mice drank only 0.4 ml ± 0.1 ml/day/mouse, as compared to 4.5 ml ± 1.5 ml/day/mouse for the other groups.

Mitogens and antigens. The mitogens and their concentrations employed in these studies were as follows: 2 μg/ml recrystallized concanavalin A (Con A, Miles); 5×10^{-5} M 2-mercaptoethanol (2-ME, Eastman Kodak Co.); 2 μg/ml phytohemagglutinin-purified (PHA,

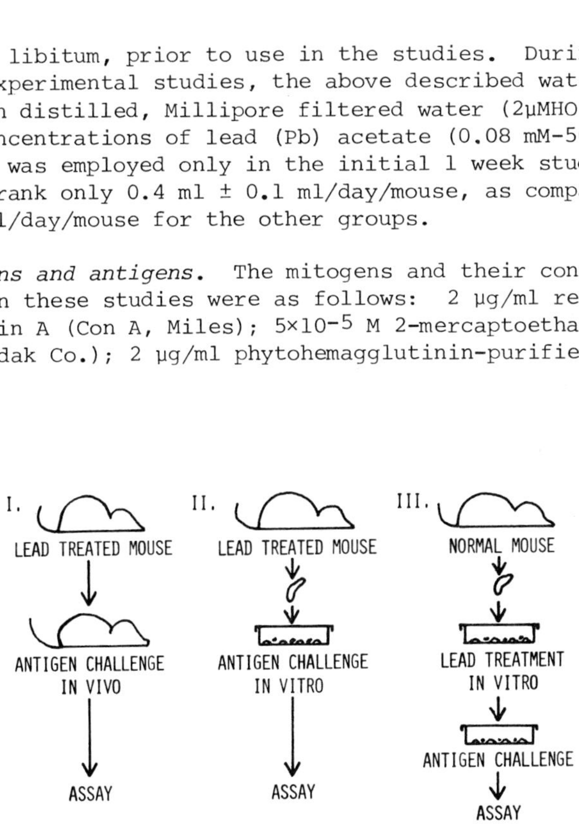

Fig. 1. Assessment of lead (Pb) effects on the immune response. The effect of Pb treatment were assayed in this study by the three protocols described: I. in vivo Pb-treatment and in vivo assessment; II. in vivo Pb-treatment and in vitro assessment; and III. in vitro Pb-treatment and assessment.

Wellcome Reagents); and 50 µg/ml *E. coli* 055:B5 lipopolysaccharide *(LPS, Difco)*. Sheep erythrocytes (SRBC), obtained from Griffin Laboratories, New York State Department of Health, were stored in Alsever's solution. The *Listeria monocytogenes* inoculum was prepared and stored as previously described (Schell and Lawrence, 1977).

Preparation of cells. Spleens were aseptically removed from the mice after exsanguination. Spleens were teased and settled to obtain sterile single cell suspensions as previously described (Lawrence *et al.*, 1978). A balanced salt solution was used for the isolation of all cells.

Assessment of in vivo humoral immunity to SRBC. Mice were injected intravenously with 10^8 SRBC and 5 days later their spleens were removed, single cell suspensions were obtained, and the cell preparations were assayed in the hemolytic plaque assay. All results are expressed as number of plaque-forming cells (PFC)/spleen. The results were relatively comparable to PFC/10^6 spleen cells.

Assessment of in vivo cell-mediated immunity to Listeria. The in vivo immune response to *Listeria* was assessed as previously described (Schell and Lawrence, 1977). Briefly, approximately 10^5 viable *Listeria* were injected intravenously, and 2-4 days later the spleens were aseptically removed, homogenized, and plated for enumeration of *Listeria* colonies.

Assay for in vitro humoral immunity to SRBC. The *in vitro* cultures were set up as previously described (Lawrence *et al.*, 1978). Briefly, 5×10^6 spleen cells were cultured in 0.5 ml medium/well by the technique described by Mishell and Dutton (1967). The number of direct hemolytic PFC in each culture well was determined at day 5. All results are expressed as the number of PFC/culture averaged from 3 cultures. Background responses to SRBC were determined from cultures lacking antigen, and background PFC values were subtracted from the experimental values.

Hemolytic plaque assay. Cells producing antibody specific for SRBC were enumerated by use of a modification (Golub *et al.*, 1968) of the Jerne plaque technique (Jerne and Nordin, 1963).

Assay for mixed lymphocyte culture (MLC) responsiveness. Responder lymphocytes were mixed with 2000 R irradiated stimulator lymphocytes in various ratios (2:1, 1:1, 1:2) and the proliferative response measured 5 days later by pulsing with ^{3}H-thymidine from day 4.5-5 (12 hrs). A total of 2×10^6 cells were cultured in 0.2 ml of RPMI 1640 medium supplemented with non-essential amino acids, Na pyruvate, $NaHCO_3$, penicillin and streptomycin (100 units/ml), 5×10^{-5}M 2-ME, and 5% heat-inactivated human AB serum. A flat bottom tissue culture microtitre plate was employed (Costar, Cambridge, Mass.).

Assay for in vitro mitogen responsiveness. Single cell suspensions of spleen cells were cultured at a cell concentration of 2×10^5/0.2 ml/well. A MEM suspension culture medium +5% fetal calf serum mitogen was employed. Cultures were pulsed with ^{3}H-thymidine for 6 hrs on day 2, 3.5, or 5.

Statistical analysis. The analysis of variance (Fryer, 1966) or the Student's t test were employed. For the Student's t test, a $p < 0.01$ was considered significant.

RESULTS

In vivo Pb Effects on Humoral and Cell-Mediated Immunity

Mice were fed Pb acetate in their drinking water in concentrations ranging from 0.08 mM-50 mM (16-10,400 ppm Pb) for 1-4 wk. After 1 (Fig. 2), 2 (Fig. 3), or 4 weeks (Fig. 4), the mice were immunized intravenously with 10^8 sheep erythrocytes (SRBC) and the primary humoral immune response to SRBC was assessed 5 days later by enumeration of the SRBC-specific PFC/spleen. As shown in Figs. 2-4, Pb did not significantly affect the response to SRBC. Additional experiments have assessed the response of mice fed Pb for 8 weeks, and Pb still did not significantly alter the *in vivo* humoral response to SRBC (data not shown).

Cell-mediated immunity was assessed in mice fed Pb for 2 weeks by determining their resistance to *Listeria*. In contrast to humoral immunity, cell-mediated immunity to *Listeria* appeared to be suppressed. As shown in Fig. 5, 10 mM Pb acetate solutions significantly reduced resistance to *Listeria*, as apparent from the enhanced number of viable *Listeria*/spleen. The 0.08-2 mM Pb acetate solutions did not significantly alter the number of viable *Listeria*/spleen; however, 0.4-10 mM solutions enhanced the mortality rate of the *Listeria* infected mice. Ten days after infection, all 0.08 mM and control mice were alive, but the 2 and 10 mM groups were all dead by day 3, and the 0.4 mM mice were all dead by day 7.

In vitro Assessment of in vivo Pb-Treated Lymphocytes

The ability of splenic lymphocytes from Pb fed mice to produce a primary *in vitro* humoral immune response to SRBC was assessed. Unlike the *in vivo* responses, substantial differences were observed *in vitro*. Lymphocytes isolated from mice drinking 0.08-0.4 mM Pb acetate for 4 weeks had a 2-fold increase in their number of SRBC-specific PFC when challenged *in vitro* (Table 2). On the other hand, lymphocytes from the 2 mM group did not produce more PFC than the control group, and the lymphocytes from the

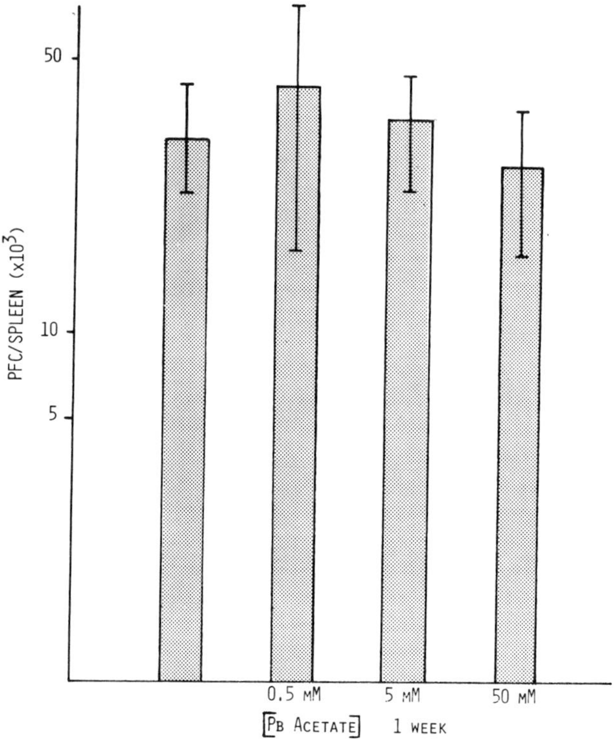

Fig. 2. In vivo assessment of the influence of Pb (1 week) on the primary humoral immune response of Nylar mice to SRBC. Mice drank water (control, —) or various amounts of Pb acetate (0.5 mM-50 mM) for 1 week prior to immunization with 10^8 SRBC, intravenously. The PFC/spleen were enumerated 5 days later. The bars represent the mean standard deviation of 5 mice/group.

10 mM group were suppressed in that only 30% of the control response was obtained.

The *in vitro* assessment of T-lymphocyte activity also differed as compared to the *in vivo* assessment of CMI (*Listeria* resistance). *In vitro* responsiveness of T-lymphocytes from Pb fed mice was investigated by MLC reactivity. Mice fed Pb for 2 or 4 weeks were assessed (Table 3 and 4). As in the *in vitro* assessment of HI, the 0.08 mM and 0.4 mM groups had enhanced activities, the 2 mM group was not affected, and the activity of the 10 mM group was slightly suppressed (10-24% inhibition). It should be noted that the Nylar strain is composed of genetically similar mice but is not inbred; therefore, 2 mM and 10 mM solutions inhibited the positive homologous response.

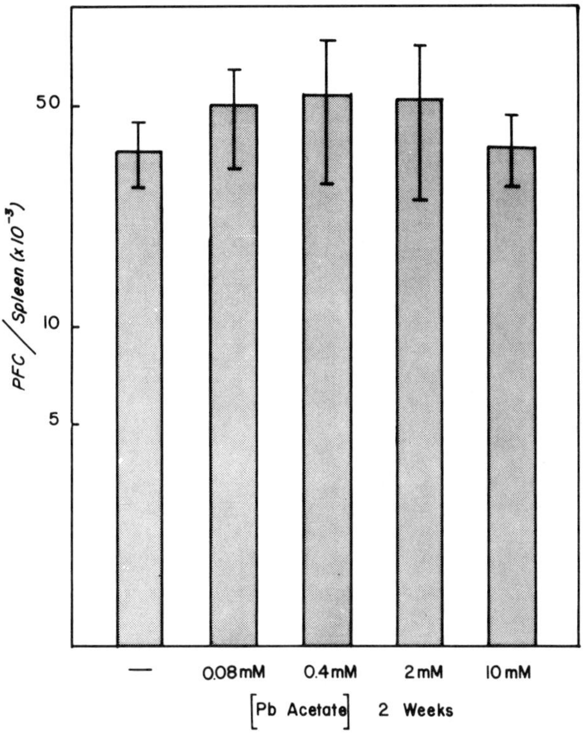

Fig. 3. In vivo assessment of the influence of Pb (2 weeks) on the primary humoral immune response of Nylar mice to SRBC. Mice drank water (control, —) or various amounts of Pb acetate (0.5 mM-50 mM) for 2 weeks prior to immunization with 10^8 SRBC, intravenously. The PFC/spleen were enumerated 5 days later. The bars represents the mean standard deviation of 5 mice/group.

The mitogenic responsiveness of the in vivo Pb-treated (4 wk) splenic lymphocytes also was determined (Table 5). The lymphocyte preparations from the Pb groups had responses to 2-ME and PHA that did not significantly differ from the control cells. The Con A-induced response was significantly enhanced in the 0.4 mM group. The LPS-induced response was significantly suppressed in the 2 mM and 10 mM groups. This suggests that intermediate *in vivo* doses of Pb enhance T-lymphocyte proliferation and suggests that intermediate *in vivo* doses of Pb enhance T-lymphocyte proliferation and that higher doses of Pb inhibit proliferation of LPS-induced B-cells but not 2-ME induced B-cells. In addition, lymphocytes from the 0.08-2 mM Pb-treated groups had significantly higher background responses (medium; no mitogens).

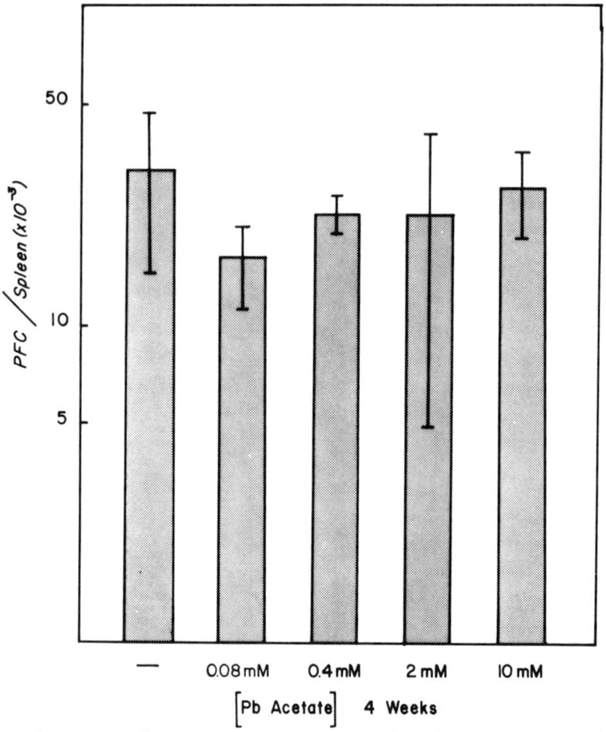

Fig. 4. In vivo assessment of the influence of Pb (4 weeks) on the primary humoral immune response of CBA/J mice to SRBC. Mice drank water (control, —) or various amounts of Pb acetate (0.08 mM-10 mM) for 4 weeks prior to immunization with 10^8 SRBC, intravenously. The PFC/spleen were enumerated 5 days later. The bars represent the mean standard deviation of 5 mice/group.

In vitro Pb-Treatment

Pb enhanced the *in vitro* primary humoral immune response to SRBC (Fig. 6). Metal chloride concentrations ranging from 10^{-4}M to 10^{-7}M were employed, and the effects of Pb, Ca, and Hg were compared. Ca produced no significant effect on the response, whereas Hg inhibited the response and Pb enhanced the response. The ability of Pb to potentiate the HI response to SRBC *in vitro* correlates with the enhanced activity of *in vivo* Pb-treated spleen cells tested *in vitro* (Table 2).

Likewise, the T-lymphocyte activities tested in *in vitro* MLC responses were enhanced by Pb (Fig. 7), which correlates with the enhanced *in vitro* MLC reactivity of *in vivo* Pb-treated cells (Table 3 and 4). The MLC reactivity in both cases was inhibited by the highest concentration of Pb (0.5 mM, *in vitro*; 10 mM, *in vivo*).

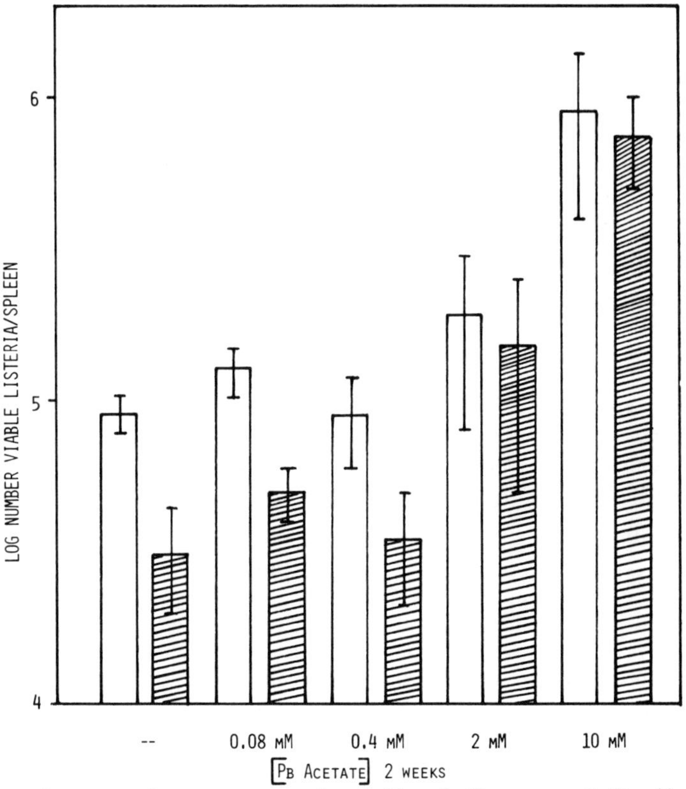

Fig. 5. In vivo assessment of the influence of Pb (2 weeks) on the cell-mediated immune response of Nylar mice to Listeria monocytogenes. Mice drank water (control, —) or various amounts of Pb acetate (0.08-10 mM) for 2 weeks prior to infection with 1.6×10^5 *viable Listeria. The log number of viable Listeria/spleen were enumerated 48 (open bars) and 72 (hatched bars) hrs. after infection. The bars represent the mean standard deviation of 5 mice/group.*

Some aspects of the mitogen responsiveness of *in vitro* Pb-treated spleen cells did not correlate with that of *in vivo* Pb-treated spleen cells. Again, Pb, itself, induced limited proliferation, produced no effect or slightly enhanced the Con A- and PHA-induced responses of T-lymphocytes, and slightly enhanced the 2-ME-induced response of B-lymphocytes (Table 6). However, the LPS-induced response was not inhibited *in vitro* by Pb-treatment even with the highest concentration (10^{-4} M). Higher concentrations of $PbCl_2$ (5×10^4M) were not used, because they were toxic. Since Hg dramatically suppressed the *in vitro* primary HI response to SRBC (Fig. 6) and some studies have indicated that Hg is mitogenic

TABLE 2. Effect of Lead (4 weeks) on in vitro Primary Humoral Immune Response to SRBC[a]

Lead Dose[b]	*PFC/Culture*[c]	*% Control*
--	378 ± 36	100
0.08 mM	768 ± 81	203
0.4 mM	725 ± 71	192
2.0 mM	473 ± 51	125
10.0 mM	112 ± 62	30

[a]*The in vitro primary humoral immune response (day 5) to SRBC was assessed with cultures established with* 5×10^6 *spleen cells (from mice treated with various doses of Pb) plus* 2×10^6 *SRBC.*

[b]*Groups of 4 mice were fed various doses of Pb acetate (0.08-10.0 mM) in their drinking water.*

[c]*The number represents the mean number of PFC from triplicate cultures of the spleen cells from 4 mice/group ± standard deviation.*

for human lymphocytes (Berger and Skinner, 1974; Caron *et al.*, 1970), the Pb effects on murine lymphocyte proliferation were compared to the effects of Zn and Hg. As shown in Table 7, the heavy metal effects are reported as % of control for day 3 responses. Pb produced results equivalent to those reported for day 2 responses (Table 6) except that on day 3 the LPS-induced response was substantially enhanced by Pb. Zn was inhibitory at a high concentration (10^{-4}M), but not at lower concentrations, whereas Hg was toxic at all concentrations.

In vitro macrophage function was not significantly affected by 10^{-4}M $PbCl_2$, although there was a slight increase in the uptake of SRBC preincubated with Pb (Table 8). Since macrophage processing of SRBC did not appear to be substantially altered, the Pb effects on *in vitro* PFC and MLC responses may be due to a direct effect on B-lymphocytes and T-lymphocytes, respectively.

DISCUSSION

Numerous reports have indicated that heavy metals can react with cells of the immune system and alter humoral (Vos, 1977; Selye *et al.*, 1966; Hemphill *et al.*, 1971; Cook *et al.*, 1974; Koller, 1973; Koller and Kovacic, 1974) and cell-mediated (Vos, 1977; Muller *et al.*, 1976) immune responses. However, many aspects of these reports are conflicting, and the types of cells affected and the mechanisms involved are unresolved. Since, to date, no studies have attempted to dissect the cellular components of the immune system, the effects of Pb and other heavy metals on

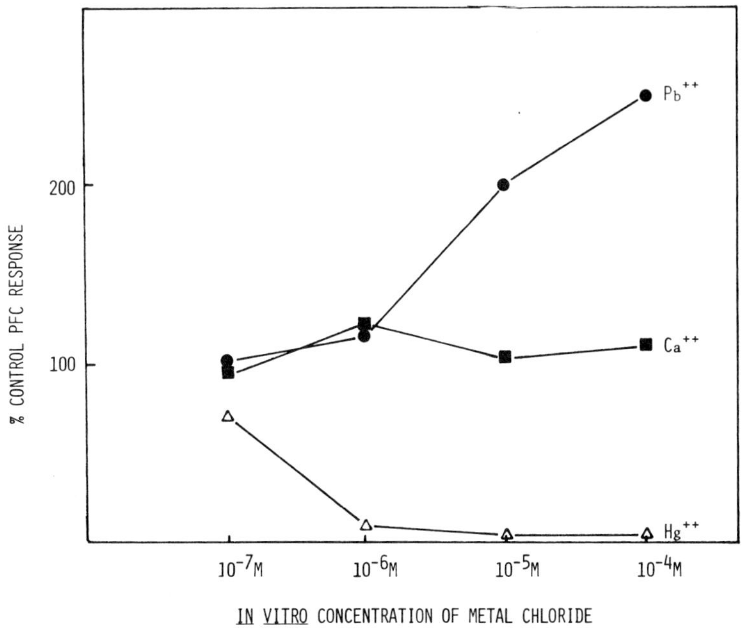

Fig. 6. In vitro assessment of the influence of metals on the humoral immune response to SRBC. Cultures of 5×10^6 CBA/J spleen cells + 2×10^6 SRBC were initiated with various concentrations of Pb, Ca, or Hg, and the development of SRBC-specific PFC were determined 5 days later. Each point was calculated from the mean number of PFC/culture from triplicate cultures. The % control value was determined by dividing the mean experimental values by the mean control (cultures without addition of metal chlorides) values. The control cultures had 553 PFC/culture.

B-lymphocyte and T-lymphocyte activities are unknown. The ability to resist infection and control neoplasia rests on the functional integrity of the immune system. The need for competent T-lymphocytes (thymus-derived) and B-lymphocytes (bone-narrow derived) can be most easily depicted when patients with immunologic deficiencies are examined. Individuals with B-lymphocyte deficiencies lack antibody production; therefore, they have recurrent pyrogenic infections and lowered respiratory and gastrointestinal resistance to viral pathogens. Patients lacking T-lymphocyte function experience extensive morbidity due to multiple viral, fungal, and bacterial infections. Untreated immune deficiencies will result in early death. B- and T-lymphocytes also can be hyperactive and loss of immuno-regulatory controls can result in autoimmune disease. Since the immune system is intimately involved in the maintenance of a healthy state, the effects of Pb on the activities of B-lymphocytes and T-lymphocytes must be evaluated.

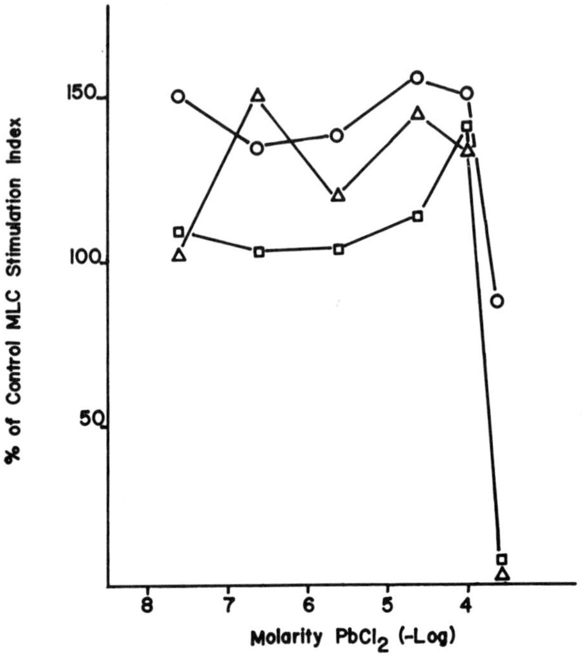

Fig. 7. Effect of various concentrations of $PbCl_2$ on a one way mixed lymphocyte culture response. CBA/J responder spleen cells were mixed with 2000 R C57Bl/6 stimulator spleen cells in ratios 2:1 (○-○), 1:1 (□-□), or 1:2 (Δ-Δ). Each point represents the percent of the control (no Pb) stimulation index (SI). The stimulation indices were determined from triplicate cultures and were calculated as follows: cpm (^{3}H-thymidine incorporation) for CBA vs. C57Bl/6 (allogeneic response) cpm for CBA vs. CBA (syngeneic response) × 100. The control SI were 1.9 (2:1), 3.6 (1:1), and 7.7 (1:2).

In this investigation, the ability of Pb to alter humoral immunity (HI) and cell-mediated immunity (CMI) has been correlated with the *in vivo* and *in vitro* effects of Pb on B-lymphocyte, T-lymphocyte, and macrophage function. T- and B-lymphocyte and to a lesser extent macrophage functions were altered by Pb; the net effects were enhancement or suppression of the *in vitro* responses dependent on the Pb dose. *In vitro*, Pb, on the whole, enhanced HI and CMI, whereas, *in vivo*, Pb did not significantly alter HI and suppressed CMI.

The *in vivo* HI response (production of antibody) to SRBC was not altered by oral administration of 0.08 mM-10 mM Pb for 1-4 weeks. Previous *in vivo* studies employed an acute Pb dose (a single intraperitoneal dose of 4 mg) or chronic Pb exposure (10

TABLE 3. Effect of Lead (2 weeks) on MLC Response of Nylar Mice to CBA/J Mice

Lead Dose on Nylar Responders	*Stimulator Cells[a]* *Nylar*	*CBA/J*	*S.I.[b]*	*% Control*
--	*11,236[c] ±1,335*	*89,313 ±4,815*	*7.95*	*100*
0.08 mM	*13,632 ±1,423*	*123,178 ±5,069*	*9.04*	*114*
0.04 mM	*13,411 ±1,420*	*128,162 ±6.609*	*9.55*	*120*
2.0 mM	*3,770 ±164*	*28,492 ±6,116*	*7.56*	*95*
10.0 mM	*4,152 ±220*	*29,561 ±4,059*	*7.12*	*90*

[a]*Stimulator cells were the homologous nylar spleen cells or the allogeneic CBA/J spleen cells; a ratio of 2 stimulators: one responder was used and the stimulator cells were irradiated with 200 R before addition to the MLC.*

[b]*S.I. = stimulation index* $\frac{(\text{cpm CBA/J:Nylar})}{(\text{cpm Nylar:Nylar})}$

[c]*The number represents the mean from triplicate cultures ± standard error.*

weeks; orally) and reported that Pb enhanced (Koller *et al.*, 1976) or suppressed (Koller, 1973; Koller and Kovacic, 1974), respectively. Our study employed intermediate exposure times (1-4 weeks), which may explain why neither enhancement nor suppression were observed. However, when the lymphocytes from Pb-treated mice were removed, placed in culture, and assessed for activity, *in vivo* doses of 0.08 mM-0.4 mM were shown to enhance HI and 10 mM suppressed. Likewise, when isolated lymphocytes were exposed to Pb *in vitro*, 10^{-5}M to 10^{-4}M Pb significantly enhanced the HI response. Only *in vitro* doses >5×10^{-4}M were toxic.

Since the HI response is a direct function of B-lymphocyte antibody production and LPS is a B-lymphocyte mitogen in the mouse (Greaves and Janossy, 1972), it is important to note that *in vivo* Pb-treated LPS-responsive cells were inhibited (Table V) by Pb concentrations that appeared to suppress or have no effect on the HI response (Table 2). A positive correlation also exists for the *in vitro* studies, because Pb enhanced LPS-induced proliferation (Table 7) and the HI response (Fig. 6). The differences between our *in vivo* and *in vitro* studies may relate to the fact that *in vitro* exposure was acute and *in vivo* expsoure was relatively chronic. This suggests that an acute exposure to Pb during an HI

TABLE 4. Effect of Lead (4 weeks) on MLC response of CBA/J Mice to C57Bl/6 Mice

Lead Dose on CBA/J Responders	*Stimulator Cells*[a] *CBA/J*	*C57Bl/6*	*S.I.*[b]	*% Control*
--	*20,685*[c] *±1,211*	*70,983 ±6,284*	*3.43*	*100*
0.08 mM	*20,463 ±5,180*	*91,116 ±3,198*	*4.45*	*130*
0.04 mM	*14,023 ±365*	*79,485 ±4,785*	*5.67*	*165*
2.0 mM	*22,025 ±2,825*	*72,928 ±2,775*	*3.31*	*97*
10.0 mM	*28,887 ±3,435*	*74,696 ±4,644*	*2.59*	*76*

[a]*Stimulator cells were the syngeneic CBA/J spleen cells or the allogeneic C57Bl/6 spleen cells; a ratio of 2 stimulators: one responder was used, and the stimulator cells were irradiated with 2000 R before addition to the MLC.*

[b]*S.I. = stimulation index* $\frac{(cpm\ C57Bl/6:CBA/J)}{(cpm\ CBA/J:CBA/J)}$

[c]*The number represents the mean from triplicate cultures ± standard error.*

response enhances, as shown in our *in vitro* study and in the *in vivo* acute Pb exposure study (Koller *et al.*, 1976), but chronic *in vivo* exposure can suppress (Koller, 1973; Koller and Kovacic, 1974; Table 2) when doses >2 mM are employed.

Long-term *in vivo* exposure to Pb may alter normal B-lymphocyte differentiation and, thus, alter the HI response. The data support this possibility, in that, *in vivo* Pb treatment enhanced 2-ME-induced B-lymphocyte proliferation (Table 5). LPS-responsive B-lymphocytes are known to be less mature than the 2-ME-induced B-lymphocytes (Goodman *et al.*, 1978). This suggests that chronic *in vivo* exposure to Pb may preferentially inhibit a young B-lymphocyte or alter B-lymphocyte differentiation. Chronic *in vivo* Pb exposure has been reported to alter the percentage of B-lymphocyte with C3b receptors (Koller and Brauner, 1977). B-lymphocytes lacking C3b receptors exist very early or late in the ontogenic developmental pathway of stem cell to B-lymphocyte to plasma cell. Pb alteration of the ontogenic development of the B-lymphocyte would explain why our *in vivo* and *in vitro* studies indicated differential Pb effects on B-lymphocyte activities.

Differential Pb effects on the subpopulations of B-lymphocytes

TABLE 5. Effect of in vivo Lead (Pb) Treatment on the in Vitro Mitogen-Reactivity of Spleen Cells[a]

	Mitogenic Response (cpm/culture)				
Mitogen	--[b]	0.08 mM	0.4 mM	2.0 mM	10 mM
--	479[c] ±144	842 ±235	1,164 ±262	1,136 ±259	658 ±88
2-ME	7,756 ±891	12,113 ±882	14,804 ±658	11,389 ±249	11,564 ±306
Con A	37,361 ±1,264	39,664 ±4,866	54,238 ±2,486	42,838 ±3,329	36,335 ±2,720
PHA	38,549 ±3,361	48,726 ±3,486	39,096 ±3,053	36,727 ±3,259	32,309 ±2,905
LPS	41,061 ±2,361	43,164 ±1,168	44,659 ±1,819	26,417 ±1,283	29,782 ±1,072

[a]*The mitogenic response of* 2×10^5 *CBA/J spleen cells from the experimental and control groups was determined with the mitogens on day 2 by pulsing for 6 hours with* ^3H*-thymidine.*

[b]*Concentrations of Pb acetate fed to the mice for 4 weeks.*

[c]*The number represents the mean cpm ± standard error for triplicate cultures with the spleen cells of 3 mice/group.*

also would indicate why B-lymphocyte responses to SRBC seem to differ from responses to endotoxin or endotoxin-yielding bacteria. Endotoxin (otherwise known as LPS) is a T-independent (TI) antigen and SRBC in a T-dependent (TD) antigen. It has been suggested that TI antigens (and possibly some TD antigens) stimulate different subpopulations of B-lymphocytes (Mosier *et al.*, 1977). Numerous reports have been shown that Pb reduced *in vivo* resistance to endotoxin (Selye *et al.*, 1966; Hemphill *et al.*, 1971; Cook *et al.*, 1975; Filkins and Buchanan, 1973; Seybarth *et al.*, 1972), which is believed to be controlled by HI, whereas in this study the *in vivo* HI response to SRBC was not suppressed (Figs. 2-4), although LPS-induced proliferation of *in vivo* Pb-treated lymphocytes was inhibited (Table V). This supports the possibility that Pb differentially affects a subpopulation of B-lymphocytes or alters the lineage of B-lymphocytes.

T-lymphocyte activities also were differentially affected by Pb treatment. A central figure of CMI is the interaction of soluble mediators (lymphokines) from an activated subset of T-lymphocytes with macrophages to enhance their nonspecific bactericidal (Nathan *et al.*, 1971; Simon and Sheagrean, 1971) or bacteriostatic (Foules *et al.*, 1973) activity. The enhancement of macrophage bactericidal activity has been shown to be mediated by T cells (North, 1973). Challenge with the facultative intracellular bac-

TABLE 6. Effect of in vitro Lead (Pb) Treatment on the Mitogen-Reactivity of Spleen Cells[a]

	Mitogenic Response (cpm/culture)			
Mitogen	*--*[b]	10^{-6}*M*	10^{-5}*M*	10^{-4}*M*
--	*477*[c] ±*110*	*646* ±*181*	*799* ±*99*	*946* ±*192*
2-ME	*16,179* ±*702*	*17,292* ±*1,948*	*18,372* ±*1,119*	*19,750* ±*801*
Con A	*52,005* ±*2,849*	*57,489* ±*1,059*	*56,214* ±*2,301*	*60,843* ±*7,020*
PHA	*44,702* ±*2,229*	*43,818* ±*2,404*	*47,754* ±*2,329*	*62,983* ±*4,706*
LPS	*71,064* ±*1,189*	*76,142* ±*713*	*85,461* ±*3,895*	*91,306* ±*2,161*

[a]*The mitogenic response of* 2×10^5 *CBA/J spleen cells incubated with various concentrations of Pb mitogen was determined on day 2 by pulsing for 6 hours with* ^3H*-thymidine.*

[b]*Final concentrations of Pb chloride in culture*

[c]*The number represents the mean cpm ± standard error for triplicate cultures.*

teria *Listeria monocytogenes* has commonly been employed to monitor macrophage activation (Balnden *et al.*, 1969). Regulation of immunity to *Listeria* has been correlated with a cell that carries the Thy-1 alloantigen (T-cell), and treatment of immune spleen cells with anti-Thy-1 serum and complement abrogates the transference of adoptive immunity to *Listeria* (North, 1973) and macrophage mobilization at the sites of infection (North, 1973).

Since activation of macrophages is a central feature of CMI (Mackaness, 1970), investigation of the suppression or enhancement of the number of *Listeria* recovered from Pb-treated mice can provide evidence for evaluating the regulatory effects of Pb on CMI. Pb significantly reduced *in vivo* resistance to *Listeria*, which indicates that CMI (or T-lymphocyte activity) was suppressed; 0.5 mM-10 mM Pb treatment enhanced *Listeria*-mediated mortality. However, by another criterion for the assessment of T-lymphocyte activity, the *in vitro* MLC response, T-lymphocyte activity was enhanced by *in vivo* treatment with 0.08 mM-0.4 mM and slightly suppressed by 10 mM Pb acetate (Table 3 and 4). Although the 10 mM *in vivo* dose slightly suppressed the reactivity of the T-lymphocytes in the MLC response, no *in vivo* Pb dose suppressed the mitogenic response of the T-lymphocytes to Con A or PHA, selective T-cell mitogens (Greaves and Janossy, 1972). All *in vitro* doses $<5\times10^{-4}$M enhanced MLC reactivity and Con A and PHA responsiveness.

TABLE 7. Effects of Heavy Metals on Lymphocyte Proliferation[a]

Metal	--	2-ME	Con A	LPS
Pb 10^{-4}M	290	120	100	419
10^{-5}M	307	105	118	ND
10^{-6}M	235	155	112	145
10^{-8}M	75	172	88	150
Zn 10^{-4}M	12	0	0	5
10^{-5}M	13	ND	30	ND
10^{-6}M	30	134	104	121
10^{-8}M	38	147	97	ND
Hg 10^{-4}M	24	1	0	10
10^{-5}M	8	ND	1	ND
10^{-6}M	15	0	0	5
10^{-8}M	15	1	20	ND

[a]*Splenic lymphocytes 2×10^5/0.2 ml/well) were cultured for 3 days and their proliferation assessed by a 6 hour pulse with ^{3}H-thymidine (72-78 hrs). The cultures were done in triplicate.*

[b]*Control responses to -- (control; medium), 2-ME, Con A, and LPS were 430 ± 36, 25,55 ± 768, 80,250 ± 5,812, and 21,156 ± 688 cpm/culture ± standard deviation, respectively.*

% control response = experimental control 100. Variance within a group was consistently less than 10%. The numbers represent the mean of duplicate experiments.

Mitogens can activate a large portion of the T-lymphocytes, and although this stimulation must be considered nonspecific (Greaves and Janossy, 1972), the activity of mitogen-induced T-lymphocytes mimics, in part, the activity of antigen-activated T-lymphocytes. However, T-lymphocytes are heterogeneous, and at least three subpopulations exist (Ly-1, Ly-2,3, and Ly-1,2,3; Cantor and Boyse, 1975). Mitogens are known to stimulate multiple subpopulations, whereas distinct activities are known to be functions of specific subpopulations. For example, the major MLC responsive T-lymphocyte resides in the Ly-1$^+$ subpopulation (Cantor and Boyse, 1975). The T-lymphocyte responsible for macrophage activation in resistance to *Listeria* may reside in a different subpopulation. Therefore, Pb may have had differential effects, because it affected different subpopulations of T-lymphocytes. *In vivo* and *in vitro* Pb treatments did not substantially differ in their ability to alter the Con A or PHA responsiveness of T-lymphocytes possibly because different subpopulations of T-lymphocytes are not involved.

Unlike B-lymphocyte differences which seemed to be a result of *in vivo* versus *in vitro* assessment of the Pb effects as well as

TABLE 8. In vitro Effect of Lead (Pb) on Macrophage Binding and Phagocytosis

Cultures	% Rosettes Exp. 1	% Rosettes Exp. 2	% Phagocytosis Exp. 1	% Phagocytosis Exp. 2
Mϕ + SRBC	0	0	0.5	1.0
Mϕ + SRBC + Pb	5.5	0	1.0	2.0
Mϕ + Pb-SRBC	ND	2.0	ND	3.5
Mϕ + AbSRBC + Pb	19.0	0	60.0	77.0
Mϕ + SRBC + Pb	0	1.0	69.0	70.0
Mϕ + Pb-AbSRBC	ND	0	ND	75.0

[a]5 10^4 noninduced, adherent peritoneal cells (Mϕ) were cultured for 1 hr in the presence of sheep erythrocytes (SRBC) anti-SRBC:SRBC complexes (AbSRBC), SRBC preincubated with 10^{-4}M $PbCl_2$, or AbSRBC preincubated with 10^{-4}M $PbCl_2$. The percent of Mϕ rosetted with ≥3 SRBC or AbSRBC or phagocytosed ≥3 SRBC or AbSRBC was calculated as previously described (Mantovani, 1975).

[b]Each number represents the mean of duplicate cultures in which >100 cells were counted.

in vivo versus *in vitro* Pb treatment, *in vivo* versus *in vitro* Pb treatment did not result in substantial differences with T-lymphocytes. The differences in T-lymphocyte activity resulting from *in vivo* versus *in vitro* assessment could be due to the involvement of different subpopulations of T-lymphocytes, as posited above, or indirect *in vivo* effects could be involved. Pb has been shown to alter hepatic clearance of foreign matter (Trego *et al.*, 1972) and the phagocytic Kupffer cells of the liver were shown to be morphologically altered by Pb (Hoffman *et al.*, 1972). Pb alteration of antigen (SRBC or *Listeria*) clearance would indirectly alter HI and CMI. Pb reduction of hepatic clearance of endotoxin has been suggested as the cause of Pb enhancement of endotoxin-induced shock (Filkins and Buchanan, 1973; Seyberth, 1972) and may be the cause for Pb increasing susceptibility to pyrogenic bacteria as well as a concomitant direct or indirect Pb-induced reduction in HI (Selye *et al.*, 1966; Hemphill *et al.*, 1971; Cook *et al.*, 1974; Cook *et al.*

These indirect effects would not be applicable in *in vitro* assessment of lymphocyte activities; therefore, *in vitro* assessment would render a more accurate assessment of the direct effects of Pb on B- and T-lymphocyte functions. In addition, the effects of Pb on macrophage functions as shown in Table 8 must be done *in vitro*, especially if the effects of Pb on macrophage interaction with lymphocytes are to be assessed. Pb did not appear to significantly alter macrophage capture of antigen, but it may affect macrophage processing of antigen (Koller and Roan, 1977), which could influence lymphocyte activation as well as the effector phase of activated lymphocytes.

Hypothetically, the results could be interpreted to indicate the following: (a) Chronic subclinical Pb exposure may enhance B-lymphocyte proliferation and/or differentiation, which could lead to immunoregulatory loss of antibody synthesis. The end result could be Pb induction of autoimmune phenomena; (b) Chronic subclinical Pb exposure may enhance T-lymphocyte subpopulations involved in graft rejection (as seen in the enhancement of the MLC response). Thus, the end results could lead to 1) enhanced tumor rejection, 2) enhanced transplant rejection, or 3) alteration of a feto-maternal relationship; (c) Acute Pb exposure may transiently enhance or suppress HI and CMI.

The three different approaches employed in this study (Fig. 1) have shown that the methods employed can affect the influence Pb may have on the immune system. Pb did produce differential effects on the immune system, and the use of different experimental systems can be manipulated to answer various questions about the ability of Pb to alter B- and T-lymphocyte activities. If assessment of direct Pb effects on the lymphocyte activities is desired, *in vitro* assays should be employed. *In vitro* and *in vivo* Pb treatments will be required to evaluate direct acute versus chronic Pb exposure, respectively. In addition, *in vivo* Pb treatment is required to assess the influence of Pb on the ontogenic development of the B- and T-lymphocytes. Evaluation of the Pb effects on the immune system must correlate the *in vitro* and *in vivo* Pb effects and the subpopulations of lymphocytes actually being assessed in the assays employed. The ability of Pb to alter the development of B- and T-lymphocyte subpopulations is currently under investigation in this laboratory.

ACKNOWLEDGMENT

The author would like to thank Randolph J. Noelle, Allison Eastman, and Susan Robbins for their assistance in various aspects of the reported experiments, and Kathy Benedetto for her assistance in the preparation of the manuscript. Tables 2, 4, 5, 6 and 8 reprinted from *Infect. Immunity 31,* 136 (1981).

REFERENCES

Berger, N. A., and Skinner, A. M. (1974). Characterization of lymphocyte transformation induced by zinc ions. *J. Cell. Biol. 61,* 45.

Blanden, R. V., Lefford, M. J., and Mackaness, G. B. (1969). Host response to Bacillus Calmette-Guerin infection in mice. *J. Exp. Med. 129,* 1079.

Cantor, H., and Goyse, E. A. (1975). Functional subclasses of T-lymphocytes bearing different Ly antigens. I. The generation of functionally distinct T-cell subclasses is a differentiative process independent of antigen. *J. Exp. Med. 141,* 1376.

Caron, G. A., Poutala, S., and Provost, T. T. (1970). Lymphocyte transformation induced by inorganic and organic mercury. *Int. Arch. Allergy Appl. Immunol. 37,* 76.

Chisolm, J. J. (1971). Lead poisoning. *Sci. Amer. 224,* 15.

Cook, J. A., Marconi, E. A., and DiLuzio, N. R. (1974). Lead, cadmium, endotoxin interaction: Effect on mortality and hepatic function. *Toxicol. Appl. Pharmacol. 28,* 292.

Cook, J. A., Hoffman, E. O., and DiLuzio, N. R. (1975). Influence of lead and cadmium on the susceptibility of rats to bacterial challenge. *Proc. Soc. Exp. Biol. Med. 150,* 741.

Filkins, J. P., and Buchanan, B. J. (1973). Effects of lead acetate on sensitivity to shock, intravascular carbon and endotoxin clearances and hepatic endotoxin detoxification. *Proc. Soc. Exp. Biol. Med. 142,* 471.

Foules, R. E., Fajardo, I. M., Liebowitch, I. M., and David, J. R. (1973). The enhancement of macrophage bacteriostasis by products of activated lymphocytes. *J. Exp. Med. 138,* 952.

Fryer, H. C. (1966). *Concepts and Methods of Experimental Statistics.* Allyn and Bacon, Boston, p. 260.

Gainer, J. H. (1974). Lead aggravates viral disease and represses the antiviral activity of interferon inducers. *Environ. Health Perspect. 7,* 113.

Gainer, J. H. (1977). Effects of heavy metals and of deficiency of zinc on mortality rates in mice infected with encephalomyocarditis virus. *Am. J. Vet. Res. 38,* 869.

Golub, E. S., Mishell, R. I., Weigle, W. O., and Dutton, R. W. (1968). A modification of the hemolytic plaque assay for use with protein antigens. *J. Immunol. 100,* 133.

Goodman, M. G., Fidler, J. M., and Weigle, W. O. (1978). Nonspecific activation of murine lymphocytes. IV. Proliferation of a distinct, late maturing lymphocyte subpopulation induced by 2-mercaptoethanol. *J. Immunol. 121,* 1905.

Goyer, R. A. (1971). Lead toxicity: A problem in environmental pathology. *Amer. J. Pathol. 64,* 167.

Goyer, R. A., and Rhyne, B. C. (1973). Pathological effects of lead. *Int. Rev. Exp. Pathol. 12,* 1.

Greaves, M. F., and Janossy, G. (1972). Elicitation of T- and B-lymphocyte responses by cell surface binding ligands. *Transplant Rev. 11,* 87.

Hemphill, F. E., Kaeberle, M. L., and Buck, W. B. (1971). Lead suppression of mouse resistance to *Salmonella typhimurium. Science 172,* 1031.

Hoffman, E. O., Trejo, R. A., DiLuzio, N. R., and Lamberty, J. (1972). Ultrastructural alterations of liver and spleen following acute lead administration in rats. *J. Exp. Molec. Pathol. 17,* 159.

Jacquet, P., Leonard, A., and Gerber, G. B. (1975). Embryonic death in mouse due to lead exposure. *Experientia 31*, 1312.

Jerne, N. K., and Nordin, A. A. (1963). Plaque formation in agar by single antibody producing cells. *Science* (Wash., D.C.) *140*, 405.

Kehoe, R. A. (1976). Chemotherapy, toxicology and metabolic inhibition: lead. *Pharmacol. Ther. A1*, 161.

Koller, L. D. (1973). Immunosuppression produced by lead, cadmium and mercury. *Am. J. Vet. Res. 34*, 1457.

Koller, L. D., and Kovacic, S. (1974). Decreased antibody formation in mice exposed to lead. *Nature 250*, 148.

Koller, L. D., Exon, J. H., and Roan, J. G. (1976). Humoral antibody response in mice after single dose exposure to lead or cadmium. *Proc. Soc. Exp. Biol. Med. 151*, 339.

Koller, L. D., and Brauner, J. A. (1977). Decreased B cell response after exposure to lead and cadmium. *Toxicol. Appl. Pharmacol. 42*, 621.

Koller, L. D., and Roan, J. G. (1977). Effects of lead and cadmium on mouse peritoneal macrophages. *J. Reticuloendothel. Soc. 21*, 7.

Lawrence, D. A., Eastman, A., and Weigle, W. O. (1978). Murine T-cell preparations: radiosensitivity of helper activity. *Cell Immunol. 36*, 97.

Mackaness, G. B. (1970). The monocyte in cellular immunity. *Sem. Hematol. 7*, 172.

Mantovani, B. (1975). Different roles of IgG and complement receptors in phagocytosis by polymorphonuclear leukocytes. *J. Immunol. 115*, 15.

Mishell, R. I., and Dutton, R. W. (1967). Immunization of dissociated spleen cell cultures for normal mice. *J. Exp. Med. 126*, 423.

Mosier, D. E., Mond, J. J., and Goldings, E. A. (1977). The ontogeny of thymic independent antibody responses *in vitro* in normal mice and mice with an X-linked B cell defect. *J. Immunol. 119*, 874.

Muller, S., Gillert, K. E., Krause, C., *et al*. (1976). Suppression of delayed type hypersensitivity of mice by lead. *Experientia 33*, 667.

Nathan, C. F., Karnovsky, M. L., and David, J. R. (1971). Alterations of macrophage functions by mediators of lymphocytes. *J. Exp. Med. 133*, 1356.

North, R. J. (1973). Importance of thymus-derived lymphocytes in cell-mediated immunity to infection. *Cell Immunol. 7*, 166.

Raffel, S. (1961). *Immunity*, Chapter 1. Appleton-Century-Crofts, New York.

Rosenthal, A. S. (1978). Determinant selection and macrophage function in genetical control of the immune response. *Immunol. Rev. 40*, 136.

Schell, R. F., and Lawerence, D. A. (1977). Differential effects of concanavalin A and phytohemagglutinin on murine immunity. Suppression and enhancement of cell-mediated immunity. *Cell Immunol. 31*, 142.

Selye, H., Tuchweber, B., and Bertok, L. (1966). Effect of lead acetate on susceptibility of rats to bacterial endotoxins. *J. Bacteriol. 91,* 884.

Seyberth, H. W., Schmidt-Gayk, H., and Hackental, E. (1972). Toxicity, clearance and distribution of endotoxin in mice as influenced by actinomycin D, cycloheximide, alpha-amanatin and lead acetate. *Toxicon. 10,* 491.

Simon, H. B., and Sheagrean, J. N. (1971). Cellular immunity *in vitro.* I. Immunologically mediated enhancement of macrophage bactericidal capacity. *J. Exp. Med. 133,* 1377.

Trego, R. A., DiLuzio, N. R., Loose, L. D., and Hoffman, E. (1972). Reticuloendothelial and hepatic functional alterations following lead acetate administration. *Exp. Mol. Pathol. 17,* 145.

Treves, A. J. (1978). *In vitro* induction of cell-mediated immunity against tumor cells by antigen-fed macrophages. *Immunol. Rev. 40,* 205.

Vos, J. G. (1977). Immune suppression as related to toxicology. *CRC Critical Rev. Toxicol. 5,* 67.

Waldron, H. A., and Stofen, D. (1974). *Sub-Clinical Lead Poisoning.* Academic Press, New York.

A STUDY OF FILTER PENETRATION BY LEAD IN NEW YORK CITY AIR

T. J. Kneip
M. T. Kleinman
J. Gorczynski
M. Lippmann

New York University Medical Center
New York, New York

INTRODUCTION

Data reported by Robinson *et al.* (1974) and Seeley *et al.* (1974) suggested that previous air sampling studies might have seriously underestimated airborne lead concentrations. Their data and more recent data of Skogerboe *et al.* (1977) indicated that a major fraction of airborne lead could pass through membrane or glass fiber filters. These investigators did not attempt to determine the species that they called "filterable" lead, but suggested that it might be either particles of molecular size or a vapor of some inorganic species.

The possible existence of lead in the ambient atmosphere in a form which penetrates "high efficiency" sampling filters is of serious concern, since it calls into doubt virtually all of the existing data on atmospheric lead concentrations. Because of our interest in airborne particulates in general, and airborne lead in particular, we undertook to address some of the questions relating to this problem. Glass fiber sampling filters were selected for the field studies, since the bulk of the available data on ambient airborne lead levels in the United States has been obtained using such filters.

In this study, we have used a cyclone pre-selector and three filters in series, followed by a split stream with a diffusion battery and charcoal trap on one side, and acid scrubbers on the other side, in order to obtain the following concentrations.

A. Total Pb (i.e., all of the Pb collected in the sampling train)

B. Particulate Pb > 1.0 μm (i.e., Pb collected in a 10 mm nylon cyclone with a 1 μm cut-size)

ISBN 0-12-460520-6

C. Particulate Pb < 1.0 μm (i.e., Pb collected by in-line filters downstream of the cyclone)

D. Particulate Pb < 0.05 μm (i.e., Pb which penetrates the filter and is collected by a diffusion battery which collects >83% of particles (<0.05 μm)

E. Vapor Pb (i.e., Pb which passes both filters and diffusion battery and is collected by a charcoal trap)

F. Total acid soluble Pb (i.e., Pb which passes the filters and is collected by an acid scrubber)

The information obtained was used to determine the extent to which Pb in ambient air can penetrate filters.

EXPERIMENTAL

Filter Media and Test Conditions

Cellulose acetate membrane filters (Millipore Type AA and Type SC), glass fiber filters (Gelman Type AE) and cellulose fiber filters (Whatman 41) were tested for retentivity of the sodium fluorescein particles of ≤0.07 μm mmad at face velocities ranging from 0.84 cm/sec. to approximately 25 cm/sec.

Gelman A, AE, E and Spectrograde A filters all have essentially the same fiber diameter, porosity and collection efficiency characteristics (Lundgren and Gunderson, 1975). The face velocities used in these tests were chosen to overlap those reported by Lockhart *et al.* (1964) and to extend the measurements to flow ranges reported by Seeley, *et al.* (1974). The percent penetration was determined from the fraction of aerosol collection on a glass fiber (Gelman AE) backup filter in-line and downstream of the filter being tested. The total aerosol was taken as the sum of the material on the test and backup filters. If there is any penetration of both filters, the efficiencies reported would be higher than the true values.

Generation of Test Aerosols

Sodium fluorescein aerosols were generated for laboratory tests of filter penetration using an Environmental Research Corporation Atomizer-Impactor Model 7300 system with a Wright nebulizer (Wright, 1958). The size of the aerosol generated is a function of the concentration of sodium fluorescein in 0.1 N ammonium hydroxide in the atomizer (Stöber and Flachsbart, 1973). A spiral centrifuge (Stöber, 1967) was used to confirm the relationships between sodium fluorescein concentrations and the aerodynamic size of the test aerosols generated.

We also used the spiral centrifuge as a preselector (with its backup filter removed) downstream from the atomizer-impactor. Particles of ≤0.07 μm pass through the spiral duct in sufficient concentrations to be used for the collection efficiency determinations in our test systems.

Sampling System Design

A diagram of the field sampler is shown in Fig. 1. A 10 mm nylon cyclone was operated at 9 L/min., where its aerodynamic cut-size is ∿1.0 μm (Blachman and Lippmann, 1974). By removing the larger particles from the air stream, the filter loading effects were reduced. Three glass fiber filters in series were used to collect the submicrometer particulate fraction (Gelman-Spectro Grade). Thoron (radon-220, $t_{1/2}$ = 55s) released from a generator containing 10 g thorium nitrate in 100 ml HNO_3, was introduced with the sample at the entrance to a mixing chamber. Thoron rapidly decays to Pb-212 atoms (10.4 h $t_{1/2}$) which are nearly instantly deposited on any available surface, including the surface of ambient aerosols. The "tagged" aerosols provide a means of assessing the efficiency of each stage of the train.

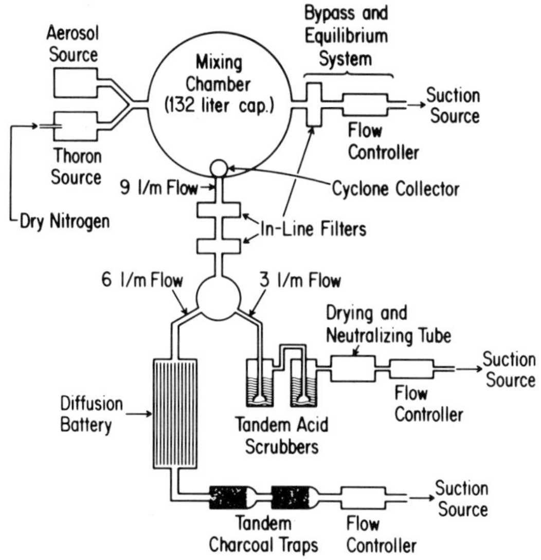

Fig. 1. Schematic of field sampler. Sampling stages and flow rates are indicated. Pressure drop measured at an orifice at each flow control and across cyclone to establish and control flow rates; bypass system was adjusted to 8.5-9.5 l ppm as required to balance the system. Dry nitrogen was adjusted to a flow of 1-2 bubbles a second through the thoron generator.

Any of the extremely small lead particles which penetrate the filters were collected in a 100 channel parallel plate diffusion battery. The battery is 50 centimeters in length and 20 centimeters in width, with 0.10 centimeters between channels. It has an efficiency of approximately 83% for particles ≤0.05 μm diameter at the flow rate used in this sampling system (Knutson, 1977, personal communication).

Vapor lead species passing the filter and diffusion battery were collected in the charcoal trap. Identification of lead species on the charcoal was not attempted. The charcoal trap was 2.5 cm in diameter and 9.5 cm long, and was operated at 6 l/min, conditions recommended by Brandt and Ter Haar (1974).

The other parallel fraction of the effluent from the filter was passed through two acid scrubbers in series for comparison with similar experiments reported by Skogerboe *et al.* (1977). The flow rates used were recommended by Skogerboe as providing good efficiency for vapors and particulate lead species. A scrubbing solution of 60 ml of 50% nitric acid was made up from redistilled nitric acid (G. F. Smith Co.) and distilled-deionized water. A summary description of flow rates and nominal collection characteristics is given in Table 1.

A high volume (Hi Vol) air sample was collected by a second sampler during several of the sampling periods. The high volume sampler is an independent unit which collects a sample at a constant 0.57 m^3/min flow rate on a 20.3 by 25.4 cm glass fiber filter (Kneip *et al.*, 1970).

TABLE 1. Operating Flow Rates

Sampler	*Flow Rate Liters per Minute*	*Retention Characteristics*	*References*
Cyclone	*9.0*	*50% at 1.0 μm*	*Blachman and Lippmann, 1974*
Filter 1-3 (Glass fiber, Gelman Type A)	*9.0*[a]	*99%*	*Lockhardt, 1964 Lippmann, 1978*
Diffusion Battery	*6.0*	*83% at ≤0.05 μm*	*Knutson, E., 1977*
Charcoal Traps	*6.0*	*100%*[b]	*Brandt and Ter Haar, 1974*
Acid Scrubbers	*3.0*	*High* [c]	*Skogerboe, R. K., et al., 1977*

[a]*Face valocity 1.5 cm/sec. Data this report.*
[b]*For tetramethyl and tetraethyl lead.*
[c]*Indicates no lead recovered in 3rd scrubber in series.*

Sampling Locations

Samples were collected at the New York University Medical Center, 32nd Street between First Avenue and the East River Drive, at rooftop (50 m) and near ground level (4 m), both at about 25 m west of the East River Drive. Samples containing freshly generated motor vehicle exhaust aerosols were also collected at a station about 9.3 m directly above and between the lanes of the State Thruway at a location 10 km north of Suffern, New York, a suburban community with essentially no industrial sources of lead nearby.

Analyses

The Pb-212 was measured by a scintillation counter with a NaI(Tl) well crystal. The sampling filters and a filter blank were oxidized in a low temperature asher for 6-1/2 hours. Each filter went through a series of three extractions with 50 ml HNO_3, 1:1 HNO_3 and deionized H_2O. The extracts were filtered and analyzed on an IL 453 atomic absorption spectrophotometer.

Each scrubber solution along with an appropriate blank was evaporated to approximately 2 to 4 ml and brought to a volume of 10 ml for atomic absorption spectrophotometry (AAS) analysis. The activated carbon from the charcoal trap was weighed and extracted sequentially with HNO_3 and HNO_3/HCl followed by a boiling water extraction. The extract was digested to destroy organics, filtered, evaporated to about 1 to 2 ml and diluted to 10 ml with 5% HNO_3. Analysis was by AAS using a carbon rod (Varian Model 63 with IL-151) using background correction. The diffusion battery was washed with water, the washings digested with hot concentrated HNO_3 and the resulting solution brought to volume in a 25 ml volumetric flask with 5% HNO_3.

Recovery of organic vapor lead species was attempted with the HNO_3 scrubbers; however, only a few percent of ambient lead concentrations can exist as the tetramethyl or tetraethyl forms (Skogerboe *et al.*, 1977). The charcoal samples were processed to recover only inorganic forms of lead.

Corrections for Radon Decay

The mixing chamber was designed to provide sufficient residence time to achieve secular equilibrium between the parent nuclide RN-220 and the daughter, Pb-212. However, there will always be a small fraction of undecayed RN-220 passing through the system which can decay at a later stage. The second filter is separated from the first by only a sheet of tea paper; thus, there is a negligible residence time between filters 1 and 2.

The ratio of activity on filter 3 to that on filter 1 is a maximum of 0.012 (Kneip *et al.*, 1979), based on a 132 L mixing volume before the first filter, 220 ml volume between filters 1 and

3, and the radioactive decay law. There may be very little aerosol available downstream of filter 1 to which the newly formed Pb-212 daughters can attach, and a significant fraction of daughter atoms formed may adhere to the walls of tubing and filter holders in this region; thus, ratios found may be artificially low. The major volume beyond the third filter is in a 5 L mixing chamber which preceeds both the diffusion battery-charcoal trap branch, and the acid scrubber branch. Decay in this region could also produce measurable activity, but the wall losses may be even more significant in these regions of the sampler than downstream of the first filter.

RESULTS

Filter Efficiency

The percent of the aerosol which penetrated the filters in the laboratory tests is shown in Figure 2. The data indicate that both cellulose acetate Type AA and SC membrane filters and Gelman Type AE glass fiber filters are better than 99% efficient for collection of ≤0.07 μm diameter particles at all face velocities tested with essentially zero filter loading. Whatman 41 exhibited considerably lower retentivities, averaging between 51 and 72%.

Total Lead Comparison

The comparison between the Hi Vol and special sampler total lead data indicates generally satisfactory agreement. A paired t-test shows no significant difference in the total lead values obtained by the two samplers for five periods, with values of 1.43±0.3 and 1.54±0.4 for the Hi-Vol and train samples, respectively. This provides assurance that the inlet systems are reasonably similar in particle size acceptance characteristics.

Error Calculations

A series of calculations have been performed in order to define the detection limits for each stage. These values are necessary to evaluate the significance of the measurements in terms of penetration of the first filter by lead containing particle or vapor species.

In order to minimize false negatives, the criteria of Currie (1973) have been used to calculate the limiting value, designated "critical limit," L_C. For a "well known blank" this is 1.64 times the standard deviation above the mean blank for $p = 0.05$. Where our blanks are not fully and clearly established we may be accepting a larger risk than necessary of false positives: however this is a conservative approach in examining the data for potential penetration of the filter by lead species.

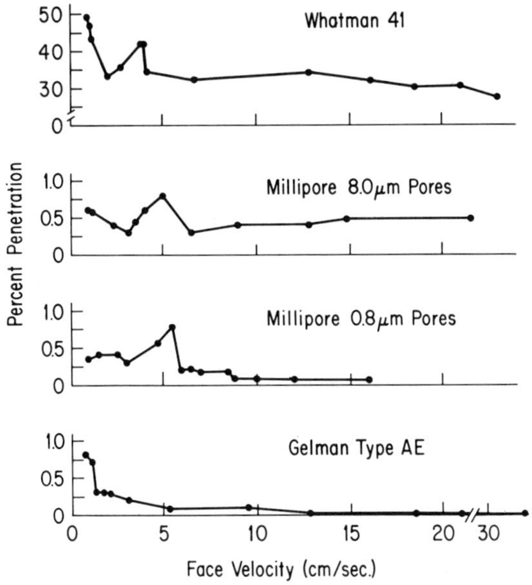

Fig. 2. Comparison of filtration efficiency of filter media commonly used in atmospheric sampling. Particles ≤0.07 mmad as described in text.

Lead-212. Errors in the measurements of Lead-212 are combinations of sampling variations and counting errors. The latter can be calculated from the total counts accumulated for sample fractions and background samples.

These results are calculated by means of the following equations:

$$S_{Sample}(\text{cpm}) = \frac{\sqrt{\text{counts}}}{\text{counting time}}$$

$$S_{blank}(\text{cpm}) = \frac{\sqrt{\text{counts}}}{\text{counting time}}$$

$$S_{tot}(\text{cpm}) = \sqrt{(S_{sample})^2 + (S_{blank})^2}$$

$$S_{tot}(\text{cpm/m}^3) = \frac{S_{tot}(\text{cpm})}{\text{Sample Volume (m}^3)}$$

The values obtained are generally small in comparison to the sum of the activity found in the cyclone and first filter. Estimates of particle penetration based on activity exceeding these limits or maximum estimates based on the limits should provide an adequate evaluation of this factor. As errors other than counting have not been accounted for, any conclusions regarding penetration will be conservative.

Stable Lead. Stable lead analyses were not possible on diffusion battery washes, as contamination originating from soldered seams caused high blanks. Among the rest of the sampler sections only the scrubber solutions caused unusual difficulty with detection limits. The data in Table 2 show that the glass scrubbers had large blank values for laboratory runs with dry nitrogen. Both laboratory and field blanks were in the range of 0.4 to 0.66 μg total lead when gas was not passed through the scrubbers. Much less lead was introduced from plastic scrubbers. One of five laboratory N_2 runs gave a value of 1.35 μm total lead which was rejected by Chauvenet's criterion (Kneip *et al.*, 1979). The overall blank for 14 determinations including four laboratory runs with dry nitrogen was then 0.53±0.1 with a critical limit of 0.69 μg.

Critical limits are given in Table 3 for typical flow rates and sampling times which gave ~6 m^3 through the cyclone and filters,

TABLE 2. Scrubber Solution Blank Values

	Lead, μg(n) 1975	Lead, μg(n) 1976
HNO_3 (Conc.)	0.27±0.02 (4)	0.27±0.03 (3)
Laboratory Blank[(a)]	0.43±0.03 (5)	-
Field Blank[(a)]	0.55±0.04 (4)	0.54±0.12 (6)
Glass Scrubbers[(b)]		
Laboratory 1	4.0±0.3 (3)	-
Laboratory 2	9.6±6.8 (3)	-
Field 1	2.7±1.3 (4)	-
Field 2	8.6±8.9 (4)	-
Plastic Scrubbers[(b)]		
Laboratory 1	-	0.66±0.4 (5)
Laboratory 2	-	0.98±0.4 (5)
Field 1	-	0.74±0.2 (5)
Field 2	-	0.90±0.3 (6)

(a) Laboratory and field banks involved filling, transfer and storage of the acid only.

(b) Scrubber measurements in the laboratory with dry N_2 gas, in the field with ambient air.

TABLE 3. Critical Limits for Stable Lead Determinations

Sampling Stage	*Lc* *Pb, μg*	*Blank* *Pb, μg*	*Detection Limit* *Pb, μg/m³* (m)
Cyclone	*0.13*	*0.02 ±0.07*	*0.002 (6)*
Filter	*2.65*	*2.14 ±0.31*	*0.008 (6)*
Charcoal 1976	*0.778*	*0.653±0.076*	*0.03 (4)*
Absorber 1975-77	*1.23*	*0.845±0.317*	*0.09 (4)*
Scrubber, HNO_3	*0.69*	*0.53 ±0.1*	*0.08 (2)*

∿4 m^3 through the diffusion battery and charcoal traps, and ∿2 m^3 through the acid scrubbers. The data for 1976 resulted in critical limits sufficiently low to provide a satisfactory test of the fractions of lead species penetrating the first filter.

DISCUSSION

Particulate Lead

Sample volumes have been used in conjunction with the detection limit data for each sampler stage for each run to calculate the penetration values for lead-212 and stable lead respectively. Sampling runs were carried out for a period of approximately 12 hours. Variations in operating time resulted in sample volumes ranging from 5.5 to 6.5 m^3. This affects the detection limits to only a slight extent. Shifts in the limits are observed for the changes in scrubbers and charcoal detection limits discussed above.

Filter efficiencies increase with increased particle loading (Davies, 1973); therefore, in order to achieve loadings as low as those reported to be associated with filter penetration by airborne lead, we operated a 10 mm cyclone with a 1 μm cut-size upstream of the first filter, as previously discussed. The cyclone should collect a significant fraction of the mass. The results shown in Table 4 confirm the success of this approach, with an average of 35.5% of the stable lead collected in the cyclone. With one exception, a much larger fraction of stable lead than of Lead-212 is found on particles exceeding 1 μm mmad.

An approximate mass removed by the cyclone or the filter can be calculated by assuming a concentration of about 1% Pb in the aerosol mass and equal removal for both lead and total mass. Calculated filter loadings for the field runs of some 7-8 μg/cm^2 compares well to Skogerboe's sample runs (1977), which can be estimated to give 2-4 μg/cm^2. Normal filter loadings for 24-hr Hi-Vol

TABLE 4. Tracer Fractions Collected by Stage

Run	1	2	3	4	5	6	7	8	9	10
Date	10/27/75	9/1/76	10/25/76	12/15/75	7/21/76	8/18/76	4/18/77	11/20/75	12/1/75	7/7/76
Location	N.Y. State Thruway				NYC - 1st Floor			NYC 14th Floor		
Pb-212 cpm/m^3	6,099	1,086	2,185	8,980	9,257	6,212	1,439	12,742	7,634	16,074
Cyclone, %	0.8	0.7	0.3	2.3	0.9	2.1	6.0	21.4	1.4	3.9
Filter 1, %	97.8	98.8	98.9	97.2	98.5	96.7	93.6	78.4	98.2	95.1
Sum	98.6	99.5	99.2	99.5	99.4	98.8	99.6	99.8	99.6	99.0
Stable Lead, μg/m^3	2.49	1.62	2.76	2.13	1.62	1.22	2.14	1.47	1.23	1.60
Cyclone, %	37.4	32.1	6.5	55.2	32.7	48.4	38.8	29.9	33.0	41.9
Filter 1, %	62.6	67.9	93.5	41.1	67.3	51.6	61.2	70.1	67.0	58.1

samples can range from 60 to 210 μg/cm^2 for 30 to 100 μg/m^3 TSP values.

The activity data were evaluated by comparison to the decay criteria. As given previously, they relate expected ingrowth of daughters to residence time in the train. Filters 1 and 2 were mounted on a sheet of tea paper separating the glass fiber filters. The tea paper was counted with filter 2, and activity was found in only 2 cases (0.2 and 0.3% of total activity). Actual ratios of $R_{3/1}$ range from 0.001 to 0.006, well below the maximum 0.012 calculated earlier.

Similar calculations can be applied to each successive stage of the sampling system. As with the $R_{3/1}$ values, these estimates demonstrate that the Pb-212 activity measured downstream of filter 1 with one exception (Run No. 10) can be attributed to the small amount of Rn-220 decaying in the sampler volume behind filter 1, and hence that determinations of Pb penetrations based on uncorrected Pb-212 data would be conservative. (Activity measurements in the diffusion battery were possible on only four occasions where available counting time permitted measurement before decay had proceeded for too long a period of time.)

Lead-212 activity on filters 2 and 3, the diffusion battery and charcoal, and the scrubbers afford the most sensitive means of evaluation of particle penetration. The measured lead-212 activity for filters 2 and 3 ranged from 0.1 to 0.6% of the total activity. Data for diffusion battery-charcoal branch also shows a maximum of 0.6% while that for the scrubbers has a maximum of 0.8%.

All of these activities are less than what is calculated for radon decay beyond filter 1 and are therefore maximum indicators of particle penetration. Summing the filters and scrubbers or filters, diffusion battery and charcoal (including less than figures at the stated values) gives a range of 0.4 to 1.6% penetration for the New York State Thruway samples and all others fall in this same range. Using only detected values, the range is 0.3 to 1.2%. These estimates are biased high by the failure to subtract the decay correction, but they fully support the many laboratory studies which demonstrate a maximum fractional penetration of a few tenths of 1% of total particles. The average uncorrected collection efficiency is 99.3% for the cyclone-filter sum (Table IV). Corrected for decay and deleting values below detection limits, this would be 100% for retention of particles, with the exception of Run No. 10 at 99.6%.

Stable Lead. The stable lead data do not provide as sensitive a means of determining penetration of the filter as was obtained with the lead-212 tag for particles. Only the scrubbers indicate the possibility of penetration by a species other than a particulate form of lead (Table 5).

The filter data give a range of detection limits of 3.3 to 6.7% with no measured penetration, supporting the conclusion based

TABLE 5. *Lead Collected on Each Stage as Represented by Stable Lead Found, Percent*

Location	Thruway (Suburban)			Ground Level (NYC)				Rooftop (NYC)		
Date	10/27/75	9/1/76	10/25/76	12/15/75	8/18/76	4/18/77	7/21/76	11/20/75	12/1/75	7/7/76
Cyclone	37.4	32.1 (28.4)[a]	6.5	55.2	48.4	29.9 (27.2)	32.7	38.8	33.0	41.9 (38.5)
Filter 1	62.6	67.9 (60.1)	93.5	41.1	51.6	70.1 (63.6)	67.3	61.2	67.0	58.1 (53.4)
2	<3.3	<4.6	<3.4	<3.7	<6.7	<5.6	<4.9	<3.7	<6.4	<4.9
3	<3.3	<4.6	<3.4	<3.7	<6.7	<5.6	<4.9	<3.7	<6.4	<4.9
Scrubber 1	-	(11.5)	Lost	-	<6.3	(9.3)	<4.6	-	-	(8.0)
Charcoal	<1.2	<1.9	<1.1	Lost	<2.5	<2.1	<1.90	<1.4	<2.4	<1.80
Total Lead g/m^3	2.49	1.62 (1.83)	2.76	2.13	1.22	1.47 (1.62)	1.62	2.14	1.23	1.60 (1.74)

[a]Percentage values shown in parentheses are based on total lead values including scrubber 1 for the three runs as shown.

on the lead-212 activity. The charcoal traps indicate a maximum of 2.5% penetration of lead past the three filters and the diffusion battery.

Vapor Lead. Both scrubber and charcoal data clearly demonstrate the difficulty of avoiding lead contamination in field experiments. Multinanogram quantities of lead are only too readily picked up from materials of construction or from the ever present ambient aerosol which contains 1 or more percent of lead.

The three measured values noted for the scrubbers (8.0, 9.3, and 11.5%) can only be a vapor form, as the lead-212 data indicate maximum penetration of particles of less than 1% especially beyond the second and third filters. The detection limit for the scrubbers is in the range of 4 to 6%, and considerable data indicate that contamination by the scrubber itself may be a problem. We believe that these results do not represent lead from the atmosphere sampled; however, they indicate a maximum penetration by vapor species of some 10% of total lead.

Despite the contamination problems, evidence from the charcoal traps is convincing that the penetration by inorganic vapor forms of lead does not reach 10% of total lead and more likely is less than 2%.

Several misconceptions or misinterpretations have appeared in the literature concerning filter collection of lead aerosols. In each case reliable data are available to correct the mistake. The principle misconceptions include:

1. Filtration mechanisms are similar in air and water filtration, i.e., air filters function as sieves.

In air filtration, the dominant mechanisms are usually impaction and diffusion (Lippmann, 1978). Other mechanisms, i.e., interception and electrostatic precipitation may also be important in special cases. The selection of 0.3 μm DOP as a test aerosol by Lockhart *et al.* (1964) was based on their appreciation that this size is representative of the size range which is most difficult to collect. The collection of smaller particles may be expected to be more efficient because of increased collection by diffusion, while the collection efficiency for larger particles may be expected to be increased because of more efficient deposition by impaction.

The data of Lockhart *et al.* (1964) indicate >99.9% retentivity by cellulose ester membrane filters with 0.8 μm pores (Millipore AA) and glass fiber filters (Gelman Type A and MSA 1106B) for 0.3 μm DOP droplets at face velocities of 7.2 to 283 cm/sec.

2. Glass fiber filters are so contaminated as to preclude their use in lead sampling.

This statement may be based on the work of Kometani *et al.* (1972), where the filter material was totally dissolved with HF. It does not apply to modern, high purity filters and analytical methods which do not dissolve the glass.

3. Cohen (1973) reported results on collected masses and concluded that the data indicated increasing penetration of Hi Vol filters with increasing face velocities.

Cohen actually measured increasing masses collected as face velocities decreased. He made no measurements of penetration. The data are readily interpreted as showing increasing large particle impaction losses on the filter shelter as flow rates increase, although this explanation is not supported by any measurements either.

4. Many past tests of filter penetration relied on assumptions of absolute collection by backup filters and thus may be in error.

Actually many workers have used photometric or particle counting systems, rather than backup filters. Several examples are:

Lockhart, L. B. *et al.*	(1964)	Photometric
Megaw, W. J. and Wiffen, R. D.	(1963)	Particle Counter Radioactivity Counter
Stenhouse, J. I. T., *et al.*	(1970)	Flame Photometer
Liu, B. Y. H. and Lee, K. W.	(1976)	Electrical Aerosol Detector

Specifically, the broad early study by Lockhart (1964) which showed efficiencies exceeding 99+% for glass fiber and high efficiency membrane filters, was based on a continuous detector capable of measuring penetrations of 0.001%.

5. Various publications show penetration of filters by particles; therefore, field data showing penetration of membrane or glass fiber filters by lead is not unusual.

Papers discussing the retention characteristics of some cellulose filters, Nuclepore filters, or large pore membranes (Lockhart *et al.*, 1964; Rimberg, 1969; Stafford and Ettinger, 1971; Liu and Lee, 1976) cannot be used to justify data indicating penetration of high retentivity membrane or glass fiber filters. No accurate studies are known which show penetration of these filters exceeding a few tenths of 1% of particle numbers or mass.

CONCLUSIONS

Our laboratory measurements show >99% collection efficiencies for solid particles of $\leq$0.07 m mmad on both membrane (Millipore AA and SC) and glass fiber (Gelman Type AE) filters at very low loadings and flow rates to less than 1 cm/sec face velocities.

Ten field samples were taken with a specially designed sampling system for periods of 10 to 12 hours at sampling sites on the median strip of a heavily traveled six-lane highway at the location of a rural rest stop, and at first floor and 14th floor levels near the East River Drive in New York City. Data were obtained for both a lead-212 tag and for stable lead. The data demonstrate that a maximum of a few tenths of 1% of total lead can penetrate a glass fiber filter in particulate form, at low filter loadings and at a low face velocity at these locations.

Contamination problems in scrubber collectors make the use of this system for ambient air samplings questionable. The maximum estimate of vapor species penetration based on the scrubber data would be 10% of total lead. It is very probable that any inorganic vapor species penetrating the filter constitute less than 2% of the total lead based on the additional data obtained in a parallel charcoal absorber system.

The historical lead data obtained by analysis of high retentivity filter samples is appropriate for the evaluation of the exposure hazard to humans. There is no convincing evidence of any additional hazard due to lead species capable of penetrating high retentivity filters.

ACKNOWLEDGMENT

This research was performed with the support of International Lead Zinc Research Organization, Inc., Grant No. LH 231, and is part of a center program supported by Grant No. ES00260 from the National Institute of Environmental Health Sciences and Grant No. CA13343 from the National Cancer Institute.

REFERENCES

Blachman, M. W., and Lippmann, M. (1974). Performance characteristics of the multicyclone aerosol sampler. *Amer. Ind. Hyg. Assoc. J. 35,* 311-326.

Brandt, M., and Ter Haar, G. (1974). An evaluation of filters and activated carbon for collection of lead in air. Presented at EPA Seminar *Analysis of Various Forms of Atmospheric Lead.* Ethyl Corporation Research Laboratories, Detroit, Michigan, September 16.

Cohen, A. L. (1973). Dependence of Hi-Vol measurements on air flow rate. *Environ. Sci. Technol. 7,* 60.

Currie, L. A. (1973). Limits for qualitative detection and quantitative determination. *Anal. Chem. 40,* 586-593.

Davies, C. N. (1973). *Air Filtration,* Academic Press, New York.

Kneip, T. J., Eisenbud, M., Strehlow, C. D., and Freudenthal, P. C. (1970). Airborne particulates in New York City. *Air Poll. Con. Assoc. J. 20*(3), 144-149.

Kneip, T. J., Kleinman, M. T., Gorczynski, J., and Lippmann, M. (1979). A study of filter penetration by lead in New York City air. Final report on Project LH-231 Air Sampling-Particulate. International Lead Research Organization, Inc., 292 Madison Avenue, New York, New York 10017.

Kometani, T. Y., Bove, J. L., Nathanson, B., Siebenberg, S., and Magyar, M. (1972). Dry ashing of airborne particulate matter on paper and glass fiber filters for trace metal analysis by atomic absorption spectrometry. *Environ. Sci. Technol. 6,* 617-620.

Knutson, E. (1977). Personal communication.

Lippmann, M. (1978). Filter media for air sampling. *Air Sampling Instruments,* p. N-10, Fifth ed.

Liu, B. Y. H., and Lee, K. W. (1976). Efficiency of membrane and nucleopore filters for submicrometer aerosols. *Environ. Sci. and Tech. 10,* 345-350.

Lockhart, L. B., Jr., Patterson, R. I., Jr. and Anderson, W. L. (1964). Characterization of air filter media used for monitoring airborne radioactivity, Naval Research Laboratory Report NRL 6054.

Lundgren, D. A., and Gunderson, T. C. (1975). Efficiency and loading characteristics of EPA's high-temperature quartz fiber filter media. *Amer. Ind. Hyg. Assoc. J. 36,* 806-872.

Megaw, W. J., and Wiffen, R. D. (1963). The efficiency of membrane filters. *Int. J. Air Wat. Poll. 7,* 501-509.

Rimberg, D. (1969). Penetration of IPC 1478, Whatman 41, and Type 5G filter paper as a function of particle size and velocity. *Amer. Ind. Hyg. Assoc. J. 30,* 394-401.

Robinson, J. W., and Wolcott, D. K. (1974). Simultaneous determination of particulate and molecular lead in the atmosphere. *Environ. Lett. 6*(4), 321-333.

Seely, R., and Skogerboe, R. L. (1974). Combined sampling analysis method for the determination of trace elements in atmospheric particles. *Anal. Chem. 46,* 415-421.

Skogerboe, R. K., Dick, D. L., and Lamothe, P. J. (1977). Evaluation of filter inefficiencies for particulate collection under low loading conditions. *Atmos. Environ. 11,* 243-249.

Stafford, R. G., and Ettinger, H. J. (1971). Comparison of filter media against liquid and solid aerosols. *Amer. Ind. Hyg. Assoc. J. 32,* 319-326.

Stenhouse, J. I. T., Harrop, J. A., and Freshwater, D. C. (1970). The mechanisms of particle capture in gas filters. *Aerosol Sci. 1,* 41-52.

Stöber, W. (1967). Design and performance of a size separating aerosol centrifuge facilitating particle size spectrometry in the submicron range. *In* "Assessment of Airborne Radioactivity, pp. 393-402, International Atomic Energy Agency, Vienna.

Stöber, W., and Flachsbart, H. (1973). An evaluation of nebulized ammonium fluorescein as a laboratory aerosol. *Atmospheric Environment (7),* 737-748.

Wright, B. M. (1958). A new nebulizer. *Lancet,* 24-25.

DISCUSSION OF PAPER BY DR. THEODORE J. KNEIP

Chamberlain: I think Dr. Kneip has put the question of particle penetration through the filters fairly well to rest. There is this other thing as to what the charcoal fraction, assuming it is organic vapor, amounts to either in percentage or in $\mu g/m^3$. Do you have anything to say on that as regards to New York atmosphere?

Kneip: Well, we actually could have extracted the charcoal or done some other procedure. If there was some there, we probably got some. What we did was a straight nitric acid dissolution. Other people have used other halides, added or mixed acids, or extracted with a solvent first. We didn't try to do those. Even in the reports by Skogerboe, it's pretty well agreed that the maximum possible vapor, organic vapor, species is probably 10 percent of the total. The actual numbers run 3 or 4 percent in both his data and in other people's data where they've looked specifically for the organic species.

Wixson: The thing I noticed on your slides is you had a number of collections made in June, July, and August, and you contrasted those with some made in January and February. Was there any seasonal differences noticeable during the field collection?

Kneip: There was the one collection where the fraction of the tagged aerosol collected in the cyclone was much higher than all the others, just on that one day. But for overall total penetration, most of the data were less than the detection limit. We've simply taken those and added up the detection limits to find out what the maximum could have been, and that doesn't seem to have anything to do with season or location. We had three runs at the thruway where we were dealing, presumably, with a fairly fresh aerosol in between two high speed lanes. We also had four runs at ground level and three at the rooftop. Those two sites are 50 meters different in height, and none of them show anything spectacular. They all look like they're one set of data.

Wixson: They don't steal our hi-vol samplers, but they shoot them up, so we have several problems.

Ter Haar: I congratulate you on the thoroughness of the approach you used in answering what I agree was a problem we all thought was solved. I think this problem you've been working on is typical of one that we are facing in analytical chemistry today, that is, as new techniques come along that can measure 1/100th or 1/1000th of the amounts that classical analytical chemistry has, some different answers are coming up. This includes, in some cases, the measurement of lead in tissue and that sort of thing. I think it is incumbent on anyone who wishes to use these new methods to demonstrate that they are comparable with the accepted methods of analytical chemistry, and, in most cases, the analytical chemists are doing that. If they have done the sort of things you've done here, they wouldn't come to these kinds of conclusions. I mean that frequently the more micro methods are not checked against the macro methods in a side by side comparison as you've made here to demonstrate that they're comparable.

Kneip: The message I tried to get across is that the problem is not so much in measuring as it is in sampling. There are problems of sampling and handling samples in the field for something like lead because if you touch anything out in the field or in the city, you've got lead on it.

Ter Haar: That's right, but if you're sampling two thousand cubic meters, your problems are trivial. Many people are now recognizing the enormous problems that they have to confront when sampling two or three liters.

EFFECTS OF LEAD ON AQUATIC LIFE

Leonard J. Warren

Commonwealth Scientific and Industrial
Research Organization (CSIRO)
Port Melbourne, Australia

In recent years there has been a major research effort into the effects of the so-called "heavy metals" on natural ecosystems. The size of the effort may be gauged by the many papers presented at conferences such as the 1973 Nashville Conference,[1] and the 1975 Toronto Conference.[2]

Clearly, the non-human aspects of heavy metals in our environment are of international concern. In the U.S., the National Science Foundation has funded at least two major projects, each of five years duration, costing several million dollars. I recently heard Professor Tim Parsons describing some of the results of the NSF Saanich Inlet Study.[3] His talk reinforced my own conviction that it is very difficult to determine ecological changes resulting from long-term, sub-lethal doses of contaminants and yet, as the Saanich Inlet work revealed, those changes can sometimes be important. Parsons showed that elevated levels of copper in seawater slowly changed the predominantly diatom ecology to a predominantly flagellate ecology, which means, in terms of the higher members of the food chain, a switch from fish to jellyfish--and that, for those of us still higher in the food chain who prefer fish, is a result of some importance.

ILZRO also recognized the need for studies of the natural environment and chose to initiate work on the effects of heavy metals on aquatic life. In view of the 1977 NSF report "Lead in the Environment" edited by Boggess and Wixson,[4] which concentrated on the terrestrial environment, the emphasis on aquatic ecosystems is timely. Furthermore, the need to abide by or challenge environmental legislation makes it imperative to accumulate sound scientific knowledge in this area, knowledge which as far as possible should be beyond dispute.

ILZRO commissioned first a state-of-the-art review on the effects of heavy metals on aquatic life. That comprehensive review, by Dr. Bob James,[5] has provided the background for the CSIRO research effort which commenced formally in January of this year.

ISBN 0-12-460520-6

CSIRO has formed a "Heavy Metals Task Force," whose chairman is Mr. Ivan Newnham, Director of the CSIRO Minerals Research Laboratories. The task force (Table 1) comprises scientists from a range of disciplines that it would be hard to find in most research organizations. However, the problem is a complex one and needs a multi-disciplinary approach. In the first year, the task force has used in particular the skills of its chemists and geochemists. In subsequent years, it is expected that the biologist and the ecologist will assume a more dominant role.

As an aid to our research planning, we used some of the techniques of technological forecasting (Fig. 1). Seven possible world scenarios were identified, the most probable selected, and three policy options flowing from that scenario identified. By proceeding in this way from the general to the specific, we identified strategies, then objectives, then projects, and finally 8000 experiments associated with those projects. A "relevance tree" was

TABLE 1. Members of the CSIRO Task Force on Heavy Metals

J. BAULD	—	MICROBIAL ECOLOGIST
B.V. BUBELA	—	BIOGEOCHEMIST
D. GARDNER	—	MARINE CHEMIST
D.J. SWAINE	—	GEOCHEMIST
K.G. TILLER	—	SOIL SCIENTIST
T.J. WARD	—	MARINE BIOLOGIST
L.J. WARREN	—	COLLOID CHEMIST

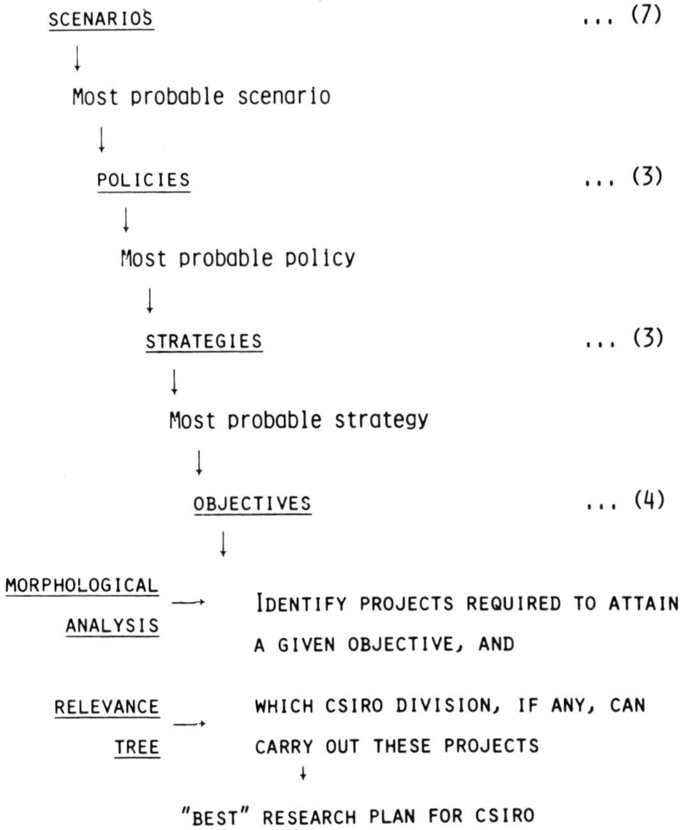

Fig. 1. Research planning by technological forecasting: an aid to CSIRO research.

used to identify CSIRO capabilities for achieving given objectives and to pinpoint possible difficulties in coordinating the work of geographically and academically separated scientists.

Our research plan indicated that assistance was required in several areas and the ILZRO grant was used to appoint a marine biologist and four technical assistants to work in the areas indicated in Table 2. The deployment of the assistants will change as the project develops.

The basic objective set by the task force was to establish sound scientific knowledge of the effect on marine life of effluents and emissions from lead and zinc smelters. The resources provided by ILZRO are being used specifically to augment a thorough study of the ecosystem around Port Pirie, South Australia (Fig. 2). This decision was taken because of the unique location of Port Pirie lead smelter[6] in a rural environment, the historical accumu-

TABLE 2. Deployment of ILZRO Appointees During 1978

RESEARCH SCIENTIST	UPTAKE OF HEAVY METALS BY AQUATIC LIFE
TECHNICAL ASSISTANTS	1. CLEAN AIR LABORATORY
	2. ANALYSIS OF SEA GRASSES AND AQUATIC LIFE
	3. AVAILABILITY OF METALS FROM SEDIMENTS
	4. SIMULATION OF TIDAL ENVIRONMENT AND ALGAL MATS

lation of heavy metals in the area, and the knowledge that, while general scientific principles underlie all environmental problems, each particular pollution study will be different according to the nature of the local ecosystem.

The smelter is situated on the eastern side of Spencer Gulf. The Gulf floor shelves gradually to a deeper channel on the western

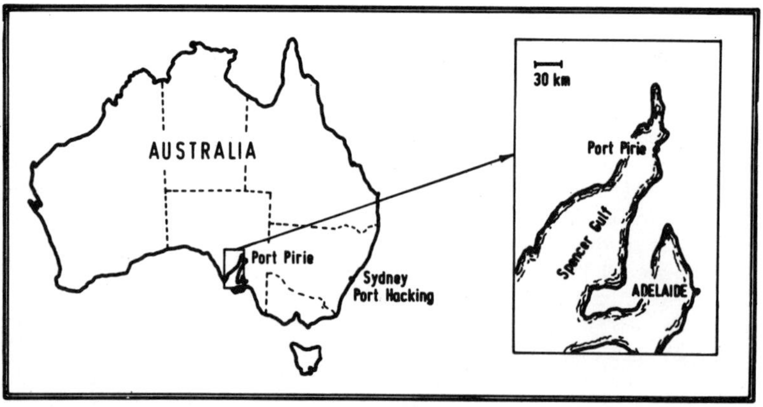

Fig. 2. Location map for Spencer Gulf and Port Pirie.

Fig. 3. View of smelter from shipping channel.

side, so that in the vicinity of the smelter ships are confined to a narrow, periodically dredged channel (Fig. 3). Contamination occurs through ship spillage, smelter stack emissions, fugitive dusts, and a liquid effluent stream.

The liquid effluent stream cuts through tidal flats which surround the smelter and cover the eastern shores of the Gulf (Fig. 4). The intertidal areas are thought to be important breeding grounds for fish and prawns.

Fig. 4. View of tidal flats near the Port Pirie smelter.

Fig. 5. Plastic sediment sampling tube in position in sea-grass bed.

On one sampling expedition we used four scuba divers and a local fisheries research vessel.[7] The divers forced plastic tubes into the surficial sediments, capped the top, withdrew the tube carefully, and then capped the bottom. Figure 5 shows a sampling tube in position in a bed of the dominant sea-grass in the area, Posidonia *australis*. Other sampling expeditions collected near-surface water samples and deep sediment cores.[8]

In many ways, the key to environmental studies is correct analysis. The reproducibility and accuracy of the analytical procedures must be properly tested. Those procedures, listed in Table 3, represent several man-years of work by different scientists in different CSIRO laboratories. For samples with higher lead contents, atomic absorption spectrometry (AAS) is suitable, providing that a correction is made for non-atomic background absorption by the calcium carbonate in the sediment. For lower concentrations, the AAS results are checked periodically against independent methods. Polarographic determinations are made on background samples. Anodic stripping voltammetry (ASV) is used for sea-waters and interstitial waters.

Sample contamination during handling and analysis was not found to be a serious problem for polluted sediments and sea grasses, but for background samples and all water samples, great care was necessary. A "class 100" clean air room, therefore was constructed. Filtered air flows through the perforated wall shown at the back of the room which houses the AAS (Fig. 6). An adjoining semi-clean room was built for the ASV to minimize mercury contamination.

TABLE 3. Analytical Methods Used for Different Concentration Ranges of Lead

SAMPLE	METHOD OF ANALYSIS FOR Pb		
	DIGESTION	PRE-CONCENTRATION	ANALYSIS
CONTAMINATED SEDIMENTS, SEA GRASSES (> 50 μg/g)	HNO_3, $HClO_4$ HNO_3, $HClO_4$, HF 0.1 M HCl	APDC/MIBK — —	AAS WITH BACKGROUND CORRECTION
OTHER SEDIMENTS, SEA GRASSES (< 50 μg/g)	"	"	CHECK AAS WITH ISOTOPE DILUTION, POLAROGRAPHY, GRAPHITE FURNACE
SEA WATERS INTERSTITIAL WATER (< 50 μg/litre)	pH 4.8 HNO_3	Hg DROP OR FILM	ASV

There are a number of possible sources of heavy metals in Spencer Gulf, but as far as lead, zinc, and cadmium are concerned, it is the smelter at Port Pirie which makes the major contribution

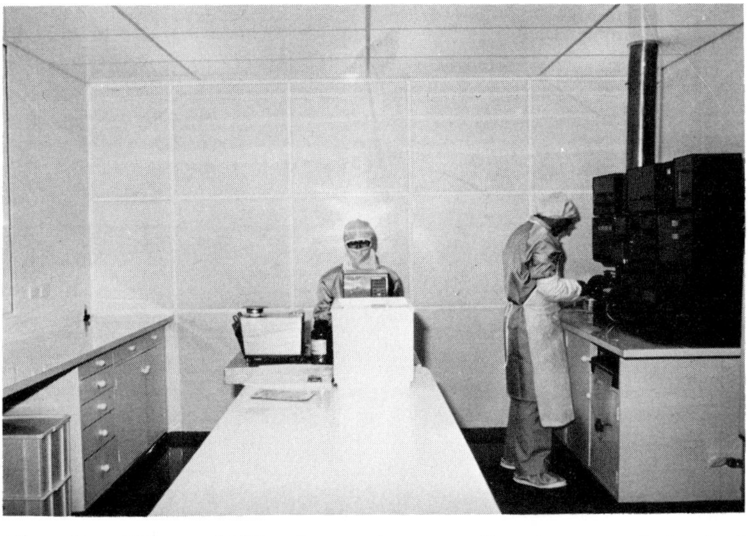

Fig. 6. "Class 100" clean air room for trace metal studies.

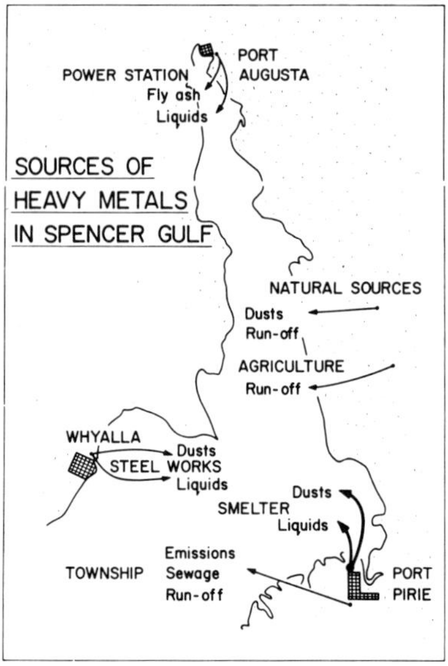

Fig. 7. Sources of heavy metals in Spencer Gulf.

(Fig. 7). The Gulf is about 30 km wide near Port Pirie, the townships are small, the countryside sparsely settled. The coal fired power station at Port Augusta is a potential source of metals through its fly ash emissions and we analyzed these in detail. However, the results showed that fly ash from the "Stack Inlet" (Table 4) contained 60 ppm Pb and at these levels the fly

TABLE 4. Trace Elements in Fly Ash from Port Augusta Power Station

(as $\mu g\ g^{-1}$)

	FLY - ASH FROM:		
	ELECTROSTATIC PRECIPITATORS	STACK INLET	COAL ASH
As	5 , 6	7·5	7·5
Cd	0·3 , 0·35	0·5	0·5
Cu	70 , 90	100	100
Hg	0·12 , 0·12	3·3	(0·4)*
Ni	40 , 40	30	40
Pb	60 , 80	60	40

* Determined in coal; calculated to ash-basis

ash is unlikely to contribute significantly to the lead concentrations some 60 km to the south, around Port Pirie.

The concentration of lead in uncontaminated areas of the Gulf is on average about 0.40 μg/l (Fig. 8). This compares favorably with the value of 0.37 μg/l determined in a pristine environment near Sydney on the east coast of Australia.[9] Contamination is most severe where the liquid effluent stream from the smelter enters the Gulf.

On the other hand, the sediments of the Gulf are much more widely contaminated. The zone of pollution extends north from the Port Pirie River roughly in an arc following the channel (Fig. 9). However, sediments with high lead were also found in shallow areas well away from the shipping channel. We are still in the process of analyzing these results, but it does appear that there is no direct relation between the concentration of lead in the sediment and that in the overlying surface water some 3-6 meters above.

Free swimming aquatic species are, therefore, likely to be affected directly only if they enter the contaminated water zone around the effluent stream. Bottom-dwelling species will be exposed to lead levels about 10 times that of the background concentration of 15 μg/g in the much more extensive zone containing the contaminated sediments.

In another sampling expedition, we used a "Vibrocore" to obtain deep sediment cores from selected locations in the Gulf.[8]

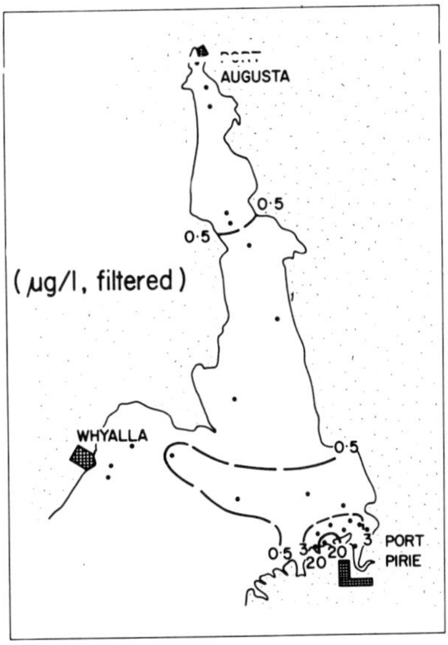

Fig. 8. Isoconcentration lines for lead in near-surface waters of Spencer Gulf.

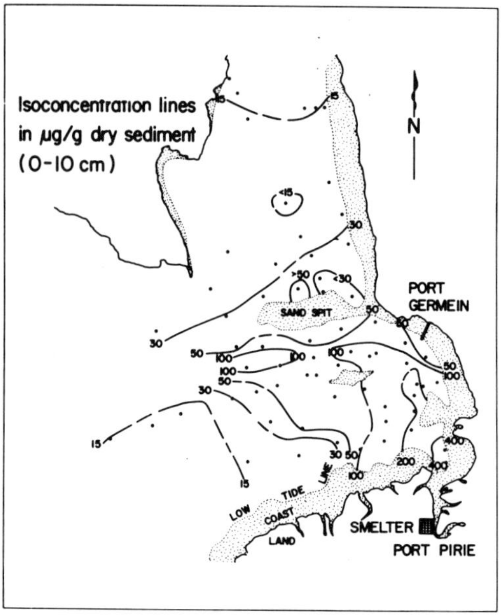

Fig. 9. Isoconcentration lines for lead in Spencer Gulf sediments.

The core from Station 1 was fairly typical with the topmost 150 cm being a sub-tidal sediment consisting mainly of shell grit, followed by an intertidal band and then red clay down to 300 cm (Fig. 10).

Cores from contaminated zones showed considerable enrichment in the top 10 cm, whereas those from uncontaminated areas, such as Station 5, had more uniform metal levels down the whole length of the core (Fig. 11). These results indicate that the high surficial lead and zinc are of recent origin, consistent with the fact that the smelter began operations 80 years ago. The results also show that the metal has not percolated downwards into the underlying sedimentary column but has remained at the water-sediment interface. Further cores are being analyzed because of the implications of these preliminary results for aquatic life. It appears that because of the low rate of natural sedimentation in the Gulf, metals accumulating on the sea floor are not automatically buried and immobilized, but remain in the top layer, where they are potentially available to aquatic species.

Lead enters Spencer Gulf either from air or in waters and may follow a variety of paths to and fro between the various geographic compartments shown in Fig. 12. One geographic compartment which has not been well-studied in the past, but which is of particular importance at Port Pirie, is that of the tidal flats or marshes.

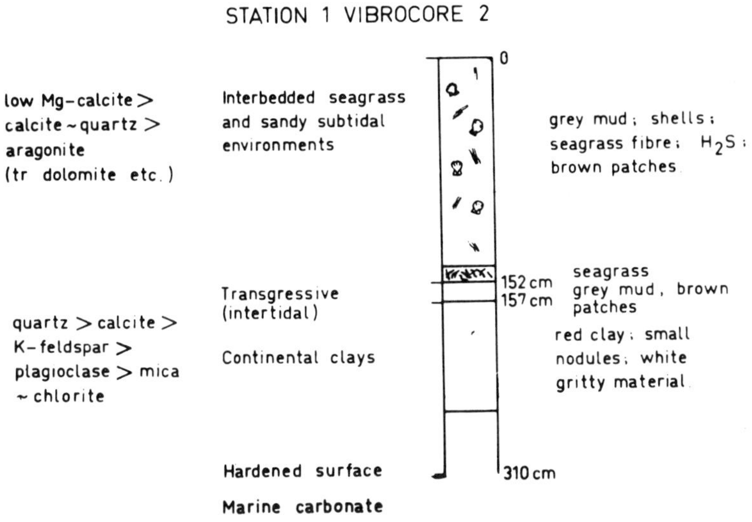

Fig. 10. Composition of a typical deep sediment core from Spencer Gulf.

These are important not only in the geochemical cycle of heavy metals but also in their biological cycle.

In many areas the tidal flats are covered by a thin surface layer of living filamentous blue-green algae (Fig. 13). Underneath the algal mat is a black zone of active sulfate reduction. The filamentous nature of the algae is evident in the electron

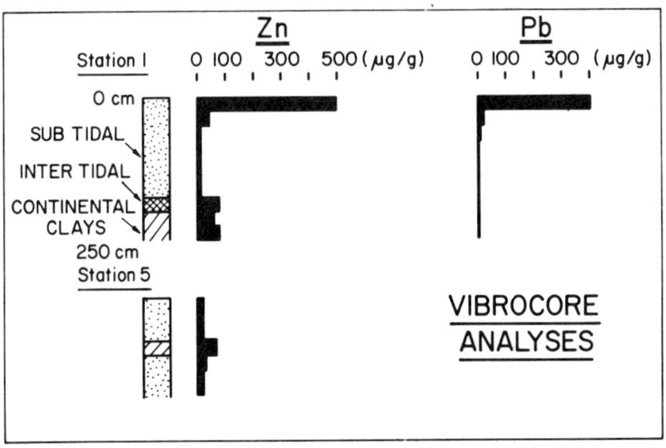

Fig. 11. Metal distribution down deep sediment cores.

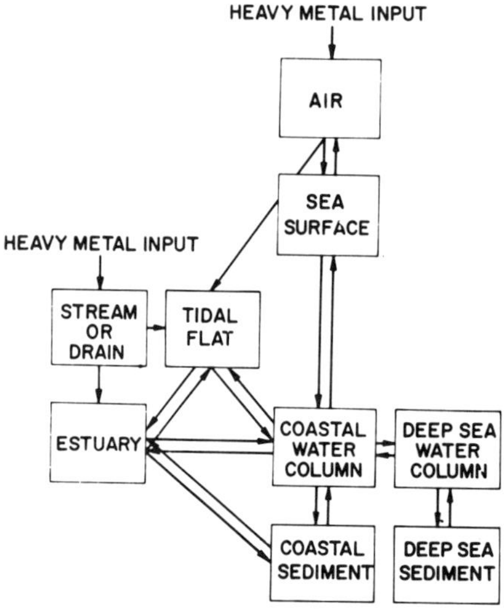

Fig. 12. Heavy metal pathways between geographic compartments in Spencer Gulf.

microscope photograph (Fig. 14), which also reveals the polysaccharide sheath around each filament. Growth of the algae depends on such things as the supply of sunlight and CO_2, the salinity of the receding tidal waters, temperature, season, and so on. The movement of tidal flows and of ground waters must also be considered (Fig. 15).

Field measurements have shown that the rate at which the algal mats fix CO_2 is about 60 m M $m^{-2}d^{-1}$. In the process, the algae releases dissolved and particulate organic carbon, which becomes the primary energy source for the sulfate-reducing bacteria "Desulfovibrio". This bacteria in turn produces H_2S at a rate which, within experimental error, is about the same as the rate at which it is supplied with organic carbon (Fig. 16). At present, the effect of lead and zinc on these carbon and sulphur cycles is under study.

The field observations were difficult to interpret because of the periodic inundations of the marsh sediments and changes in temperature and algal growth during diurnal cycles. A laboratory apparatus was constructed to simulate the periodic movement of waters through a tidal zone. Figure 17 shows the tank we have used to simulate evaporative sedimentary environments near Port Pirie. The simulation tank is automatically controlled (Fig. 18) and is designed to operate unattended for long periods. A particular ad-

Fig. 13. View of tidal flats covered by blue-green algae.

vantage has been the suitability of the system for measuring permeability and porosity of sediments with and without algal growths.

The first six-month of a two-year simulation trial have shown that in many respects the algal mat can be considered to act as a semipermeable membrane (Fig. 19). It is an effective gas barrier trapping H_2S below the mat and excluding air from above. Preliminary experiments with copper and zinc have indicated that metal ions can pass freely through the algal mat only if they are not complexed by organic ligands such as EDTA. Further work is in progress on the biology of the marsh sediments and the role of permeability in the transport of nutrients and metals.

From the overall picture which we have built up of the aquatic environment, it is clear that aquatic life will be exposed to two main sources of non-biological lead, lead in the water and lead from the sediments (Fig. 20). We know these sources exist and we have quantified them to a certain degree, whereas we are not yet sure of the importance of metal transfer and accumulation through the food chain, which is why the biological pathways are not indicated in Fig. 20. Even if the present emissions of lead to the

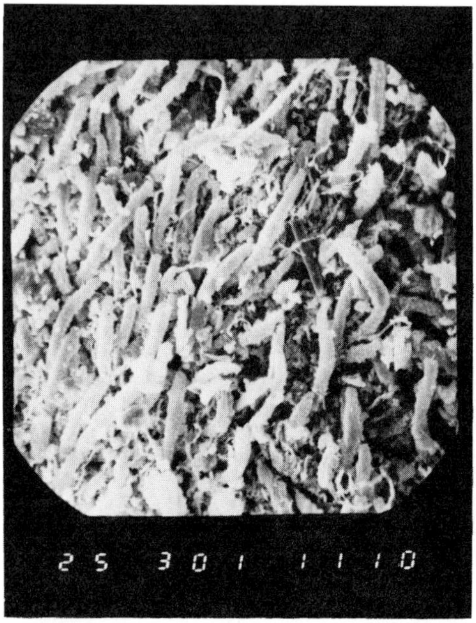

Fig. 14. Scanning electron micrograph of blue-green algae collected near Port Pirie (×30).

water column were reduced to negligible values, the reservoir of lead in the sediments would remain, and, at least in the Port Pirie area would remain exposed on the top of the sediment.

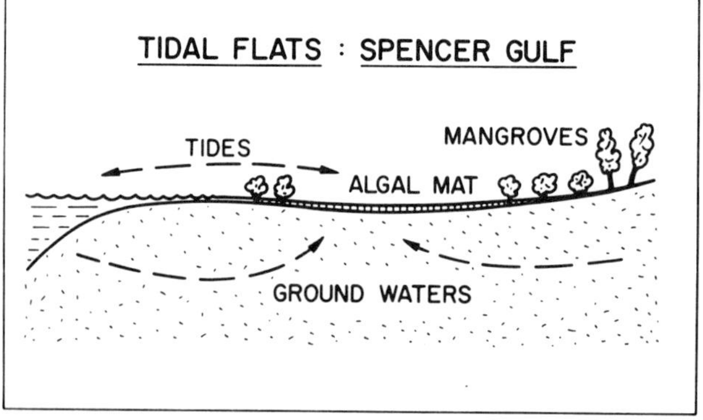

Fig. 15. Diagrammatic cross-section of tidal flats near Port Pirie.

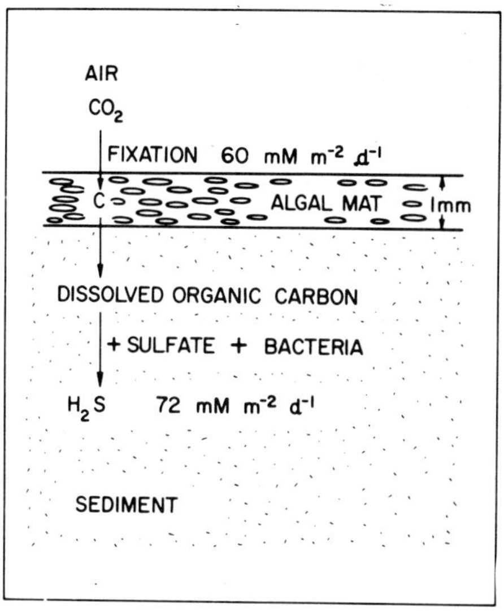

Fig. 16. Carbon and sulfur production in algal mats.

Figure 21 shows a metal balance diagram for fish. Those fish which are bottom feeders will take up particulate lead from the sediment by ingestion. Lead is also absorbed through the gills and body surface and ingested with food. The balance of uptake, storage and excretion will determine the steady state level of lead in the fish. The same principles will apply to other aquatic species. Because of the widespread contamination of the Gulf sediments and the size of this potential source of lead, we have investigated in some detail the location of lead in the components of the sediment and its biological availability.

We have developed a new method for looking at sediments, the basis of which is that the sediment components differing in specific gravity will separate into distinct bands in a density gradient in a suitable heavy liquid. After centrifugation bands of particles of similar density collect at the interfaces between the heavy liquid layers (Fig. 22).

The technique gives a clean separation of most of the sediment components without changing them chemically in any way. For example, the topmost layer consists mostly of fibrous organic debris (Fig. 23), presumably derived from the seagrass Posidonia. Then comes a layer of conglomerate particles, each a mixture of fine magnesian calcite, mica, kaolin, quartz, feldspar, and goethite. The next layer was mainly quartz with some hexagonal calcitic rods (Fig. 23 shows a calcitic rod) and shell fragments. The fourth layer consisted of whole microshells and shell fragments

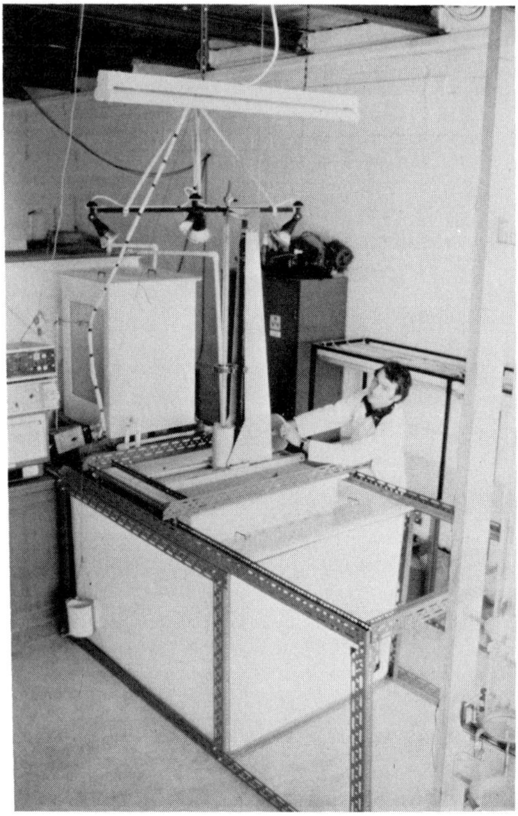

Fig. 17. View of tank constructed to simulate evaporative sedimentary environments near Port Pirie.

of magnesian calcite, while the fifth layer was entirely aragonitic shell pieces. Detrital heavy minerals were found in the dense particles which settled to the bottom of the tube.

Each density band was then analyzed for its total content of Zn, Pb and Cd, and significant differences were found in the metal levels of the various sediment components. The sample shown in Fig. 24 was taken from a sea grass bed and had 6-10 times more heavy metal in the organic debris than in the shell fragments and whole microshells. There was also a preferential rejection of heavy metal by those shells composed of aragonite, which had 2½ times less metal than those made of magnesian calcite. It appears that Pb, Zn and Cd are incorporated into the shell structure as the animal grows.

It could be argued that those sediment components with the highest metal content are the most important to the local ecosystem.

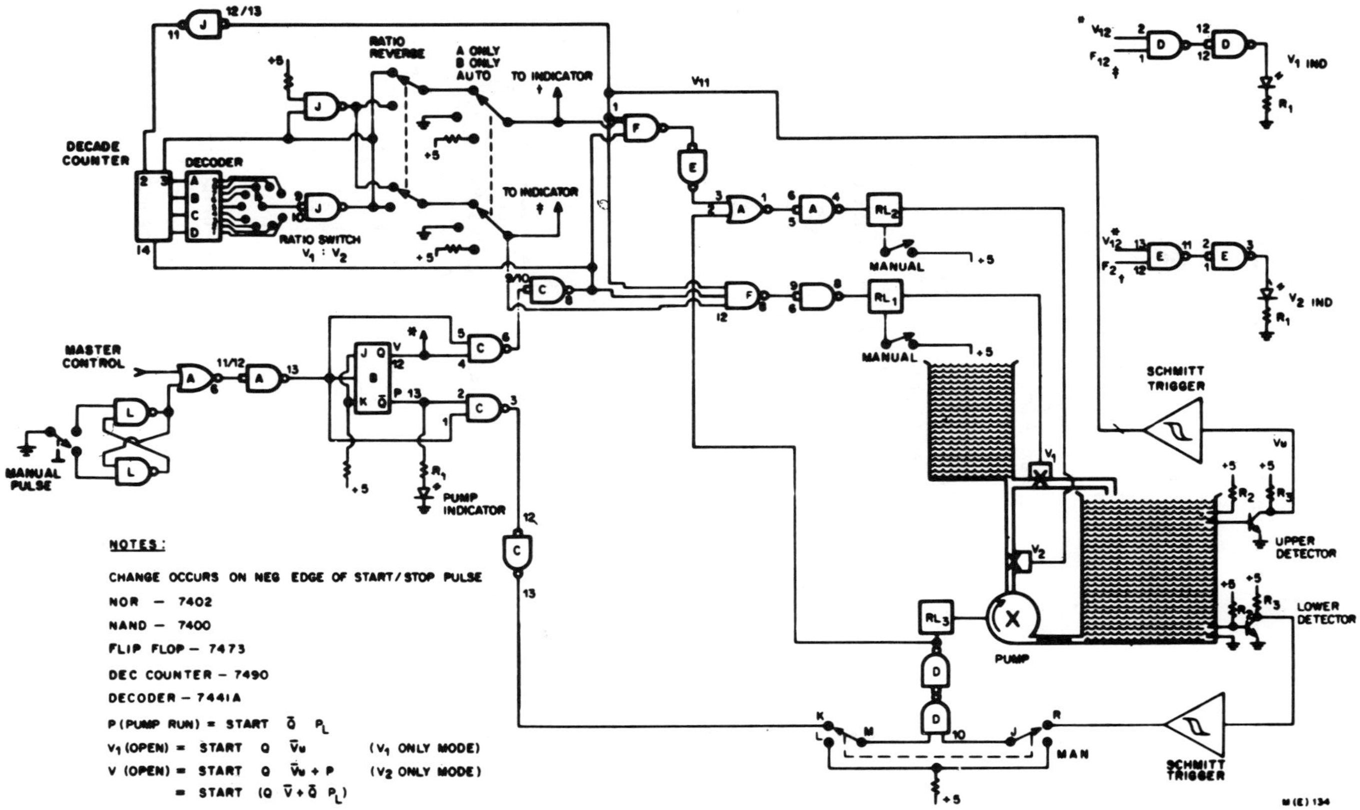

Fig. 18. Automatic hydraulic controls for simulation tank.

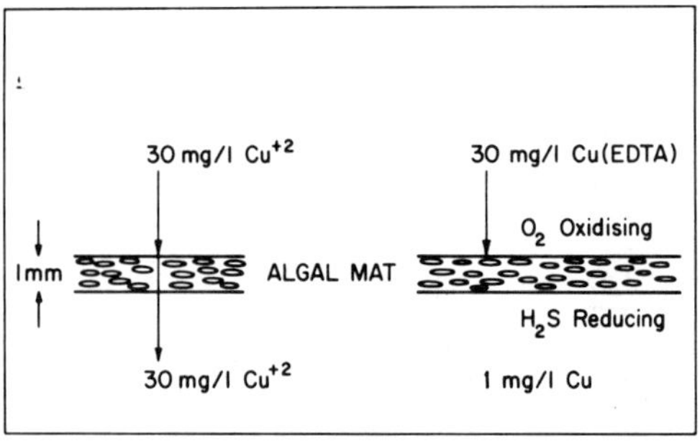

Fig. 19. Algal mat as a semi-permeable membrane.

However, this ignores the ease with which heavy metals may be extracted and the fact that the total amount of heavy metal contained in one component may be less than the total amount in another more abundant component with a lower concentration of heavy metal.

In sediment ES 35, for example, taken from a barren, shallow trench, the heavy minerals sub-fraction contained 790 ppm of lead, but contributed only 1.5% of the total lead content of the sediment (Fig. 25). The environmental significance of the heavy mineral lead would be reduced still further if it could be shown that

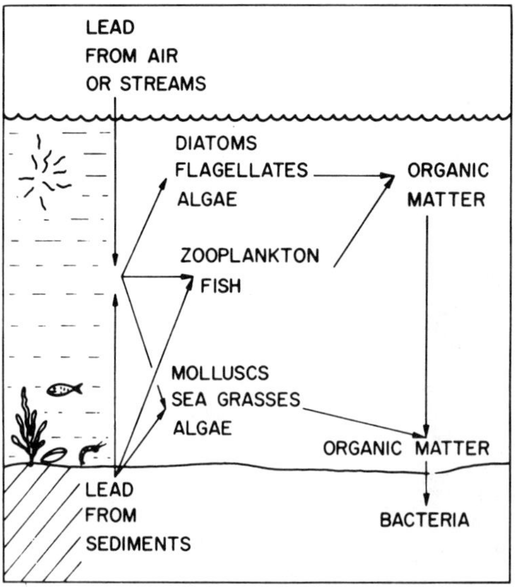

Fig. 20. Primary sources of lead in the marine ecosystem.

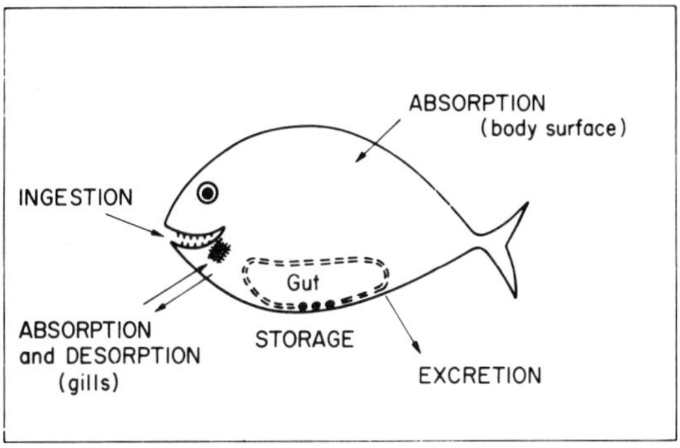

Fig. 21. Uptake of lead by fish.

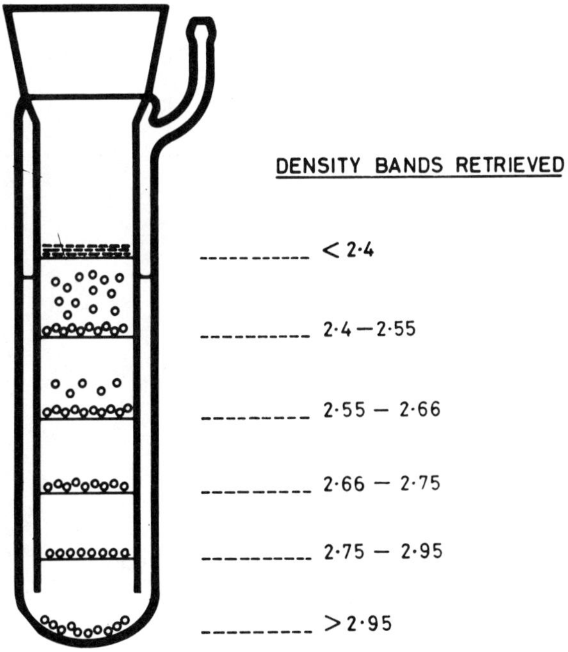

Fig. 22. Apparatus used to separate sediment components.

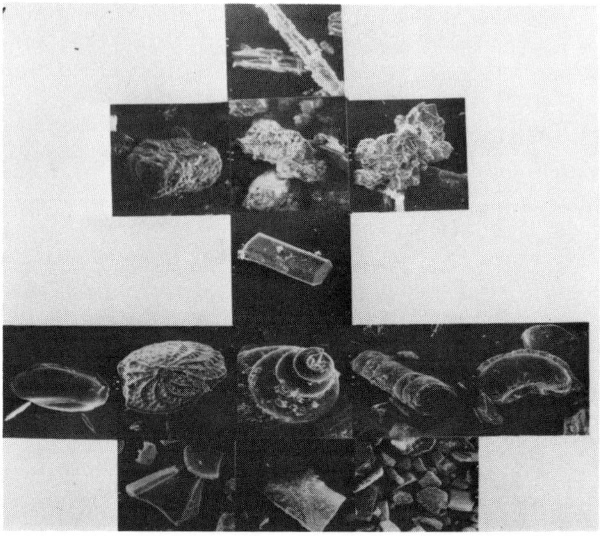

Fig. 23. SEM photographs of particles isolated in five density bands of a Spencer Gulf sediment.

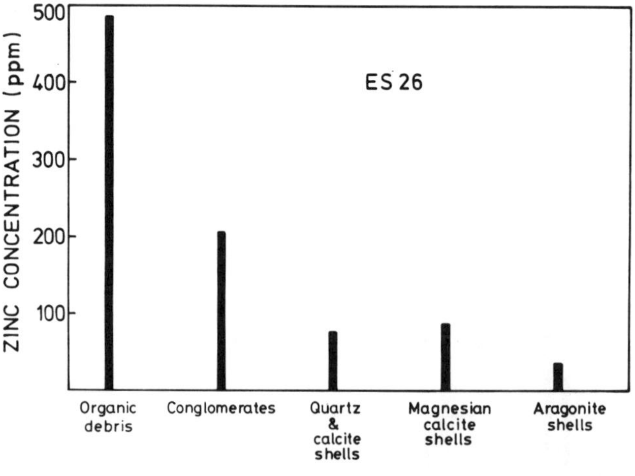

Fig. 24. Zinc concentrations in density bands separated from sediment ES 26.

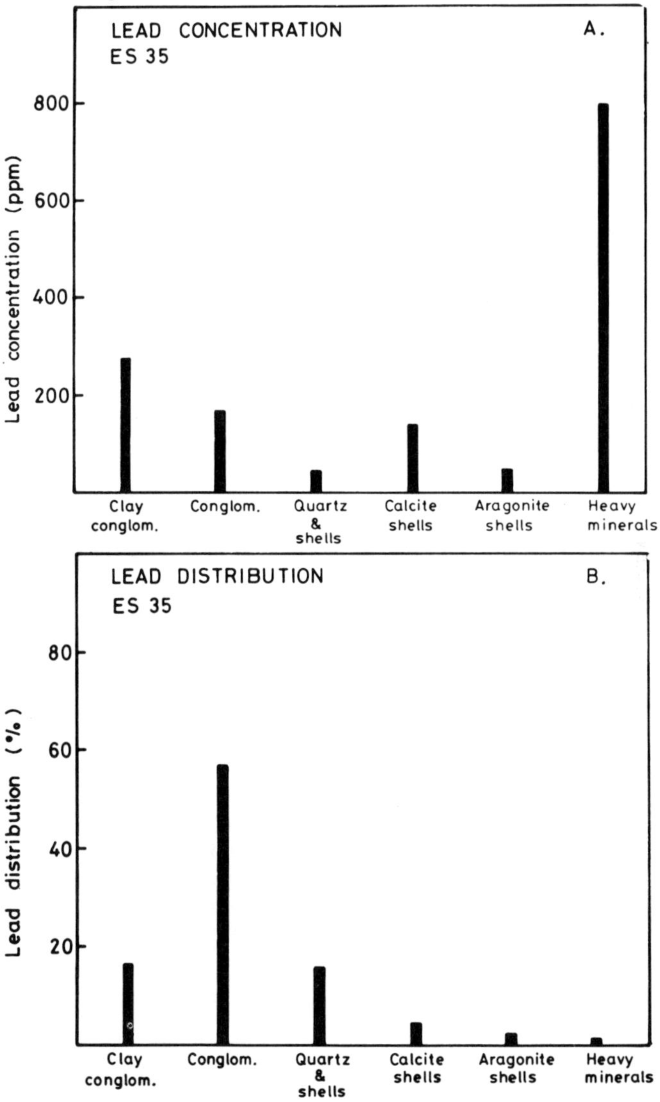

Fig. 25. Concentration and distribution of lead in density bands of sediment ES 35.

it was present as insoluble particulate matter, although the findings of Dr. Chamberlain[10] throw doubt on the assumption that less soluble compounds such as PbS are less biologically available than more soluble salts such as $PbCl_2$. Next year we plan to measure not only the total lead in each of the sediment components but also the biologically available lead, that is, the lead which would be extracted by molluscs and fish ingesting sediment particles.

In closing, I would like to emphasize two things. First, I

have had time to present only general conclusions and have not done justice to the tremendous amount of careful experimentation by all the members of the CSIRO Task Force on Heavy Metals. What we have tried to do in these first nine months is to lay the foundations for a thorough research program. Secondly, I would emphasize that we believe it is essential to define the physical parameters of the environment before the biologist and ecologist begin their detailed studies. The scene is now set for a meaningful interpretation of the levels of lead in selected fish and molluscs and the abundance and diversity of aquatic species near to and far from the smelter at Port Pirie.

ACKNOWLEDGMENTS

I would like to thank Drs. R. Arnold, J. Bauld, B. V. Bubela, D. Gardner, G. W. Skyring, D. J. Swaine and K. G. Tiller for their assistance in preparing this paper. It has been very much a joint effort.

REFERENCES

1. "Heavy Metals in the Aquatic Environment," *Proc. Int. Conf.*, Nashville, 1973, ed. P. A. Krenkel, Pergamon, 1975.
2. Int. Conf. on Heavy Metals in the Environment, Symp. Proc., Vols. I, II, III, ed. T. C. Hutchinson, Toronto, 1975.
3. Parsons, T. R., Harrison, P. J., and Waters, R. (1978). *J. Exp. Mar. Biol. Ecol. 32*, 285-294.
4. "Lead in the Environment," National Science Foundation Report RA-770214, eds. W. R. Boggess and B. G. Wixson, Washington, 1977.
5. James, R. O. (1978). "Effects of Heavy Metals on Aquatic Life," CSIRO, Melbourne. (ILZRO Venture Analysis LH-230/ZH-212).
6. The lead smelter and zinc refinery at Port Pirie is operated by Broken Hill Associated Smelters Pty. Ltd.
7. The assistance of the South Australian Government Department of Fisheries and Wildlife is gratefully acknowledged.
8. Deep cores were taken during a joint expedition with a University of Adelaide team led by J. R. Hails.
9. Batley, G. E., and Gardner, D. (1978). *Estuarine Coast. Mar. Sci. 7*, 59-70.
10. Chamberlain, A. C., Heard, M. J., Little, P., Newton, D., Wells, A. C., and Wiffen, R. D. (1978). Investigation into lead from motor vehicles. *Int. Symp. Environ. Lead Res.*, 2nd, Cincinnati.

DISCUSSION OF PAPER BY LEONARD J. WARREN

Chamberlain: Are there any data for aquatic animals, similar to that, for example, of Dr. Barry for humans, of the distribution of lead between various tissues in the mature animal?

Dr. Leonard J. Warren, Commonwealth Scientific Research Organization: Yes, there is information. The highest metal concentrations are usually in the viscera, the organs, although the distributions varies form one animal to another. I've never seen measurements where they've actually separated the bones. Usually the procedure is to remove the organs and perhaps the gills, look in detail for surface accumulations, and then analyze the rest of the animal.

Cole: I seem to recall some data which would suggest that, as far as fish are concerned, most of the lead winds up in the bone, similarly to the mammal. It seems that that might have appeared in the National Academy of Science document, *Airborne Lead in Perspectives,* some years ago. That's speculation on my part.

Shapiro: Have you, or are you, planning to use indicator organisms, those that maybe are accumulators?

Warren: Yes, I think we'll probably do that. In the Port Pirie area there are a number of shell fish we could use. There's one called the razor fish, which is not actually a fish, it's a shell. There aren't many mussels, as such, nor many oysters. But there's a lot of razor fish. An indicator organism is one aspect of monitoring the levels over a number of years. But there are other aspects of transfer through the various trophic levels in which we would look at a wider range of organisms, and at their abundance and diversity rather than merely their metal levels.

Hammond: What influence does the higher concentration of chloride in sea water have on the solubility of lead as compared to fresh water. As I recall, lead will form a complex anion with chloride. Is that a significant differential factor in seawater versus fresh water?

Warren: This is a subject that's intrigued me and is presently controversial. If you look at lead levels in fresh waters and you follow a stream into an estuary, there have been some reports indicating that lead in the estuary is released because of the solubility effect. Lead is presumably released from suspended particles. But on the other hand, the level of lead in sea water, especially in deep ocean water, is incredibly low. I gave values of .3 μg/l in a near shore environment. We're pretty confident that's right. Patterson's work, in the deep ocean environment, shows lead levels 10 times below that. The found .03 μg/l or .03 ppb, which is lower than is normally found in fresh waters. So this is an interesting situation yet to be settled.

Grandjean: I was puzzled by a report a year ago by Canadians that flounder and lobster and other fish products contained a lot of lipid soluble lead, and I wondered if you separated any lead in higher aquatic organisms in lipid soluble and non-lipid soluble lead.

Warren: No, we haven't. The next stage of the project, in which we will collect a variety of organisms for detailed analysis, is just beginning.

THE BIOLOGICAL METHYLATION OF LEAD:
AN ASSESSMENT OF THE PRESENT POSITION

P. J. Craig and J. M. Wood

University of Minnesota
Navarre, Minnesota

INTRODUCTION

It is now well-known that inorganic mercury may be converted under natural environmental conditions to methyl or dimethyl mercury (Wood *et al.*, 1968; Jensen and Jernelöv, 1969). Since, in most cases, alkyl derivatives of metals are more toxic than their ionic or neutral counterparts (Fanchiang *et al.*, 1978), attention is now being focused on the generality of such biological methylation reactions. The chemical properties of an element are rarely unique to that element and it might perhaps seem unlikely that only mercury, of all the elements, may be converted to the methyl form under environmental conditions. Lead and tin, for example, as elements much used by man, are prime candidates for study in this respect. There is, in fact, good evidence for methylation of tin (Huey *et al.*, 1974). In addition, the related metalloidal elements selenium and arsenic have been demonstrated to undergo biological methylation; for these elements the methylation process was demonstrated over 40 years ago by Challenger (Challenger, 1935; Challenger, 1945).

In the case of lead, there is much circumstantial, but no conclusive evidence, that biomethylation has occurred. This review will examine the experimental evidence so far presented for biomethylation of lead species. In addition, the inorganic, organometallic, and redox chemistry of lead will be discussed in so far as it relates to methylation prospects. Finally, suggestions for future research in this field will be made.

One point that should be emphasized here is that there is an important contrast in the background situation between lead and mercury. At the present time it can be said that any methyl mercury analyzed in the environment will have arisen by environmental methylation of inorganic mercury forms introduced into the environment. Methyl mercury is no longer directly introduced into the en-

ISBN 0-12-460520-6

vironment. By contrast, organic forms of lead are introduced in significant quantities into the environment, principally in gasoline fuel. Therefore, any biological methylation demonstrated would merely augment the existing environmental loading for alkyl lead compounds.

EXPERIMENTAL EVIDENCE FOR ENVIRONMENTAL METHYLATION OF LEAD

(1) "Biological" Experiments

In this section, both direct environmental observations ("field work") and simplified, nonbiological, chemical model experiments will be discussed. It is characteristic of this field that contributions have arisen from laboratories operating in diverse areas. Useful results have arisen from pollution, analytical, inorganic, and organometallic investigators.

The first evidence for microbiological methylation of lead was presented by Wong *et al.* in 1975. Polluted Great Lakes sediments were incubated in an anaerobic environment with water and nutrients; tetramethyl lead (TML) was detected in the gas phase above the sediment water system. It might be thought that this TML may not have been produced in the sediments; rather, it might have been already present as pollutant which was degassed into the system atmosphere. However, addition of trimethyl lead (IV) acetate (TMLA) or lead (II) nitrate greatly increased TML production. Control experiments suggested a microbiological origin for the observed TML. Not all sediments produced this effect; in not every case was lead (II) nitrate or chloride effective, although TMLA [i.e., lead (IV)] did produce TML consistently. The significant point is that an organic lead (IV) compound seemed to have been produced from a divalent inorganic lead (II) salt. The conversion for lead (IV), however, seemed quite straightforward: TMLA was converted to TML by several pure species of bacterial isolates without the presence of sediments. Such a conversion did not occur for Pb(II).

Using different sediments, Jarvie *et al.* (1975) produced another interpretation of the Wong *et al.* findings. With their sediments, Jarvie *et al.* were unable to observe TML production from lead (II) species. Following from the implication of methyl cobalamin as the methylating species for mercury methylation, they were also unable to detect any TML production from aqueous solutions of methyl cobalamin and TMLA, $(CH_3)_3PbCl$, $(CH_3)_2PbCl_2$ or $Pb(NO_3)_2$. Using both a sulphide ion-containing model and natural sediment systems, they suggested TML production was chemical and indirect rather than biological. It was proposed that TMLA was first converted to $[(CH_3)_3Pb]_2S$ by sulphide ion or sulphur species in the sediment, and that this species decomposed to give TML.

Lack of an original methylating capacity in the sediments was demonstrated by incubating $(C_2H_5)_3PbCl$; tetraethyl lead (TEL) only was formed, rather than $(C_2H_5)_3PbCH_3$. An analogous sulfur compound to $[(CH_3)_3Pb]_2S$ was again suggested as an intermediate. A similar chemical mechanism has been established for the conversion of CH_3Hg^+ to $(CH_3)_2Hg$ in the presence of H_2S; $(CH_3Hg)_2S$ is formed, which decomposes to $(CH_3)_2Hg$ and HgS (Craig and Barlett, 1978). The theoretical background for this conversion has also been discussed (Deacon, 1978). Jarvie *et al.*, therefore, concluded that a direct biological methylation of lead (IV) had not occurred and that in any case cobalamin-mediated methylation of lead (II) in water did not occur.

Notwithstanding this work, a further paper reported the production of TML from Pb(II) (lead (II) acetate) (Schmidt and Huber, 1976). The methylation was claimed to be biological by using microorganisms from an aerated aquarium under anaerobic conditions in aqueous solution. TML was detected from the gas phase above the solution. It seems unlikely that the sulfide route of the previous paper (Jarvie *et al.*, 1975) could also be operating here, as the biological material used resembled a pure culture rather than a bulk sediment sample. In addition, this German group noted that concentrations of $(CH_3)_3Pb^+$ and $(CH_3)_2Pb^{2+}$ salts in water when incubated in this way decreased rather more rapidly than would be expected from knowledge or the known chemical disproportionation routes for these compounds (which also lead to TML), i.e., $2(CH_3)_3Pb^{IV}Cl \rightarrow (CH_3)_2Pb^{IV}Cl_2 + (CH_3)_4Pb^{IV}$. The finding of excess (5 to 10 fold in a week) amounts of TML suggested an additional route to its formation, and a biological methylation by the mixed culture of microorganisms was suggested. The lack of TML production noted in the previous paper from Pb(II) was also ascribed to precipitation of non-reactive PbS from sulfur species in the sediments used.

Circumstantial evidence for the biomethylation of lead (II) rose from the demonstration under anaerobic conditions that a methylation for thallium (I) occurs (Huber and Kirchmann, 1978). thallium (I) is iso-electronic with lead (II). The product analyzed was $(CH_3)_2Tl^+$, i.e., Tl(III). A cautionary note should be made here--when acetate salts are used in methylation studies, control experiments should be undertaken to ensure that the methyl group does not originate with decomposition of the acetate moiety. This can happen under certain conditions, especially in the presence of sulfides and/or light (Akagi *et al.*, 1975).

A recent paper discussing methylation of lead by biological species is reported by Dumas *et al.* (1977). This paper reinvestigated the role of sulfide ions in the production of TML from TMLA; up to ten times as much TML was produced from a biological sediment-containing system as from an inorganic sulfide-containing aqueous medium. Also, autoclaving of the sediment system reduced H_2S production to zero (at 37°, H_2S was produced). This suggested TML production in the sediment system to be caused partially by biological activity. Alternately, the sediment could have contained high concentrations of sulfide ion. This important control was not tested experimentally (Dumas *et al.*, 1977).

More significantly, TML was produced from sediment systems to which Pb(II) as $Pb(NO_3)_2$ was added. The amounts were quite small compared to such TML production from TMLA, but larger than the amounts from the sulfide-containing inorganic system. This experiment would tend to confirm biological methylation of Pb(II), but unfortunately no control blank experiments with sediments in the absence of Pb(II) were reported.

Additional circumstantial evidence exists for the biomethylation of lead. Using an analytical method which separated between airborne particulate lead and tetraalkyl lead, Harrison and Laxen (1978) drew important conclusions from observed abnormally high alkyl lead to total lead ratios in atmospheric samples. The source of the excess alkyl lead was deduced, from backward air movement projections, to be associated with extensive coastal/estuarine intertidal mud flats to the west of the sampling cities in Cumbria, U.K. Inorganic lead is known to be present in these mud flats, and the authors conclude that the excess tetraalkyl lead can only be explained by a natural conversion of inorganic to methyl lead. Direct laboratory measurement of TML evolution from these sediment samples was not reported.

Paralleling the mercury case, it has been reported that 10-24% of the total lead in some cod samples, and 39% in mackerel muscle, is in the tetraalkyl form (Sirota and Uthe, 1977). This suggests either a biological methylation process is occurring or that strong selective concentration of the alkyl-forms are taking place in fish.

(2) Abiotic, Chemical Model Experiments

Attempts at direct methylation of Pb(II) using methyl cobalamin, as carried out for example on numerous occasions with mercury (II) salts, have been unsuccessful (Jarvie *et al.*, 1975; Taylor and Hanna, 1976; Wood, 1974; Agnes *et al.*, 1971; Lewis *et al.*, 1973). This might suggest a non-cobalamin route for any methylation observed for Pb(II) in the environment.

Methyl cobalamin has been observed to be demethylated by various Pb(IV) compounds, i.e., $Pb(OAc)_4$, Pb_3O_4, PbO_2. Using ^{14}C-labelled methyl cobalamin, loss of radioactivity, presumably as a volatile product, was observed in parallel with aquo-cobalamin formation (Taylor and Hanna, 1976). No attempts to identify the volatile products were reported, but it was suggested that any methyl lead product would decompose in the aqueous system.

Weber and Whitman (1978) have shown that dimethyl cobaloxime, a model compound for methyl cobalamin, reacts with Pb(II) in a protic solvent (isopropanol) to give a stable, insoluble methyl lead product which was not fully characterized. The initial reaction pathway was postulated as follows:

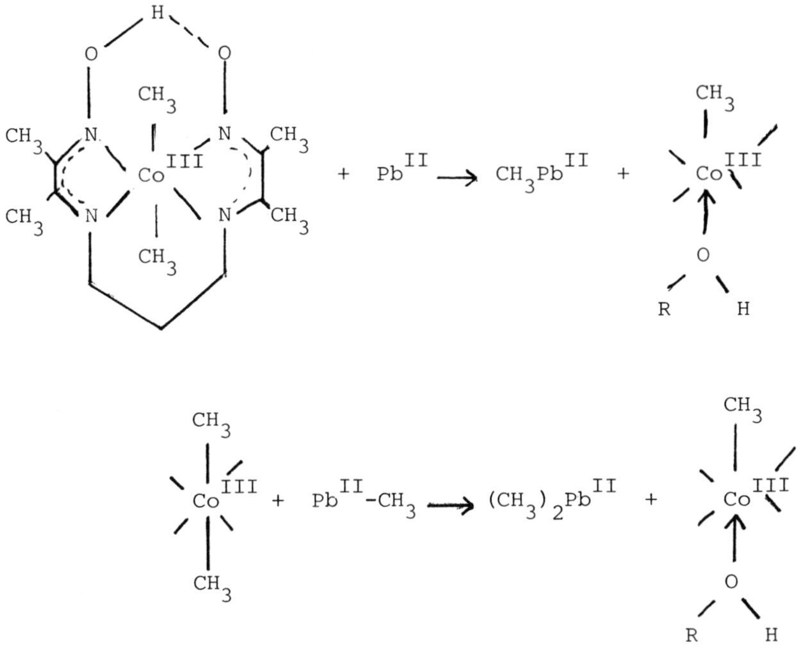

Without further evidence as to the structure of the final product, it is difficult to suggest its identity. Some kind of condensed species (e.g., $(CH_3)_6Pb_2$ mp 38°) or a lead oxygen polymer might be possible, e.g., Pb_2PbO appears to be polymeric; various polymeric $(Pb\text{-}O)_n$ systems are known. It is thought that $(C_2H_5)_2Pb$ decomposes at room temperature to $(C_2H_5)_6Pb_2$ and lead metal (Shapiro and Frey, 1968). A polymeric dimethyl lead (IV) species is unlikely to exist, owing to weak Pb-Pb bonds, i.e., Pb-Pb bond strength is 15 or 20 Kcals (Shapiro and Frey, 1968). Early formation of a $(CH_3)_6Pb_2$ species could also produce other solid lead oxygen species (Shapiro and Frey, 1968).

In parallel with the observed biological methylation of trimethyl lead salts, methylation of dialkyl lead (IV) halides by methyl cobalamin, resulting in volatile tetra alkyl lead compounds, has been observed (Ridley *et al.*, 1977). This confirms that if the very unstable monoalkyl lead compounds could be generated and exist for sufficient time to be converted to dimethyl lead (IV) species, subsequent methylation steps could occur by a cobalamin-dependent mechanism, i.e.,

$$R_2Pb^{2+}(IV) + CH_3CoB_{12} \longrightarrow R_2(CH_3)Pb^{+}(IV) \xrightarrow{CH_3CoB_{12}} R_2Pb(CH_3)_2$$

(3) Conclusions

The biomethylation of organic lead (IV) species seems well-established and from model experiments a cobalamin route is feasible. Evidence that Pb(II) salts are methylated to TML under environmental conditions exists but is not conclusive. Model experiments with Pb(II) have been negative or inconclusive. Model experiments with inorganic Pb(IV) species have demonstrated demethylation of methyl cobalamin with no production of TML.

THEORETICAL CONSIDERATIONS ON THE BIOMETHYLATION OF LEAD

This section is mainly concerned with the Pb(II) case. There is little in the way of preventing our understanding of how cobalamin-based methylation of dialkyl Pb(IV) species occurs in an aqueous medium. $(CH_3)_2Pb^{2+}$ and $(CH_3)_3Pb^+$ are sufficiently stable in water to allow further methylation to take place, leading to TML production.

Methylation by electrophilic attack of Pb(II) and inorganic (Pb(IV) presents some difficulty in that the initial methyl species $(CH_3)Pb^+$(II) or CH_3Pb^{3+}(IV) are very unstable in aqueous media. The second methylation step producing "stable" $(CH_3)_2Pb$ or $(CH_3)_2Pb^{2+}$ respectively would have to take place faster than the decomposition step. [$(CH_3)_2Pb$ when formed would most likely disproportionate to TML and Pb^o. This is the basis of most of the chemical preparations of TML.] In the case of electrophilic attack of Hg^{2+}(II) on methyl cobalamin leading to methyl mercury, the second methylation step is about 6000 times slower than the first (DeSimone *et al.*, 1973; Chu and Gruenwedel, 1976):

$$CH_3CoB_{12} \xrightarrow[k_1]{Hg^{2+}} CH_3Hg^+ \qquad CH_3CoB_{12} \quad \overset{CH_3Hg^+}{\underset{k_2}{}} \quad (CH_3)_2Hg$$

k_1 is 6000 times faster than k_2. If such a relationship exists in the case of lead, then this argues against a methylation taking place in water leading to TML. Certainly this is in accord with the cobalamin work reported so far on the kinetics of reactions with Pb(IV) species.

However, it is known that chemical alkylation of inorganic Pb(II) species to TML is possible in aqueous solution. In this case, the alkylating agent used was $(C_2H_5)_3Pb$ and subsequent disproportionation to TEL and Pb metal. This parallels the classic organometallic preparations of tetraalkyl lead species in non-aqueous solvents, e.g., using a Grignard reagent.

$$2\ RMgX + Pb^{II}X_2 \xrightarrow[\text{solvent}]{\text{ether}} R_2Pb^{II} + 2\ MgX_2$$

$$2\ R_2Pb^{II} \xrightarrow{\text{ether}} R_4Pb^{IV} + Pb^{O}$$

$(C_2H_5)_3B$ and $(nC_6H_{11})_3B$ were reported to alkylate lead in water. No report of using $(CH_3)_3B$ in this process was made, although it was mentioned in the patents; however, use of $HaB(CH_3)_3(C_2H_5$ was noted and $NaB(CH_3)_4$ was claimed. Nucleophilic attack on Pb(II) halides, therefore, is not inherently impossible in water (and, conversely, electrophilic attack on e.g. methyl cobalamin by Pb(II), also).

Further light on such electrophilic attack by metallic species comes from a comparison of the electrode potentials of lead and mercury, respectively. (The role of standard electrode potentials and the methylation mechanism has been discussed recently (Ridley *et al.*, 1977).) Assuming an initial reaction with a Pb(IV) species, electrophilic attack leading to transfer of a methyl carbanion is likely on standard electrode potential grounds, as the value of E° (+1.46) for Pb(IV)/Pb(II) is greater than that for the Hg^{II}/Hg^{O} couple (+0.854).

For the case of Pb(II), the value of the relevant standard electrode potential (E° for Pb(II)/PbO = -0.1205 volts) might suggest a reductive homolytic cleavage mechanism leading to transfer of a methyl radical, as has been suggested for tin methylation (DeSimone *et al.*, 1973). With Fe^{III} present as a single electron oxidant, the following route has been suggested (Ridley *et al.*, 1977):

$$Fe^{III} + Sn^{II} \rightleftharpoons \dot{S}n^{III} + Fe^{II}$$

$$CH_3Co^{III}B_{12} + Sn^{III} \longrightarrow CH_3Sn^{IV} + Co^{II}B_{12}$$

An analogous mechanism could be suggested for lead. Certainly, lead metal may react with methyl radicals (Shapiro and Frey, 1968).

If an initial electrophilic attack occurred, the free radical route could supplement TML formation by reaction with the lead metal generated viz:

$$6PbX_2 + 12CH_3CoB_{12} \longrightarrow 6(CH_3)_2Pb^{II} + 12Co^{III}B_{12}(H_2O) + 12X^-$$

$$6(CH_3)_2Pb^{II} \longrightarrow 2R_6Pb^{IV}_2 + 2Pb$$

$$2R_6Pb^{IV}_2 \longrightarrow 3R_4Rb^{IV}(TML) + Pb$$

In any case, since Pb^{II} undergoes electrophilic attack by CH_3^- from Grignard reactions, the value of the electrode potentials alone should not be considered to debar consideration of such possibilities for CH_3CoB_{12}. However, such a mechanism has been searched for but not observed by a number of workers. In any case, parallels of CH_3CoB_{12} with Grignard reagents should be interpreted with caution.

It should also be considered that if TML [or $(CH_3)_2Pb^{2+}$ or $(CH_3)_3Pb^+$] were to be generated in the environment, its stability under environmental conditions is less than that of methyl mercury. For example, TML is unstable in sea water and undergoes progressive dealkylation, eventually forming inorganic lead (Maddock and Taylor, 1977). Although this might suggest pollution problems with the di and trialkyl lead derivatives, those are less toxic with TML and the overall process might be considered a detoxification mechanism. It has been concluded that the major environmental impact of TML release into the marine environment would be due to acute toxic effects rather than the result of gradual bioaccumulation (Maddock and Taylor, 1977).

SUGGESTIONS FOR FUTURE RESEARCH

At this stage, it would be helpful to have more examples from different parts of the world demonstrating the biosynthesis of TML by sediment samples. TML evolution demonstrated from both doped or "natural" (i.e., lead-containing by pollution input) environments would be a valuable encouragement to workers trying to demonstrate TML production from model systems.

Where reactions of CH_3CoB_{12} with inorganic lead compounds have been shown (using labelled carbon in the CH_3 group), it would be interesting to ascertain the identity of the (presumably) radioactive volatile product. Is it $^{14}CH_4$, $^{14}C_2H_6$, or perhaps ^{14}TML?

Similarly, where methylation of lead compounds by sediments has been claimed, further experiments using labelled lead isotopes as substrates and a search for a radioactive TML product would be illustrative.

Demonstration of TML production in less polar, nonaqueous solvents would also not be irrelevant in this context. (Many biological reactions in aqueous systems take place in locally nonaqueous environments.) In a sense, the latter information already exists in the well-known Grignard preparations of TML in nonaqueous solvents.

In the final analysis, we cannot rule out the possibility of nucleophilic attack by Pb^o on biological methylating agents such as S-adenosylmethionine. Furthermore, the coordination of thiols to Pb^{II} salts could produce strong enough nucleophiles to accept a carbonium ion in a B_{12}-independent mechanism.

REFERENCES

Agnes, G., Bendle, S., Hill, H. A. O., Williams, F. R., and Williams, R. J. P. (1971). Methylation by methyl vitamin B_{12}. *Chem. Comm.* 850-851.

Akagi, H., Fujita, Y., and Takabatake, E. (1975). Photochemical methylation of inorganic mercury in the presence of mercuric sulfide. *Chem. Lett. 2,* 171-176.

Challenger, F. (1935). Biological methylation of compounds of arsenic and selenium. *Chem. Ind. (London) 54,* 657-662.

Challenger, F. (1945). Biological methylation. *Chem. Rev. 36,* 315-361.

Chu, V. C. W., and Gruenwedel, D. W. (1976). On the reaction of methylmercuric hydroxide with methylcobalamin. *Z. Naturforsch C: Biosci 31C(11-12),* 753-755.

Craig, P. J., and Barlett, P. D. (1978). The role of hydrogen sulfide in environmental transport of mercury. *Nature 275,* 635.

Deacon, G. B. (1978). Volatilisation of methylmercuric chloride by hydrogen sulphide. *Nature 275,* 344.

DeSimone, R. E., Penley, M. W., Charbonneau, L., *et al.* (1973). The kinetics and mechanism of cobalamin-dependent methyl and ethyl transfer to mercuric ion. *Acta Biochim. Biophys. 304,* 851-863.

Dumas, J. P., Pazdernik, L. R., Belloncik, S., *et al.* (1977). PNC 12th Canadian Symp. Water Pollution Research, Canada. (Translated from the French)

Fanchiang, Y.-T., Ridley, W. P., and Wood, J. M. (1978). *Organometals and Organometalloids.* ACS Symposium Series *82,* 54-65.

Harrison, R. M., and Laxen, D. P. H. (1978). Natural source of tetraalkyllead in air. *Nature 275,* 738-739.

Honeycutt, J. B., Jr., and Riddle, J. M. (1960). Triorganoboranes as alkylating agents. *J. Amer. Chem. Soc. 82,* 3051-3052.

Honeycutt, J. B., Jr., and Riddle, J. M. (1961). Preparation and Reactions of sodium tetraethylboron and related compounds. *J. Amer. Chem. Soc. 83,* 369-373.

Huber, F., and Kirchmann, H. (1978). Biomethylation of thallium (I) compounds. *Inorg. Chim. Acta 29,* L249-250.

Huey, C., Brinkman, F. E., Grim, S., and Iverson, W. P. (1974). *Proc. Intern. Conf. on Transport of Persistent Chemicals in Aquatic Ecosystems,* p. 73. ON Lettam, ed. NRC Canada, Ottawa.

Jarvie, A. W. P., Markall, R. N., and Potter, H. R. (1975). Chemical alkylation of lead. *Nature 255,* 217-218.

Jensen, S., and Jernelöv, A. (1969). Biological methylation of mercury in aquatic organisms. *Nature 223,* 753-754.

Lewis, J., Prince, R. H., and Stotter, D. A. (1973). Recent developments in the bioinorganic chemistry of [vitamin] B_{12}. II. Nonenzymatic transalkylation. *J. Org. Nucl. Chem. 35,* 341-351.

Maddock, B. G., and Taylor, D. (1977). *Proceedings of an International Conference on Lead in the Marine Environment,* M. Branica, ed. Rovinj, Yugoslavia.
Riddle, J. M. (1960). U.S. Patent 2,950,301.
Riddle, J. M. (1960a). U.S. Patent 2,950,302.
Ridley, W. P., Dizikes, L. J., and Wood, J. M. (1977). Biomethylation of toxic elements in the environment. *Science 197,* 329-332. (See also Ref. 9 therein)
Schmidt, U., and Huber, F. (1976). Methylation of organolead and lead (II) compounds to $(CH_3)_4Pb$ by microorganisms. *Nature 259,* 157-158.
Shapiro, H., and Frey, F. W. (1968). *The Organic Compounds of Lead.* Interscience, New York.
Sirota, G. R., and Uthe, J. F. (1977). Determination of tetraalkyllead compounds in biological materials. *Anal. Chem. 49,* 823-825.
Taylor, R. T., and Hanna, M. L. (1976). Methylcobalamin: methylation of platinum and demethylation with lead. *J. Environ. Sci. Health Part A A11(3),* 201-211.
Weber, J. H., and Whitman, M. W. (1978). *Organometals and Organometalloids.* ACS Symposium Series *82,* 247-264.
Wong, P. T. S., Chau, Y. K., and Luxon, P. L. (1975). Methylation of lead in the environment. *Nature 253,* 263-264.
Wood, J. M. (1974). Biological cycles for toxic elements in the environment. *Science 183,* 1049-1052.
Wood, J. M., Kennedy, F. S., and Rosen, C. G. (1968). Synthesis of methyl-mercury compounds by extracts of a methanogenic bacterium. *Nature 220,* 173-174.

DISCUSSION OF PAPER BY DR. JOHN M. WOOD

Grandjean: We have recently been concerned about the occurrence of organic lead in humans, and I was very interested to hear your paper. Our concern continues, and I would like your opinion as to what the sources of human exposure would be. We would, of course, think of air pollution, and we know that about 10 percent of the lead in the atmosphere in the cities is probably organic lead, but how much impact would the microbial production be? How much would that mean for people? How much should we be concerned about it?

Dr. John M. Wood, Fresh Water Biological Institute: I think probably the most important point that I hope you get is that organo lead compounds, just as organo mercury compounds, are transported through cell membranes at an enormous rate. So the kinetics for uptake of organo lead compounds into cell, of course, are favored. The uptake of inorganic lead complexes probably occurs but not very well. So if you want to look at kinetics, we're talking about transport rates for organo lead compounds in the nano second time range, and you're talking about uptake of organo lead on a yearly sort of time scale, so it is clear that we should be very much concerned about the small contaminating levels of organo lead compounds. A recent paper in "Nature" from a Danish group shows that. You're familiar with that are you?

Grandjean: I'm a coauthor of it.

Wood: Oh, you're a coauthor of it. It's very interesting because it turns out that people who are exposed to the city environment have more trimethyl lead chloride in the central nervous system than people who are not. And the problem is whether this trimethyl lead chloride results from biological methylation or as a result of rapid transport of trimethyl lead, uncombusted trimethyl lead, due to the kinetic aspects of this. We don't know the answer to that question, and I think it's important that we find out the answers.

Grandjean: Recently I saw a paper that mercury can be methylated in the rat intestine, and I wonder if it's possible for lead to be methylated in the intestine as well.

Wood: Well, that's an interesting question, and it's something that should be looked at. Tin is certainly methylated in the human intestine. When you look at methylation of lead species, the only inorganic lead species where there is some evidence, and I say it's very, very shaky evidence at this time, I'm not satisfied that it's hard scientific evidence. But the only evidence that we have is under anaerobic conditions starting out with plumbous with lead two as the substrate. And I think one can say from all the work that I've done in the last two years that it's highly unlikely that you'll get any methylation of plumbous in aerobic conditions that is where you have oxygen available, but under anaerobic conditions I say there's a possibility at this stage.

Ter Haar: It seems clear that this particular question is not about to go away. I believe that the fundamental difficulty in answering the question of whether it is a sediment or an alkyl lead already in the sediment type problem is the fundamental analytical question of identifying the actual species that is in the body. Methods that use extraction and so forth, in my mind, are not satisfactory to prove that the particular species are present, that is, the trialkyl or tetraalkyl. We have to use mass spec methods to definitely, unequivocally, pin down the identity of these species before we are going to begin to try to answer the question of how much of these materials are present. The situation in Canada is a typical example. A paper appeared in the literature about two years ago where the figures of 80 and 90 percent appeared. I'm not sure that the author is convinced today that the number is zero, but I know he is convinced that the number is far,

far lower than appeared in the original paper. These extractions are all fine when you're worrying about macro amounts of material. But the extraction techniques are no longer satisfactory when we're talking about a few nanograms in the presence of fairly large amounts of inorganic lead present as either the ion or perhaps associated with a protein or something like that. So we need definitive methods before answering these questions.

Wood: I couldn't agree with you more. I think there are two things that have to be done. First, we have to start using isotopes. They are much better. One could, in theory, do double labeling experiments with inorganic lead gamma emitters and carbon-14 beta emitters and really find out whether we've got biological methylation in a relatively complicated situation. I know you can't do that in humans which is what everybody's concerned about, but you can certainly do it in fish and other organism that are at the top of the food chain. I think that's important. The second thing is, you are echoing my frustrations on the tin problem. Two years ago we clearly demonstrated a biochemical basis for the synthesis of methyl tin compounds, and we published a coupld of papers and suggested that the biological methylation of tin would occur. At that time, we had no analytical methods to analyze for methyl tin compounds in tissues, in urine, and blood. Those methods have been developed, and methyl tin compounds are now being found in urine and blood and tissues. Another major thing is that the chemical species is incredibly important. We have to develop very, very sensitive analytical methods to look for these methylated metals in a complex environment. I agree with you entirely.

Ter Haar: Do you have any feeling for the amount of tin normally present in the diet?

Wood: Well, it depends on how much diet soda you drink. It depends on how much acidic tin vegetables you eat and how long the tins have been on the grocery store shelf.

Ter Haar: Do you find the alkylated tin in every situation?

Wood: In every situation looked at so far, it's there.

Ter Haar: Whether they drink diet soda or not.

Wood: No, if people do drink large quantities of diet soda there's an increase, a significant increase. It is added as an antioxidant. You can't avoid drinking it, as you know, because it's in solution. Vitamin C is better, I think, and is slightly more expensive, a good antioxidant. So, in terms of canned vegetables, you have to have acidic vegetables which slowly leach the can into the vegetables before you're going to get sizeable dose of methyl tin. There's another aspect of the tin problem, and that is the analytic methods to distinguish between monomethyl, dimethyl, and trimethyl tin have not yet been developed satisfactorily. So we have another problem to face because trimethyl tin is much more toxic than monomethyl tin in the system. So that problem isn't over by any means.

Patrick S. I. Barry, The Associated Octel Co., Ltd.: I'd just like to make a comment on Dr. Grandjean's comment in which he indicated that ten percent of the lead in the air was of organo form. I don't think this is generally accepted. Work recently done in London by Perry and his group suggest that this is nearer two percent, although you could get 10 percent in garage forecourts. But this is not the general street. Another point that I'd like to mention, again in respect to Dr. Grandjean's paper, I think it was Neilson's paper, Dr. Grandjean was associated with it, refers to high concentrations of organo lead in brain relative to total lead among people who lived in low floor areas compared to people who lived in upper stories. What was interesting to us, quite apart from the fact that Neilson, *et al.*, claims to find this organo lead, was that people living in the low stories had considerably lower inorganic lead concentrations than the people who lived in the upper stories. I think this needs a bit of explanation. I wonder, in relation to the Neilson study, whether they were really extracting organo lead as such in terms of the *in vivo* situation or whether in terms of the extraction methods used which resulted in converting an inorganic lead to organo lead.

Wood: Well, I'm afraid I didn't do the experiments, and I can't discuss that. But there's one point that I want to get across again. Even if you have very, very small concentrations of organo lead present in the atmosphere or anywhere, the rate of transport through cells is very fast. It may very well be much more significant than slow biological methylation of lead in a sediment environment, and it's time we got a picture of that problem, because I'm not enormously concerned about inorganic lead complexes which are tied up in humic material. I was in Los Angeles recently, where there are 39 million cars, and the grass and the trees are growing at the side of the freeways. If grasses, trees, and things can grow at the side of the highway, a lot of inorganic lead which is precipitated out, or very strongly bound, is not really available unless you had extreme acidic conditions to increase solubility and uptake by the biota. I'm more or less convinced at that, but I'm very much concerned about the uptake of trace amounts of organo lead compounds.

Warren: I would like to get a feeling for the relative toxicity of these metal compounds compared to the inorganic species. Reports I've seen show that the levels of the methylated species in sediments, for instance, are generally less than one percent, is that correct?

Wood: Yes.

Warren: So that means that they have to be at least 100 times more toxic if you're going to attribute importance to them compared to the inorganic lead that's already there.

Wood: I should point out the data that I showed you. Those data are taken from short term toxicity studies where the experimental animals are loaded up with the metal versus the organo metal, and you do the classic LD_{50} study. This is obviously not the way to do it. You see about an order of magnitude difference in toxicity. It turns out that an accumulation of metal alkyls is transported through the blood brain barrier, through the placenta, and partitioned into the central nervous system. The reaction in the central nervous system is an extremely rapid process relative to transport of inorganic lead. So it's very, very difficult to extrapolate from a short term chronic toxicity experiment. There must be toxicologists here. If anybody doesn't agree with me, stand up and say so. But it is difficult to extrapolate from that to what might be happening in the real world.

Ter Haar: We have a population at which we can look. This population has been exposed to extremely high concentrations, relatively speaking, of organic lead for 50 years, and that data is published. These people have been exposed to an average concentration of from 50 to 100 $\mu g/m^3$ for 8 hours a day for their lives, and the exposure has no detectable effect on either morbidity or mortality. So from a chronic point of view we don't see a lot of difference between organic lead and inorganic lead.

Wood: Yes, I know, but how many post mortem examinations have you done to look for alkyl lead compounds in the central nervous system of these people? I understand your point, but have you done it, or can you do it?

Ter Haar: Well, it certainly has been done in animals.

Wood: You were telling me earlier that you don't yet have the analytic methods.

Ter Haar: No, that is not so. In this case we're talking about macro amounts. I don't have any problem with macro amounts for the analytical procedures, because in those cases the extraction procedure does at least a 95 percent job of differentiating. It's when there's nanogram or picogram amounts that I worry about the analytical methods. When the levels are micrograms, it's not a problem. But even if there is a considerable amount in these tissues, the effect on the workers is not detectable.

Wood: You mean the workers don't become sick?

Ter Haar: They don't become sick, and they don't have shorter life spans. Not only do they not become physically sick, but they don't have milder symptoms either. We previously discussed the so-called healthy worker effect. Well, in this case, the lead exposed worker population was compared to another set of workers.

Wood: Yes, but how do you feel about people walking around having some methyl mercury in their central nervous system, some methyl tin compounds in their central nervous system, and potentially some methyl lead compounds in the central nervous system, with no real fundamental understanding of the neurotoxicity of these three highly dangerous neurotoxins. All of these three neurotoxins can be the consequence of the environment you and I bring our children up in. Aren't you just a little bit curious to find out what really is happening?

Ter Haar: Well, I'd be curious from a scientific point of view, but I'm not curious from a toxicological point of view as far as the health of my children is concerned. This is because I see these concentrations, a thousand fold higher than found in the general environment, that people have lived in their whole working life with no adverse health effects. That's the basis of toxicology. So I'll look at the high levels and say, all right, I've got a factor here of a thousand fold compared to my child, and so I don't have a concern about it. I'm interested, I'm obviously interested in what you're doing. I don't mean to downgrade it at all.

Wood: I get the point, but I'm very much concerned about the additive effects of neurotoxins which give you similar clinical symptoms and which move at 20 nanoseconds through a cell membrane into your central nervous system. I'd like to know what those things are doing from every point of view. I'd like to know what the biochemical lesion is, because there has to be one.

Cole: If you're concerned about the organic metals such as organic lead, in tissues, isn't it probable that this is not a new phenomenon. These metals that you're talking about have been around for eons, and man has been exposed to them. You're just now getting to the point where you can look at them and study them, but it doesn't necessarily mean that we're dealing with anything that we haven't been dealing with for centuries.

Wood: Yes, I agree with you entirely. These biological processes can look at metals. Mercury, for example, is a natural biological process where you get the methylation of inorganic mercury to methyl mercury. It is accumulated in fish, look at tuna fish and sword fish as classical examples of this, and you can say, fine, all right, these are natural biological processes that have evolved with organic systems. What I'm concerned about is not that these processes occur, but the <u>rate</u> at which these processes occur and any perturbation which is put on a rate increase as a consequence of increasing the concentration of the substrate. If you are moving 70 times more lead per year around the world and spreading it around than would occur by natural geological processes, you are increasing the fundamental substrate concentrations for natural reactions. If these natural reactions occur, and if we decide to ignore this concentration effect, and if we decide to ignore this kinetic effect, I think we're making a serious error.

Cole: I agree with Dr. Ter Haar that it's not something we ought to ignore, but by the same token, there's a tendency sometimes to overreact.

Grandjean: I think that ILZRO has really put its money in an extremely important area by helping Dr. Wood do this study. I liked his talk very much. Dr. Barry made some comments to me, and I have discussed his comments with him before, but let me just mention that Neilson and I made a review on organic lead for the Swedish government a year or two ago, and we're putting this out in English. We have published an unpublished information showing that about 10 percent of lead in the atmosphere is organic. There's just a few studies that show that it's much less, but the majority of studies show that it's somewhat higher. I don't know if it's influenced by microbial production. Dr. Ter Haar's comment on Dr. Robinson's study on the health of tetramethyl lead workers was based on measurements of blood pressure, height, and sickness and absence rates and things like that. I don't consider that sensitive measures of insidious central nervous system effects. We did some studies on cells in Copenhagen, and it's now been published by Paul Biorem of the Institute of BioPhysics. We found that very low concentrations of trialkyl lead is amphophyllic, and it acts as a ferry in the cell membrane and transports chloride ions and exchanges with hydroxyl ions. Therefore, it could possibly interfere with the electric phenomenon occurring in the nerve cell membranes, especially since nerve cells have a very low natural chloride transport and would be very sensitive to such an influence by trialkyl lead.

Wood: Well, I must say that we don't understand why these things are neurotoxic yet. So there's another very important fundamental area for research. We know that they react extremely well with sulfhydryl groups. We know that they will catalytically hydrate and hydrolyze plasmalogens, which are vinyl-ether lipids which are found primarily in the central nervous system. So you can rationalize biochemical mechanisms for the lysis of cell membranes in the central nervous system on the basis of a couple of reactions that have been known to occur with metal alkyls. But whether these are the correct sites for the neurotoxicity of these things or not remains to be determined. So this is another area where we need to do a lot of work. There is a great deal of similarity in the clinical symptoms and possibly in the mechanisms of action of all of these metal alkyls in the central nervous system.

Kneip: I just want to make a short comment on organo lead in the atmosphere. I'll be very interested to see the review by Dr. Grandjean, but I don't agree with what he just said. I think somebody has to take a careful look at where the sampling is done before we come to the conclusion that there's a general level of ten percent organo vapor in the atmosphere. Regarding sampling, Skogerboe specifically sampled next to the road, trying to find organo lead, and he concluded that the concentration was three to four percent at that position. Well, we know it's in filling stations. You know you can go there and find higher organo lead in the atmosphere. But if you go to the general atmosphere, the well mixed atmosphere away from the immediate source, I don't believe 10 percent is the right number, and I don't believe 10 percent is the right number for the exposure in the general population. Ten percent may be the correct figure for the traffic policeman in the middle of the intersection at Times Square.

Wood: Well, you have put your finger on one of the issues that has always troubled me, and that is, in many of the field experiments involving sampling, in many of them, you could always stand up and make very strong statistical arguments against what people have been doing. I don't know how we get around that problem because the expense involved in taking the correct number of samples properly, where you could fit a decent statistical model, is an incredibly expensive exercise. This is one of the troubles that I always have. We have a number of limnologists in the University that I'm associated with who take two phosphate samples and make a decision on the state of a lake. Well, I think that is really stretching things. It doesn't mean very much to me, and I see this often. If only we could begin to get some sort of cooperation between groups to get together and do statistically sound things using the same quantity of money. If people would not fall out, would cooperate with each other and would do these things properly, we would probably get some information.

**Since this manuscript was submitted over 2 years ago other contributions have been made in this field. Ahmad, I., Chan, Y. K., Wong, P. T. S., Corty, A. J., and Taylor, L. have published a paper in Nature 287, 716-717 (1980) showing that Pb^{II} is methylated by CH_3^+-donors in aqueous solution. Also, Reisinger, K., Stoeppler, M., and Nurnberg, H. W. have shown that micro-organisms do* <u>*not*</u> *methylate Pb^{II} salts (Nature, 1980, in press).*

CLOSING REMARKS BY DR. PAUL B. HAMMOND

I want to thank ILZRO for having made this meeting possible. I have enjoyed it thoroughly, and from all of the comments I've heard, my reaction is typical.

In this concluding discussion, I will not attempt to review the individual papers that were presented at this Symposium. Rather, I will try to give a personal reaction to what I have heard and to how it either contradicts, supplements, or in any way modifies my previous views about the problem of lead from the health standpoint.

The picture that has evolved over the years and which, basically, has been reinforced in the course of this meeting, is that the dose-response curve for lead is very, very shallow. Let me cite a few examples from what was presented.

We heard, for example, from Drs. Fugas and Saric concerning the characteristics of a population residing in a highly polluted valley in Yugoslavia. I was somewhat amazed at the relationship between the external dose and the internal dose of lead in this population. We were told that these people ingest on the average 2.9 mg of lead per day in their diet. We were also told that on the average they inhale 17 $\mu g/m^3$ of lead.

Now, if one were to use the approximations that have been made to date concerning the relationship between lead in blood and lead in air, and between lead in blood and lead in food, it would be predicted that these people would have a concentration of lead in their blood in excess of 200 μg/dl. But, as it turns out, these people have a concentration of lead in their blood of approximately 48-65 μg/dl.

While I was amazed at the magnitude of the external dose in these people, I was really not surprised at the discrepancy between what one would predict from the kinds of data that were quoted in the recent WHO document and what actually was observed. This is because not only is the dose-response curve for lead very shallow, but it is also true that the relationship between external dose and internal dose is not linear. To make linear extrapolations from observations made at low levels of lead exposure is, I think, totally falacious.

I wouldn't even bother making the point if it weren't that I am aware of the fact that this is a concept which is not generally appreciated. I suspect that most of the people in this audience appreciate it, but I really wonder how widely appreciated it is by those who make decisions concerning tolerable levels of lead in the environment.

In this connection, I would like to add a word about the discussion we heard following the last presentation. It was, I thought, quite interesting in the sense that it epitomizes the difference between toxicity and hazard. To discuss the methylated metallic compounds in terms of toxicity is a necessary exercise,

ISBN 0-12-460520-6

but to infer that there is a hazard because they are highly toxic is to ignore the fact that hazard exists only when the opportunity exists for human exposure to toxic amounts of a poison. The most toxic substances are not hazardous if exposure is inconsequential.

A few more of my comments center around this question of "how much is toxic?" The update given us by Dr. Cooper concerning mortality profiles in lead workers is helpful. It demonstrates rather clearly, I think, that if lead, indeed, is a cause of excess mortality in any category of disease, the incidence must be quite low.

This is the limitation of any kind of epidemiological study. Regardless of whether you study 1,000 men or whether you study 10,000 men, there will always be those who come back to you and say that if you had studied 100,000 men, would you not, perhaps, have shown that the statistically-insignificant elevation of cancer deaths may, in fact, not be real?

From an epidemiological standpoint, so far as mortality statistics are concerned, we are at a serious disadvantage as compared, for example, to mortality statistics regarding asbestos. There, at least, we do have something to hang on to--a mortality rate. And from a mortality rate you can construct, at least theoretically, an extrapolation downward at lower doses to get some feel for dose-response relationship.

With regard to the kidney, while it would appear that perhaps some toxic effects occur at the levels of exposure in industry today, there is no evidence that the results of this is a foreshortening of life due to nephritis. This is evident from the update by Dr. Cooper regarding the statistics on death from nephrosclerotic disease.

A few other minor points are to be made in regard to what needs to be done or is being done to resolve the question of the contribution of airborne lead to the total body burden. While perhaps a substantial amount of information is now available in the range of exposure in the general population, I think that it might be useful to institute an Azar-type study in an industrial setting. It seems to me that the studies conducted in the past have not been sufficiently assiduous in the matter of assuring that the lead in blood used as a reflection of internal dose is really the result of inhalation, as distinguished from inhalation plus poor hygienic practices, particularly smoking on the job.

In regard to the sources of lead in children, there still is work that needs to be done. Dr. Sayre told us about the use of adhesive paper to sample housedust; we heard that there was a difference in regard to the concentration of lead in housedust among children with high PbB's as compared to moderate PbB's. This is entirely understandable, and suggests that housedust can be a problem. But, a critical question is as to where is it coming from.

As you are aware, these studies generally are done in the inner cities where there is a substantial amount of old paint. The amount of patina/oxidized paint that comes down as a dust and contributes to total dust is something else that certainly needs to be more clearly understood. ILZRO would be well-advised to see that studies designed to identify the sources of and the significance of the sources of lead get further support.

I might add that we at the University of Cincinnati toyed with the idea of doing polonium-210 analyses in the feces of children. We have a large bank of feces in case any of you has any use for it.

In closing, I'd like to thank you all for coming here, and compliment ILZRO, once again, for having made and hopefully continuing to make a significant contribution to the large effort which is ongoing with regard to the seemingly endless problem of lead exposure.

INDEX

U

W

Y

Z